Parallel Algorithms for Irregular Problems:
State of the Art

Parallel Algorithms for Irregular Problems: State of the Art

Edited by

Afonso Ferreira
LIP, Lyon

and

José D. P. Rolim
University of Geneva

KLUWER ACADEMIC PUBLISHERS
DORDRECHT / BOSTON / LONDON

Library of Congress Cataloging-in-Publication Data

Parallel algorithms for irregular problems : state of the art / edited by Afonso Ferreira, José D.P. Rolim.
p. cm.
Papers from a workshop and summer school held in Geneva in 1994.
Includes bibliographical references.

1. Parallel processing (Electronic computers) 2. Computer algorithms. I. Ferreira, Afonso. II. Rolim, José D. P.
QA76.58.P363 1995
005.2--dc20 95-30322

ISBN 978-1-4419-4747-5

Published by Kluwer Academic Publishers,
P.O. Box 17, 3300 AA Dordrecht, The Netherlands.

Kluwer Academic Publishers incorporates
the publishing programmes of
D. Reidel, Martinus Nijhoff, Dr W. Junk and MTP Press.

Sold and distributed in the U.S.A. and Canada
by Kluwer Academic Publishers,
101 Philip Drive, Norwell, MA 02061, U.S.A.

In all other countries, sold and distributed
by Kluwer Academic Publishers Group,
P.O. Box 322, 3300 AH Dordrecht, The Netherlands.

Printed on acid-free paper

CONTENTS

PREFACE

Efficient parallel solutions have been found to many problems. Some of these solutions can be automatically obtained from sequential programs using compilers. However, there still exists a large class of problems, known as irregular problems, that lack efficient solutions.

Aiming at fostering the cooperation among practitioners and theoreticians of the field, the workshop and summer school on Parallel Algorithms for Irregularly Structured Problems - *IRREGULAR 94* - was organized in Geneva in 1994, addressing issues related to deriving efficient solutions to irregular problems.

Evidently, the first question to be answered is "what are irregular problems?". Therefore, during *IRREGULAR 94* we brought up all participants in a joint effort to answer it. What follows is an informal synthesis of what was said.

A regular (or iso-ergotic) algorithm is such that its dependence (or task) graph is independent of the instance data, meaning that the work accomplished by the algorithm is always the same, up to a (small) constant. On the other hand, the dependence graph of non-regular algorithms (also proposed was ataxidi algorithms) varies according to the input data, implying that the task scheduling cannot be evenly distributed beforehand in a parallel setting. An irregular problem is then characterized by the fact that there are ataxidi algorithms that execute faster than iso-ergotic ones for the same instances in parallel.

Of course, the quality of ataxidi algorithms is closely connected to its non-regularity. Complex choices are made concerning data-structuring techniques used to improve their sequential complexity, making their parallel versions very difficult implement without advanced dynamic load balancing strategies.

Interest in irregular problems is growing throughout the world. There is a new SIG from IFIP, whose main activity is to coordinate the IRREGULAR series. Also, one of the main research topics within the HCM project *SCOOP - Solving Combinatorial Optimization Problems in Parallel* is how to cope with the irregularity in such problems. Finally, some French laboratories are involved in a project, named *Stratageme*, which addresses related problems.

We were glad to see that attendance at *IRREGULAR 94* was larger than expected, showing that we were not alone considering irregular problems as the key to massive parallelism. Hence, we decided to ask speakers to contribute to the present book, in order to lay a solid foundation where researchers world wide could build the theory and practice of parallel irregular applications. We tried to give the reader an overview of the main issues arising in the field. Therefore, this book is divided into three parts. The first one deals with scientific computation, covering different topics ranging from finite element methods to computer vision. The second part gathers research results on discrete optimization, mainly based on new techniques for parallel branch and bound. Last but not least, we show some important steps towards the development of tools for automatic parallelization.

We hope that one day end-users will have at their disposal such tools that could efficiently parallelize even irregular problems . For more on this issue, *rendez-vous a* Lyon, France, in September 1995, for IRREGULAR 95.

Afonso Ferreira, José Rolim

PART I

SCIENTIFIC COMPUTATION

1

PARALLELIZATION STRATEGIES FOR MATRIX ASSEMBLY IN FINITE ELEMENT METHODS

Jorn Behrens

Alfred-Wegener-Institut
Institute for Polar and Marine Research
27515 Bremerhaven, Germany

ABSTRACT

Adaptive finite element methods include unstructured discretization meshes. Most algorithms for the matrix assembly follow the ordering of the arising meshes. Time-dependent problems resulting in rapidly changing locally refined (i.e. unstructured) discretization grids need a matrix assembly in each timestep. This demands a considerable amount of computing time which in turn calls for parallelization.

Domain decomposition techniques often suffer from insufficient load balancing. Especially in adaptive methods balancing the computational workload is difficult, because the number of unknowns in each domain is not predictable and even may change during the computation.

Analysing data dependencies in the matrix assembly, one obtains the problem to synchronize access to one node (matrix element) from neighbouring elements. An efficient algorithm based on coloring the elements following a very simple tabulated indexing scheme has been found. Within each color, calculation is free of data dependencies. Load balancing becomes easy, because data can be distributed uniformly among the processors. This method works well for shared memory architectures.

For distributed memory computers, however, the color indexing method fails, because logical and physical ordering differ, thus physical data access is not structured. A second method for parallelizing the same problem (i.e. matrix assembly in finite element methods) has been developed. Using index-arrays for mapping the logical ordering to a suitable physical (storage-) ordering, one obtains an almost optimally load balanced and efficient algorithm.

Timing results for an Alliant FX/2800 (color indexing) and a Kendall Square Research KSR-1 show the efficiency of both algorithms.

A. Ferreira and J. D. P. Rolim (eds.), Parallel Algorithms for Irregular Problems: State of the Art, 3–24.

1 INTRODUCTION

One of the "Grand Challenges" of the 1990's is the modeling of world ocean circulations. This is a research topic at Alfred-Wegener-Institute (AWI). A small working group has been established to develop new numerical methods and improve the existing ones in currently used ocean models.

One approach to improve accuracy and ease of use, especially concerning boundary conditions and complicated computational domains, is the introduction of finite element methods (FEM). These methods, however, suffer from higher computational costs compared to finite difference schemes used commonly. Therefore adaptive meshing is desirable in order to reduce the number of nodes to a minimum and save computing time while preserving accuracy.

As timescales of oceanic processes as well as computational domains are large and the spatial extent of the interresting phenomena on the other hand is rather small, a large number of nodes and timesteps occurs in ocean models. This is why supercomputers have traditionally been used by the ocean modeling community. Recently parallel computers became workhorses for these numerically intense applications.

A special problem with adaptive methods and parallelization is the load balancing issue. As grids are unstructured and even may change while the computation proceeds, domain decompositioning methods may fail. Two new approaches for the parallelization of matrix assembly in FEM are proposed.

The first one works well for shared memory architectures. It is based on coloring the elements in order to synchronize data access. This method, the color indexing method, was originally developed for an Alliant FX/2800 which served as a development system at AWI until 1993. The color indexing scheme will be described in section 2.

Due to the financial break down of Alliant this computer was exchanged by a KSR-1 in 1993. The color indexing method is not efficient on the later architecture. A second method based on index-arrays is proposed for distributed memory architectures in section 3.

Section 4 summarizes experimental results for both methods while section 5 gives a conclusion. An appendix is added which makes plausible that the color indexing method works for almost arbitrarily shaped meshes.

2 PARALLELIZATION BY COLOR INDEXING

In this section a parallelization strategy based on coloring the elements of a triangulation is proposed. We start with analyzing the data dependency which prevents parallelization, then the algorithm will be developed, and finally some of the properties of this strategy will be discussed.

A typical situation in a regularly refined grid looks like in figure 2. The algorithm for the matrix assembly in FEM proceeds as follows. Each line of the matrix consists of one diagonal element (corresponding to one node) and as many off diagonal entries as neighbouring nodes exist. Each entry is calculated by summing up all contributions from triangles containing the corresponding node as the procedure loops through the elements of the mesh (as illustrated in figure 1). If elements A and B in figure 2 belong to different processors, then a data access conflict can occur, when both processors try to add their values to the same matrix entry at the same time.

Thus the idea of this parallelization strategy is to give all triangles around each node a distinct color while keeping the total number of colors small. Then parallelization within each group of triangles of the same color is possible. With a large number of triangles one color will be present many times yielding coarse granularity. A simple regularly refined colored mesh is shown in figure 3.

This strategy for globally and regularly refined triangulations was implemented already in [2] and [3]. Here an extension for locally refined triangulations is presented. While the coloring algorithm there was based on inheritance of color patterns from parent triangles, the approach here is to use a simple table of color patterns. Choosing colors from the table, however, still depends on

```
do ie = 1,#elements
   do i2 = 1,3
      ind2= node(i2,ie)
      do i1 = 1,3
         ind1= node(i1,ie)
         find corresponding index ik in the element stiffnes matrix A_k
         A(ind1,ind2) = A(ind1,ind2)+ A_k(ik,ie)
      end do
   end do
end do
```

Fig. 1. Pseudo code of the original algorithm for matrix assembly in FEM

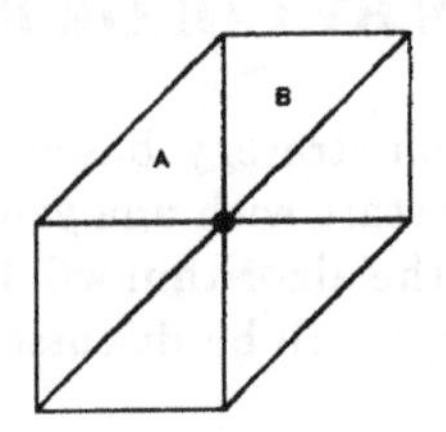

Fig. 2. A typical situation in a regularly refined triangulation

certain properties of the parent triangle. Additionally some assumptions on the initial triangulation have to be made.

Let a coarse triangulation consist of triangles of four types, where every two triangle types belong together as shown in figure 4. These types, although numbering only four, provide a sufficiently flexible means of triangulating the domain, when rotated or streched triangles are used.

Coloring depends on triangle type, level of refinement, and – for green closures – position of the triangle (for a more detailed description and the tables containing color indices see appendix 5). In this manner we obtain a colored triangulation where triangles of the same color contain only distinct nodes or, equivalently, each node has surrounding triangles of distinct colors.

Fig. 3. Colored regularly refined triangulation

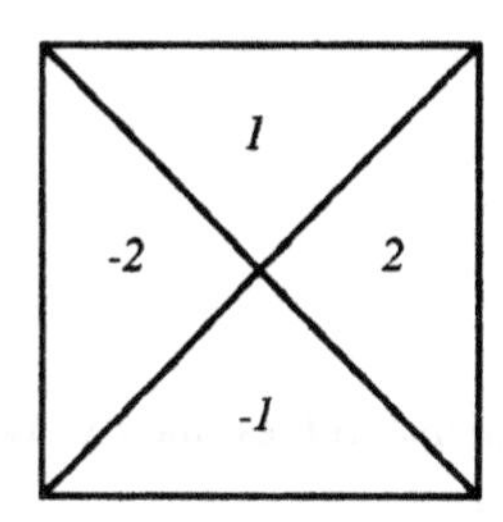

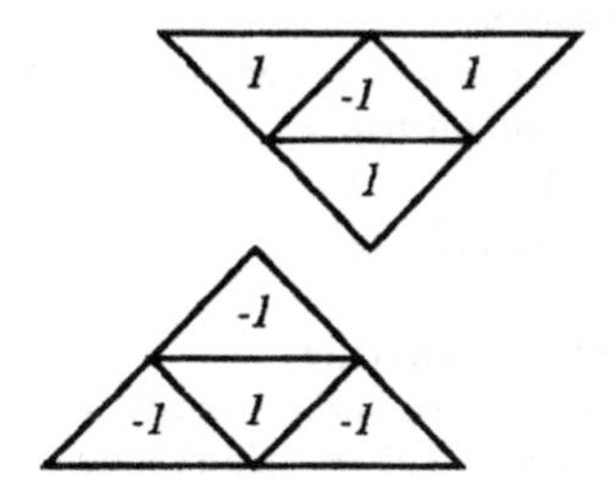

Fig. 4. Coarse grid triangle types

Parallelizing within each color leads to an almost optimally load balanced algorithm with as many synchronizations as colors are present. In the implementation discussed in section 4 up to 48 colors are used. For a regularly refined triangulation with four initial triangles, seven refinement levels, and an additional local refinement level there are approx. $32K$ triangles on the finest grid. Computation corresponding to one triangle consists of at least 9 floating point operations and a considerable number of integer index calculations (i.e. $\mathcal{O}(10)$ - $\mathcal{O}(10^2)$ instructions per element). Assuming $\mathcal{O}(10^3)$ triangles per processor and a moderate synchronization overhead of $\mathcal{O}(10^2)$ clockcycles shows that time for (in the worst case) 48 synchronizations is small compared to the computational workload. The pseudo code segment in figure 5 illustrates the algorithm.

The algorithm described above is efficient on a shared memory architecture such as the Alliant FX/2800. The structure of data access is changed slightly compared to the original algorithm. Still no knowledge about the physical location of data is collected. However the structure of the algorithm itself (loop structure, calculation, etc.) remains almost unchanged and only minor code modifications are necessary. Explicitly, the outer loop is exchanged by a pair of nested loops such that the elements of each color are processed in parallel.

Therefore data access is not predictable which makes this algorithm usable only for shared memory architectures. On the KSR-1 the algorithm still works (without changes) but efficiency drops significantly (see more detailed timing results in section 4). Lacking knowledge of data location also prohibits optimization for hierarchical memory architectures with caches.

```
do ic = 1,#colors
   do parallel ip = 1,#elements_of_color(ic)
      ie = elemnt_pntr(ip,ic)
      do i2 = 1,3
         ind2= node(i2,ie)
         do i1 = 1,3
            ind1= node(i1,ie)
            find corresponding index ik in the element stiffnes matrix A_k
            A(ind1,ind2) = A(ind1,ind2)+ A_k(ik,ie)
         end do
      end do
   end do parallel
end do
```

Fig. 5. Pseudo code of the coloring algorithm

The proposed coloring algorithm is computationally very cheap. It consists of only one assignment per element in the refinement procedure. This simplicity is punished with the relatively large number of colors resulting from this strategy. A different approach, where the smaller number of colors is purchased by a more complicated coloring algorithm, is given in [5].

3 PARALLELIZATION BY INDEX-ARRAY MAPPING

Whereas the parallelization strategy in the last section was especially taylored for shared memory architectures, a second strategy for distributed memory machines is proposed here. Beginning with the observation that only nodal data is accessed for writing, an idea of fixing these data to the processors is developed. Finally some properties are discussed.

The aim is to decouple the data dependencies and to distribute data such that a minimum of data movement occurs. Once the contributions from adjacent triangles of a node are known, the matrix entries corresponding to this node can be calculated independently from all other matrix entry calculations.

Thus changing the loop structure in matrix assembly from a loop over all triangles (adding to all nodes contained in the actual triangle) to a loop over nodes (adding contributions from all the adjacent triangles) eliminates data dependencies and makes it possible to fix data to a determined processor.

```
do parallel inod = 1,#nodes
   each processor operates on "his" nodes, indices of
   which are stored in an index-array
   do ic = 1,#elements_next_to_node(inod)
      inei = adjcnt_elmnt(ic,inod)
      find corresponding matrix indices im for the global matrix
      A and ik for the element stiffnes matrix A_k
      A(im,inod) = A(im,inod)+ A_k(ik,inei)
   end do
end do parallel
```

Fig. 6. Pseudo code of the index-array algorithm

All triangle numbers corresponding to a certain node are stored in an index-array. A second index-array is used to fix nodal data to one processor over the whole time of the calculation. Different orders of access in different routines are captured by different mapping arrays. This is the reason why we call this strategy the index-array mapping.

The algorithm proceeds as follows: Loop through all nodes of the triangulation in parallel gathering the node numbers for the corresponding processor from an index-array. Add the contributions from all adjacent triangles (which numbers are stored in an other index-array) to the corresponding matrix entries. A pseudo code example is given in figure 6.

A different approach can be found in [8]. The element-by-element ordering is maintained there while parallelization is achieved by clustering the elements and distributing clusters to the processors.

Storage requirements for the index-array containing the triangle numbers for each node depend on the initial triangulation. The following formula describes the maximum number N_t of triangles containing one and the same node:

$$N_t = max\{(K_0 \cdot K_{green}), (K_{regular} \cdot K_{green})\} \tag{1.1}$$

where K_0 is the maximum number of triangles containing one node in the initial triangulation, $K_{green} = 2$ is the number of green triangles dividing a regular triangle, and $K_{regular} = 6$ is the number of triangles surrounding one node with regular refinement. The index-array for N_{Nodes} nodes has size

Fig. 7. Locally refined triangulation with five regular refinement levels as used in the experiments

$$M_{index-array} = N_t \cdot N_{Nodes} \cdot M_{SizeOfInt}. \tag{1.2}$$

Again we obtain an algorithm with nearly optimal load balancing (imbalance occurs only due to the fact that the number of nodes might not be a multiple of the number of processors). Write access to data is carried out on the same processor for the whole computation. That means only read requests are sent to remote processors, all writing remains within local memory. Very high efficiency is achieved on the KSR-1 with this parallelization strategy.

4 EXPERIMENTAL RESULTS

Timing results from an Alliant FX/2800 (with eight processors) and a KSR-1 (with 32 processors) are presented in this section. The algorithm used for the tests is a finite element scheme using continuous piecewise linear basisfunctions and a hierarchical bases preconditioned conjugate gradient method for the solution of the arising system of linear equations (see [10] for theoretical background and [2] for the algorithm). The Poisson-Problem with Dirichlet boundary conditions on the unit square is taken as a model problem. When local refinement is applied, approximately one third of the triangles on the given finest grid level is refined once more (see figure 7 for an example). The configurations (number of refinements and resulting number of triangles on the finest grid) used for the tests are listed in table 1.

Tests on the Alliant have been performed with up to four processors, because this was the configured maximum number of processors in a computational cluster (cf. [1]). Similarly the KSR-1 tests are run on up to 26 processors, because this is the largest allocatable processor set (cf. [6]). All measurements are based on user times of the matrix assembly routine measured with the delivered timing routines.

Different performance parameters are given. ***Speedup*** measures the acceleration ratio of execution time of the algorithm on p processors over execution of the same algorithm on 1 processor:

$$s = \frac{t_1}{t_p} \tag{1.3}$$

The *efficiency parameter* is a very similar measure. It is defined as a speedup scaled by the number of processors:

$$f_e = \frac{t_1}{p \cdot t_p}. \tag{1.4}$$

Efficiency would be 1, if no overhead were imposed by parallelization. (see [4] for parametization of performance).

The *scaleup* indicates the property of the algorithm to scale with the number of processors and the problem size. It is defined:

$$S = \frac{t_p(N)}{t_{l \cdot p}(l \cdot N)}, \tag{1.5}$$

where p is the number of processors, N is the number of unknowns or generally an indicator for the problem size. Therefore $t_p(N)$ is the execution time for the model problem with N unknowns on p processors. l is a positive integer. Doubling the problem size and at the same time doubling the number of processors should keep the execution time constant.

reference	no. of refinements	no. of nodes	no. of triangles (fine grid)
A	7 (g)	8,321	16,384
B	7 (l)	16,261	32,152
C	8 (g)	33,025	65,536
D	8 (l)	64,433	128,120

Table 1. Configurations for the tests, (g): global refinement, (l): additional local refinement

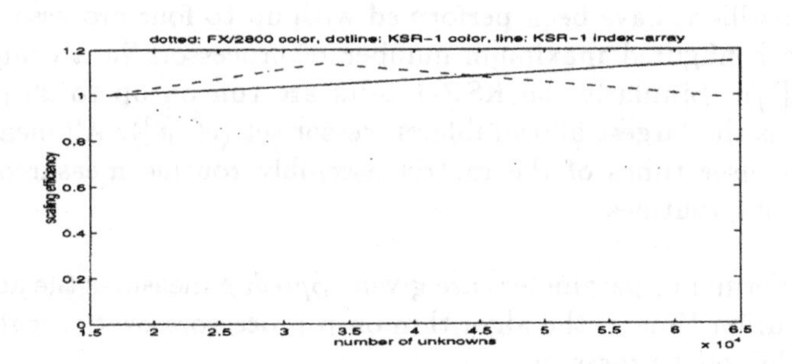

Fig. 8. Scaleup with 1, 2, and 4 processors for the Alliant FX/2800, with 6, 12, and 24 processors for the KSR-1

While the Alliant achieved a parallel efficiency of around $f_e = 0.95$ for triadic operations (cf. [3]) parallel performance in real life applications was lower. Our tests show values for f_e of around 0.70 (cases B, C, and D).

The KSR-1 numbers of f_e for the color indexing parallelization strategy drop to less than 0.50 (case B). In contrast to that the index-array mapping yields an efficiency around 0.85 in each case. It should be noted, however, that the color indexing method still works and parallelizes for a moderate number of processors, in spite of the fact that no care is taken about the location of data. All data movement is performed by the virtual shared memory mechanism, the *Allcache Engine* (cf. [6]).

Figures 9 through 14 illustrate the results. Tables of timing results, efficiency and speedup can be found in appendix 5. Note that high efficiency means that both algorithms scale well with the number of processors which is illustrated in figure 8[1].

[1] Scaleup values over one are due to cache size problems with small processor numbers.

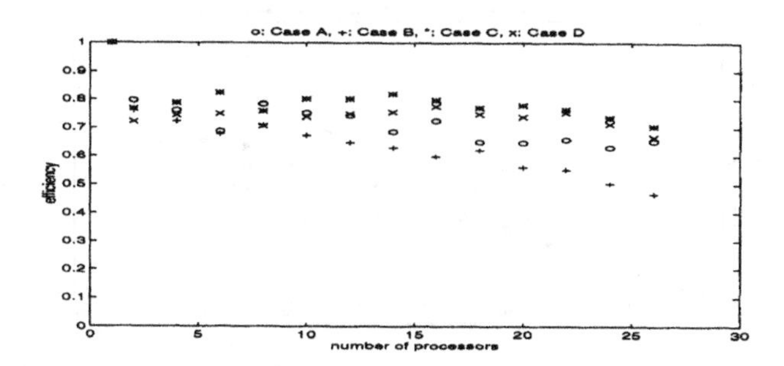

Fig. 9. Efficiency Parameter for the coloring method on KSR-1

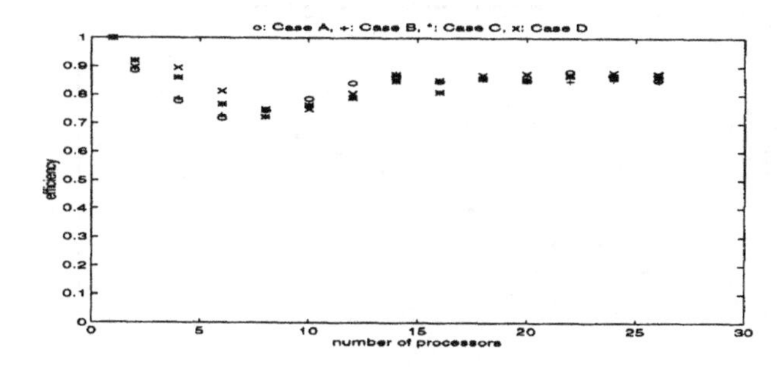

Fig. 10. Efficiency Parameter for the index-array method on KSR-1

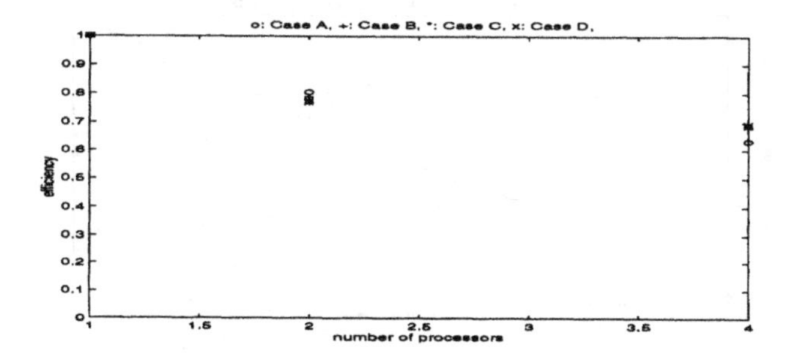

Fig. 11. Efficiency Parameter for the coloring method on Alliant FX/80

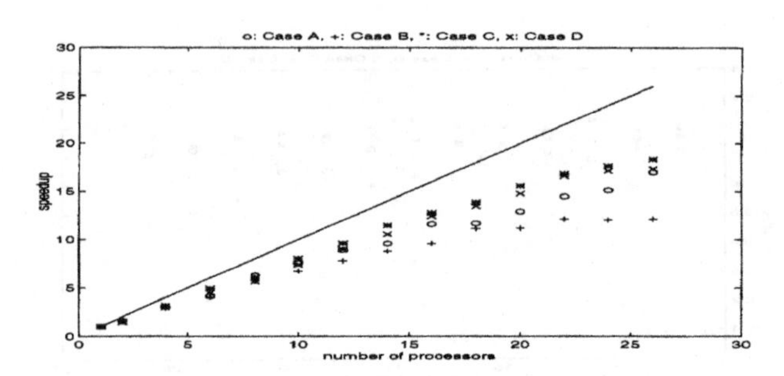

Fig. 12. Speedup for the coloring method on KSR-1 (line represents theoretical speedup)

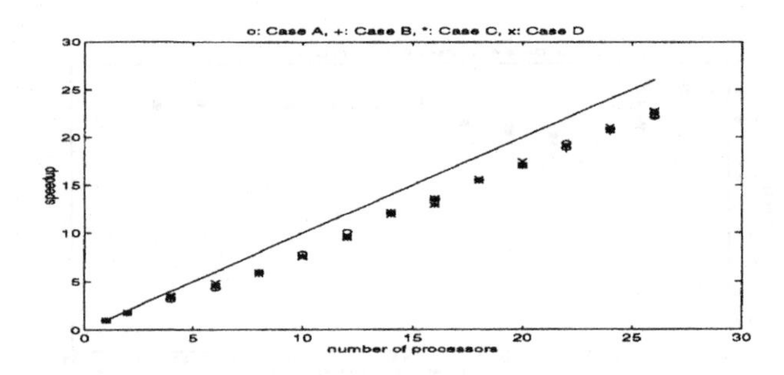

Fig. 13. Speedup for the index-array method on KSR-1 (line represents theoretical speedup)

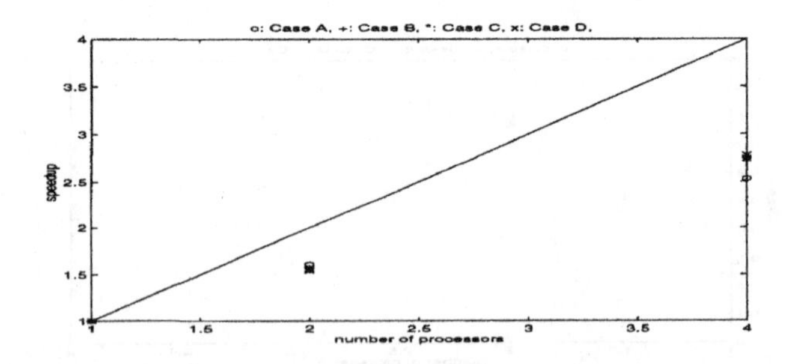

Fig. 14. Speedup for the coloring method on Alliant FX/2800 (line represents theoretical speedup)

5 CONCLUSION

Two strategies for parallelizing matrix assembly in FEM have been described. One algorithm, originally developed for a shared memory architecture (Alliant FX/2800), shows good performance on this kind of systems. This algorithm is based on coloring the elements of the triangulation such that data access to the nodal values can be synchronized. The logical structure of the algorithm remains almost unchanged, therefore only minor changes to the code are necessary.

On distributed memory architectures (such as the KSR-1) performance of this algorithm drops significantly. Therefore a second algorithm has been proposed. It is based on the employment of index-arrays for gathering information from the surrounding elements when calculating nodal matrix entries. Additionally index-arrays are used to map the logical structure of the algorithm to a determined data distribution. By this means data can remain at fixed positions in local storage areas, yielding low communication requirements.

Both algorithms are asymptotically optimally load balanced even with locally refined and unstructured grids. Both algorithms are efficient on their corresponding architectures.

The color indexing method prohibits optimization of data movement, which makes this strategy inferior for distributed memory machines and also for hierarchical memory architectures with caches. On the other hand the index array method consumes a significantly bigger amount of memory.

However both algorithms scale with the number of processors. This is an important feature when large problems are considered as is the case in ocean modeling.

A global address space is utilized. This makes coding much easier, because only directives have to be included to parallelize certain code segments. Additionally a small number of data access conflicts can be ignored, because the operating system cares for synchronization.

The color indexing method has been used in real life applications at the Bundesanstalt für Wasserbau, a state department for hydroengineering in Hamburg [7]. Modelling of tidal and dynamic transport of water of the rivers Elbe and Weser was implemented on an Alliant FX/8.

REFERENCES

[1] Alliant Computer Systems Corp., Littleton (Massachusetts). *FX/2800 System Description*, mar 1991. Part no. 300-00500.

[2] J. Behrens. Optimierung eines Mehrgitterverfahrens und eines Hierarchische Basen Verfahrens auf einer Alliant FX/80. Diplomarbeit, Rheinische Friedrich-Wilhelms-Universität, Bonn, 1990. Erstellt am Alfred-Wegener-Institut Bremerhaven.

[3] W. Hiller and J. Behrens. Parallelisierung von Mehrgitteralgorithmen auf der Alliant FX/80. In H. W. Meuer, editor, *Parallelisierung komplexer Probleme*, pages 37–82, Berlin, 1991. Springer-Verlag.

[4] R. W. Hockney and C. R. Jesshope. *Parallel Computers 2, Architecture, Programming and Algorithms*. Adam Hilger, Bristol Philadelphia, 2nd edition, 1988.

[5] M. T. Jones and P. E Plassmann. A parallel graph coloring heuristic. *SIAM J. Sci. Comput.*, 14(3):654–669, 1993.

[6] Kendall Square Research Corp., Waltham. *KSR1 System Administration*, Dec. 1992.

[7] G. Lang. Kurzbeschreibung des Programmes PFRG02. Technical report, Bundesanstalt für Wasserbau, Hamburg, 1992.

[8] J. Liou and T. E. Tezduyar. Clustered element-by-element computations for fluid flow. In H. D. Simon, editor, *Parallel Computational Fluid Dynamics: Implementations and Results*, Scientific and Engineering Computation, pages 167–187, Cambridge Massachusetts, 1992. The MIT Press.

[9] W. F. Mitchell. A comparison of adaptive refinement techniques for elliptic problems. *ACM Trans. in Math. Softw.*, 15(4):326–347, 1989.

[10] H. Yserentant. On the multi-level splitting of finite element spaces. *Numer. Math.*, 49:379–412, 1986.

APPENDIX A

DETAILED DESCRIPTION AND PLAUSIBILITY OF THE COLOR INDEXING METHOD

In this section a detailed description of the color indexing method is given. Assumptions on the initial grid are stated. The coloring scheme is described in some detail which includes listing of tables with color values. Consistency of the coloring is made plausible.

Assumption A.0.1 (Refinement Strategy) *Following assumptions to the refinement procedure are made:*

1. *Regular refinement is performed like in [9]: centerpoints of each side are connected pairwise giving four new triangles dividing the meshsize by two.*
2. *With local refinement some unallowed nodes occur. These are eliminated by green closures (i.e. triangles containing unallowed nodes are divided by a side connecting the unallowed node with the opposite triangle node giving two new triangles).*
3. *Green closures must not be refined further. In cases where refinement is desired first a coarsening step to the original triangle has to be performed, then a regular refinement is possible.*
4. *Triangle nodes (and sides) are numbered counterclockwise as shown in figure A.1.*

Note A.0.2 *Global regular refinement of an allowed triangulation yields an allowed triangulation.*

Assumption A.0.3 (Initial Triangulation) *Following assumptions to the initial grid are made:*

1. *We assume that the initial triangulation consists of only four types of triangles with type indicators {1,-1,2,-2}. A triangle of type* $i \in \{1, -1, 2, -2\}$ *has the following neighbouring triangle types:*

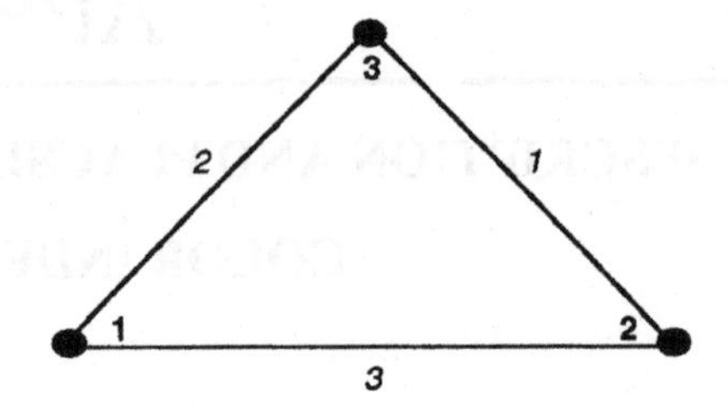

Fig. A.1. Local triangle node and side numbering

$$\begin{array}{ll} \textit{Side 1:} & (-1)^i \cdot sign(i) \cdot j \\ \textit{Side 2:} & (-1)^i \cdot sign(i) \cdot -j \\ \textit{Side 3:} & -i \end{array}$$

where $j = |i| \bmod 2 + 1$.

2. *Not more than two triangles of the same type are allowed to meet in one node.*

Note A.0.4 *A triangle of type i is refined to three peripheral triangles of type i and one central triangle of type $-i$ by a regular refinement step. This is shown in figure 4.*

Most two-dimensional domains may be triangulated corresponding to assumption A.0.3 (see the following example). However there are some restrictions. To illustrate this fact let an initial triangulation of the three-dimensional sphere

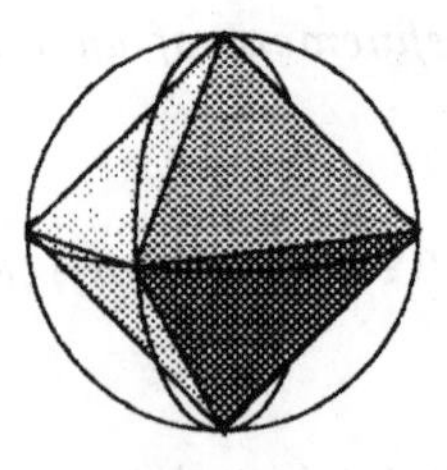

Fig. A.2. Initial triangulation of the sphere

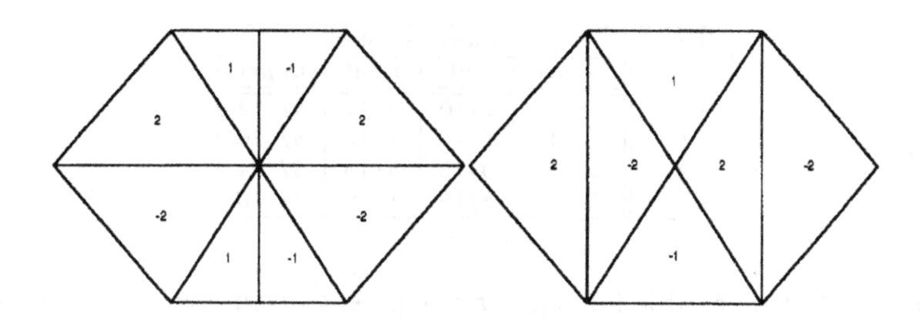

Fig. A.3. Triangletypes in the initial triangulation of a hexagon

consist of the surface of an octahedron (see figure A.2). Each semisphere is triangulated by four triangles put together to a square like in figure 4. But connecting both semispheres to conform with assumption A.0.3 is not possible, because orientation changes at the equator.

Example A.0.5 *The initial triangulation of a hexagon cannot be constructed in the obvious way without contradicting assumption A.0.3. Either constructing it from a square with two attached triangles, or inserting two additional triangles like in figure A.3 helps to solve the problem.*

Propostion A.0.6 (Coloring Algorithm) *One can construct a color pattern with these assumptions in a triangulation such that each node is surrounded by differently colored triangles, provided that*

1. *triangles from a regular refinement are colored as in table A.1;*
2. *the coloring for regular triangles is permuted every second refinement level as indicated in table A.1;*
3. *triangles from a green closure are colored according to table A.2;*
4. *a permutation is performed every second refinement level as indicated in table A.2.*

The statement of proposition A.0.6 is equivalent to the following statement:

Type	children's colors			
	centre	left	right	upper
+1	1/ 9	2/10	3/11	4/ 12
−1	5/ 13	6/14	7/15	8/ 16
+2	3/ 11	4/12	5/13	6/ 14
−2	7/ 15	8/16	1/ 9	2/ 10

Table A.1. Color table for regular refinement (even level/ odd level)

Note A.0.7 *For all colors in a triangulation constructed as in A.0.6 each node appears only once within all triangles of one color.*

Making the result of proposition A.0.6 plausible will proceed as follows: First we will show that the proposition holds within one refined triangle with arbitrarily many levels (lemma A.0.8 and A.0.9). Then interaction at the interfaces of initial triangles is considered (lemma A.0.10). These two parts together show the correctness of the proposed method.

Lemma A.0.8 *Refining one single triangle globally according to assumption A.0.1 and coloring as in proposition A.0.6 1. and 2. gives a triangulation with triangles meeting in one node or at an edge with pairwise different colors.*

PROOF: Within one triangle only refined triangles of the original type and the "inverse type" can occur. By construction type i and $-i$ triangles are colored by distinct colors under refinement. Therefore at the interfaces no triangles of the same color can meet.

Type	children's colors					
	side 1 divided		side 2 divided		side 3 divided	
	left	right	left	right	left	right
+1	17/33	18/34	19/35	20/36	21/37	22/38
−1	23/39	24/40	25/41	26/42	27/43	28/44
+2	17/33	24/40	19/35	30/46	27/43	31/47
−2	23/39	18/34	25/41	32/48	21/37	29/45

Table A.2. Color table for green closure (even level/ odd level)

Triangles of the same type meet at locally different nodes (e.g. the upper trianlge's node 1 meets with the right triangle's node 3, the upper triangle's node 2 meets with the left triangle's node 3). Thus colors adjacent to the corresponding nodes never meet.

As this holds for the first refinement step, by induction it holds for each refinement step, because triangles constructed by a regular refinement are similar to their parents. □

Lemma A.0.9 *Given a single triangle with a globally regularly refined mesh of subtriangles. When refining locally according to assumption A.0.1 and coloring according to conditions 1. through 4. in proposition A.0.6, one obtains a new refinement, where no elements of the same color touch each other.*

PROOF: We will proceed in two steps: first we will show that if one additional refinement step is allowed, then the lemma holds, the second step generalizes to arbitrarily many refinement steps.

First we look at the situation where a green closure meets with regularly refined triangles. This situation will cause no problems, because colors of green closures differ from those of regularly refined triangles.

If several green closures meet, then looking at table A.1 confirms that a conflict can only occur, when two triangles of the same type, divided at the same side meet. This situation though cannot occur in reality, because if a triangle of type i is green refined, then a neighbouring triangle of type $-i$ is regularly refined. But then the other two surrounding type i triangles (meeting with the above type i triangle at one of their nodes) are either regularly refined (which causes no problem) or are green refined but with different sides divided. That means equal colors do not meet.

Now if more than one additional refinement level is involved we observe that the difference between the refinement levels of two meeting triangles is less than 2. As colors are permuted every second refinement step, no triangles of the same color can meet. □

We have shown the consistency of the color indexing method for one initial triangle. Actually we did not need the assumption A.0.3 up to now. An efficient parallelization strategy could be implemented already, if additional synchronizations would be applied (handle each initial triangle separately). Constructing the domain of only a few initial triangles and refining sufficiently often would add only a negligible number of synchronizations.

Lemma A.0.10 *Given an initial triangulation with assumption A.0.3. With refinement and coloring as in A.0.1, A.0.6 resp., no triangles of the same color meet at the interfaces of the initial triangles.*

PROOF: We proceed in two steps: first we show that for globally refined triangulations no inconsistencies concerning coloring can occur. Then we enhance this to local refinement.

Let us check which triangle colors touch the sides of the different parent triangle types provided only global regular refinement is allowed. When the mother is of type i then the centre, right, and upper child triangles of a type-i-triangle and the left child triangle of a type-$-i$-triangle touch the first side. Analogously the centre, left and upper triangles of a type-i-triangle and the right triangle of a type-$-i$-triangle touch the second side. Finally left, right, and centre triangles of a type-i-triangle and the upper triangle of a type-$-i$-triangle touch side 3.

Now a triangle of type 1 either meets at side 1 with a triangle of type -2, at side 2 with a triangle of type 2, or with a triangle of type -1 at side 3. the later situation corresponds with the situation within one initial triangle (which is consistent as shown in lemma A.0.8). For the first situation colors 1, 3, 4, and 6 meet with colors 7, 8, 2, and 5 which is consistent with the statement of A.0.6. It is easy to check this for all possible combinations.

It remains to be shown that even with local refinement no coloring inconsistencies can occur at the interfaces of the initial triangulation.

We proceed in the same way as above. Looking at a triangle of type i we observe that all subtriangles resulting from a green closure of a type-i-triangle touch the edges of the initial triangle whereas for the type-$-i$-triangle one has to distiguish. For simplicity we will refer to the entries of table A.2 counting from left to right.

Triangles corresponding to entries 1, 2, 4, and 5 of a green closure of a type-$-i$-triangle touch edge number one. The entries for the second edge are 1, 3, 4, and 6 and for edge three 2, 3, 5, and 6.

It is an easy task to go through the table entries and check all possible combinations of triangle colors meeting at the different edges. □

We have shown in a quite heuristic way the plausibility of proposition A.0.6. Parallelization by color indexing therefore is a good possibility for a wide range of problems when the targeted computer architecture is of shared memory type.

APPENDIX B

TABLES OF TIMING RESULTS

The results of testscases A–D (table 1) are given in the tables below. Execution times represent the user times (in seconds) for the matrix assembly measured with the timing routines given on the corresponding system (**etime** with a resolution of 10^{-3}sec. on the Alliant, **user_seconds** with a resolution of 8 clock cycles ($\approx 10^{-6}$sec.) on the KSR-1). On the KSR-1 a cycle of Cases A through D was executed ten times. The given timings represent the minima of these executions. This procedure has been chosen in order to obtain smooth timing functions and to minimize the effects of interfering system activity.

Execution time														
	number of processors													
	1	2	4	6	8	10	12	14	16	18	20	22	24	26
Case A	3.04	1.91	0.99	0.74	0.49	0.40	0.34	0.32	0.26	0.26	0.24	0.21	0.20	0.18
Case B	6.01	3.95	2.08	1.48	1.06	0.89	0.77	0.68	0.63	0.54	0.54	0.50	0.50	0.50
Case C	12.50	8.15	3.97	2.53	2.06	1.56	1.30	1.09	0.98	0.90	0.80	0.74	0.71	0.68
Case D	25.70	17.80	8.65	5.72	4.55	3.48	2.86	2.43	2.07	1.91	1.74	1.55	1.50	1.48

Efficiency														
	number of processors													
	1	2	4	6	8	10	12	14	16	18	20	22	24	26
Case A	1.00	0.80	0.76	0.69	0.78	0.75	0.74	0.69	0.73	0.65	0.65	0.66	0.63	0.65
Case B	1.00	0.76	0.72	0.68	0.71	0.67	0.65	0.63	0.60	0.62	0.56	0.55	0.50	0.47
Case C	1.00	0.77	0.79	0.82	0.76	0.80	0.80	0.82	0.80	0.77	0.78	0.76	0.74	0.71
Case D	1.00	0.72	0.74	0.75	0.71	0.74	0.75	0.76	0.78	0.75	0.74	0.75	0.72	0.67

Speedup														
	number of processors													
	1	2	4	6	8	10	12	14	16	18	20	22	24	26
Case A	1.00	1.59	3.04	4.12	6.27	7.52	8.92	9.59	11.60	11.69	12.94	14.48	15.12	16.98
Case B	1.00	1.52	2.89	4.06	5.67	6.72	7.76	8.80	9.57	11.15	11.19	12.14	12.04	12.14
Case C	1.00	1.53	3.15	4.94	6.07	8.01	9.62	11.47	12.77	13.87	15.57	16.82	17.66	18.36
Case D	1.00	1.44	2.97	4.49	5.65	7.39	8.99	10.57	12.42	13.46	14.77	16.58	17.13	17.36

Table B.1. Results for matrix assembly on KSR-1 (color indexing)

Execution time

	number of processors		
	1	2	4
Case A	1.38	0.86	0.55
Case B	2.60	1.65	0.96
Case C	5.52	3.59	2.01
Case D	10.90	7.04	3.92

Efficiency

	number of processors		
	1	2	4
Case A	1.00	0.80	0.63
Case B	1.00	0.79	0.68
Case C	1.00	0.77	0.69
Case D	1.00	0.77	0.70

Speedup

	number of processors		
	1	2	4
Case A	1.00	1.61	2.53
Case B	1.00	1.59	2.72
Case C	1.00	1.54	2.75
Case D	1.00	1.55	2.78

Table B.2. Results for matrix assembly on Alliant FX/2800 (color indexing)

Execution time

	number of processors													
	1	2	4	6	8	10	12	14	16	18	20	22	24	26
Case A	3.08	1.74	0.99	0.72	0.52	0.40	0.31	0.25	0.23	0.20	0.18	0.16	0.15	0.14
Case B	6.08	3.40	1.93	1.39	1.02	0.80	0.64	0.50	0.45	0.39	0.35	0.33	0.30	0.28
Case C	12.60	6.84	3.66	2.74	2.18	1.65	1.33	1.06	0.97	0.82	0.74	0.66	0.61	0.56
Case D	25.50	14.20	7.13	5.23	4.26	3.40	2.64	2.09	1.87	1.63	1.46	1.33	1.21	1.12

Efficiency

	number of processors													
	1	2	4	6	8	10	12	14	16	18	20	22	24	26
Case A	1.00	0.89	0.78	0.72	0.75	0.79	0.84	0.87	0.85	0.86	0.85	0.88	0.87	0.85
Case B	1.00	0.89	0.79	0.73	0.75	0.76	0.80	0.87	0.85	0.86	0.86	0.85	0.85	0.85
Case C	1.00	0.92	0.86	0.77	0.72	0.76	0.79	0.85	0.81	0.86	0.85	0.87	0.87	0.87
Case D	1.00	0.90	0.89	0.81	0.75	0.75	0.80	0.87	0.85	0.87	0.87	0.87	0.88	0.88

Speedup

	number of processors													
	1	2	4	6	8	10	12	14	16	18	20	22	24	26
Case A	1.00	1.77	3.11	4.31	6.00	7.86	10.10	12.13	13.57	15.48	17.02	19.37	20.81	22.16
Case B	1.00	1.79	3.15	4.37	5.96	7.62	9.57	12.23	13.60	15.55	17.18	18.65	20.47	22.11
Case C	1.00	1.84	3.44	4.60	5.78	7.64	9.47	11.89	12.95	15.42	17.03	19.06	20.79	22.66
Case D	1.00	1.80	3.58	4.88	5.99	7.50	9.66	12.20	13.64	15.64	17.47	19.17	21.07	22.77

Table B.3. Results for matrix assembly on KSR-1 (index-array)

2

A PARALLELISABLE ALGORITHM FOR PARTITIONING UNSTRUCTURED MESHES

C. Walshaw, M. Cross, M. Everett and S. Johnson

School of Mathematics, Statistics & Scientific Computing,
University of Greenwich,
London, SE18 6PF, UK.
e-mail: *C.Walshaw@gre.ac.uk*

ABSTRACT

A new method is described for solving the graph-partitioning problem which arises in mapping unstructured mesh calculations to parallel computers. The method, encapsulated in a software tool, JOSTLE, employs a combination of techniques including the Greedy algorithm to give an initial partition together with some powerful optimisation heuristics. A clustering technique is additionally employed to speed up the whole process. The resulting partitioning method is designed to work efficiently in parallel as well as sequentially and can be applied to both static and dynamically refined meshes. Experiments, on graphs with up to a million nodes, indicate that the JOSTLE procedure is up to an order of magnitude faster than existing state-of-the-art techniques such as Multilevel Recursive Spectral Bisection.

1 INTRODUCTION

The use of unstructured mesh codes on parallel machines is one of the most efficient ways to solve large Computational Fluid Dynamics (CFD) and Computational Mechanics (CM) problems. Completely general geometries and complex behaviour can be readily modelled and, in principle, the inherent sparsity of many such problems can be exploited to obtain excellent parallel efficiencies. However, unlike their structured counterparts, one must carefully address the problem of distributing the mesh across the memory of the machine at runtime so that the computational load is evenly balanced and the amount of interprocessor communication is minimised. It is well known that this problem is NP

A. Ferreira and J. D. P. Rolim (eds.), Parallel Algorithms for Irregular Problems: State of the Art, 25–46.

complete, [7], so in recent years much attention has been focused on developing suitable heuristics, and some powerful methods, many based on a graph corresponding to the communication requirements of the mesh, have been devised, e.g. [5]. In this paper we discuss the mesh partitioning problem in the light of the coming generation of massively parallel machines and the resulting implications for such algorithms.

1.1 Why parallel mesh partitioning?

As mesh and machine sizes grow the need for parallel mesh partitioning becomes increasingly acute. For small meshes and machines, an $O(N)$ overhead for the mesh partitioning may be considered reasonable. However, for large N, this order of overhead will rapidly become unacceptable if the solver is running at $O(N/P)$.

In addition, it is often the case that the mesh is already distributed across the memory of the parallel machine. For example, parallel mesh-generation codes or solvers which use parallel adaptive refinement give rise to such distributed meshes, and in these cases it is extremely expensive to transfer the whole mesh back to a single processor for sequential load-balancing, if indeed the memory of that processor allows it. Löhner *et al.* have advanced some powerful arguments in support of this proposition, [13].

The strategy developed here to tackle these issues efficiently is to derive a partition as quickly and cheaply as possible, distribute the data and then to *optimise* the partition *in parallel.* Of course, if the mesh is already distributed then the existing partition is used and optimisation can commence immediately.

1.2 Dynamic load-balancing

An extremely important aspect of partitioning arises from time-dependent unstructured mesh codes which adaptively refine the mesh. This is a very efficient way to track phenomena which traverse the solution domain but means that the position and density of the mesh points may vary dramatically over the course of an integration.

From the point of view of partitioning this has three major influences on possible solution techniques; cost, reuse and parallelism. Firstly, the unstructured mesh may be modified every few time-steps and so the load-balancing must have a low cost relative to that of the solution algorithm in between remeshing. This may seem to restrict us to computationally cheap algorithms but fortunately help is at hand if the mesh has not changed too much, for in this case it is a simple matter to interpolate the existing partition from the old mesh to the new and use this as the starting point for repartitioning, [23]. In fact, not only is the load-balancing likely to be unnecessarily computationally expensive if it fails to use this information, but also the mesh elements will be redistributed without any reference to their previous 'home processor' and heavy node migration may result. The third factor to take into account is that the data is distributed and so should be repartitioned *in situ* rather than transferred back to some host processor for load-balancing. Again, this calls for a parallel algorithm.

1.3 Overview

In the rest of this paper we present several algorithms which combined together form a powerful and flexible tool for partitioning unstructured meshes. In particular we describe a parallel meta-heuristic for optimising partitions, §4, and a method for coarsening the communication graph arising from the mesh in order to substantially reduce the workload, §5.

The software tool written at Greenwich to implement these ideas is known as JOSTLE. This is not an acronym; rather it reflects the way the subdomains jostle one another to reach a steady-state. It is modular in nature and consists of three parts; the initial partitioning module, the partition optimisation module and the graph coarsening module. These modules are designed for interchangeable use. Currently the code runs sequentially and in this case one might coarsen the graph, partition the reduced graph, optimise the partition, interpolate onto the full graph and reoptimise. With parallel code one might partition the full graph sequentially, distribute the graph, coarsen each subdomain, optimise the partition of the reduced graph, interpolate and reoptimise.

This multi-stage approach is similar to the work of Vanderstraeten, [19], although the techniques vary in that we employ deterministic heuristics to optimise the partition. The philosophy behind the optimisation algorithm is close to that of Löhner *et al.*, [13], although in addition we employ a heuristic which attempts to improve the 'shape' of the subdomains.

The initial partitioning code uses a sequential algorithm and is only employed if the input data is stored or generated on a single processor. Any cheap algorithm could be used, including one of those based solely on the geometrical mesh data (e.g. Inertial Recursive Bisection [5]), but here we use Farhat's Greedy algorithm [4]. For certain codes (such as those employing explicit solvers) or certain machines, this may be all that is required. If not, the optimisation techniques can be employed to minimise the number and volume of interprocessor communications whilst retaining the load-balance.

The optimisation code consists of three complementary techniques grouped together to form a meta-heuristic. Collectively they take as input a graph and partition $(G, \mathcal{P}_1)$ and iterate to try and output a better partition $(\mathcal{P}_2)$. The graph reduction code takes as input a graph (G) and outputs another reduced size graph (g). Ideally the reduced graph should be considerably smaller than the original, but still capture the essential geometric or topological features.

Due to the modular nature of the code and the fact that the optimisation and graph coarsening use parallel algorithms, all the issues raised by adaptive refinement are addressed by our mesh-partitioning techniques. They do not rely on the load being balanced, they can reuse an existing partition and the methods are designed to work efficiently in parallel.

2 NOTATION AND DEFINITIONS

We use a graph to represent the data dependencies in the mesh arising from the discretisation of the domain. Thus, let $G = G(N, E)$ be an undirected graph of N nodes & E edges and P be a set of processors. In a slight abuse of notation we use N, E & P to represent both the *sets* of nodes, edges & processors and the *number* of nodes, edges & processors, with the meaning clear from the context. We assume that both nodes and edges are weighted (with positive integer values) and that $|n|$ denotes the weight of a node n; $|(n, m)|$ the weight of an edge; $|S| := \sum_{n \in S} |n|$ the weight of a subset $S \subset N$; and $|T| := \sum_{(n,m) \in T} |(n, m)|$ the weight of a subset $T \subset E$. We also assume that the graph is connected (i.e. for all $n, m \in N$ there exists a path of edges from n to m). We define $\mathcal{P} : N \to P$ to be a partition of N and denote the resulting subdomains by S_p, for $p \in P$. The optimal subdomain weight is given by $W := \lceil |N|/P \rceil$. We denote the set of cut (or inter-subdomain) edges by E_c and the border of each subdomain, B_p, is defined as the set of nodes in S_p which have an edge in E_c.

We shall use the notation $\leftrightarrow$ to mean 'is adjacent to'. For example, for $n, m \in N$, $n \leftrightarrow m$ if $\exists (n, m) \in E$. Also if $n \in N$ and $S, T \subset N$, $n \leftrightarrow S$ if $\exists (n, m) \in E$

Notation	Name	Definition
$\lvert.\rvert$	weight function	
$\lceil.\rceil$	ceiling function	$\lceil x \rceil :=$ smallest integer $\geq x$
$\leftrightarrow$	adjacency operator	$n \leftrightarrow m \Leftrightarrow \exists(n,m) \in E$
$\Gamma(n)$	neighbourhood function	$\{m \in N : m \leftrightarrow n\}$
$\mathcal{P} : N \rightarrow P$	partition function	
E_c	cut edges	$\{(n,m) \in E : \mathcal{P}(n) \neq \mathcal{P}(m)\}$
S_p	subdomain p	$\{n \in N : \mathcal{P}(n) = p\}$
B_p	border of subdomain p	$\{n \in S_p : \exists(n,m) \in E_c\}$
H_p	halo of subdomain p	$\{m \in N - S_p : m \leftrightarrow S_p\}$
W	optimal subdomain weight	$\lceil \lvert N \rvert / P \rceil$

Table 1. Definitions

with $m \in S$ and similarly $S \leftrightarrow T$ if $\exists(n,m) \in E$ with $n \in S$ & $m \in T$. We then denote the neighbourhood of a node n (the nodes adjacent to n) by $\Gamma(n)$ and the halo of subdomain S_p by H_p. Table 1 gives a summary of these definitions.

2.1 The Graph Partitioning Problem

The definition of the problem is to find a partition which evenly balances the load or node weight on each processor whilst minimising the communications cost. More precisely we seek $\mathcal{P}$ such that $S_p \leq W$ for $p \in P$ and such that $|E_c|$ is approximately minimised. It is a matter of contention whether $|E_c|$ is the most important metric for minimisation but see §6.1 for a further discussion on this topic.

Note that in general, for graphs arising from finite element or finite volume discretisations, the degree of each node is low. As a result the surface of solutions of the mesh-partitioning problem tends to have many local minima and can have several optimal solutions which are far apart.

2.2 Localisation and the processor graph

An important aim for any parallel algorithm is to keep communication as localised as possible to avoid contention and expensive global operations; this issue becomes increasingly important as machine sizes grow. Throughout the

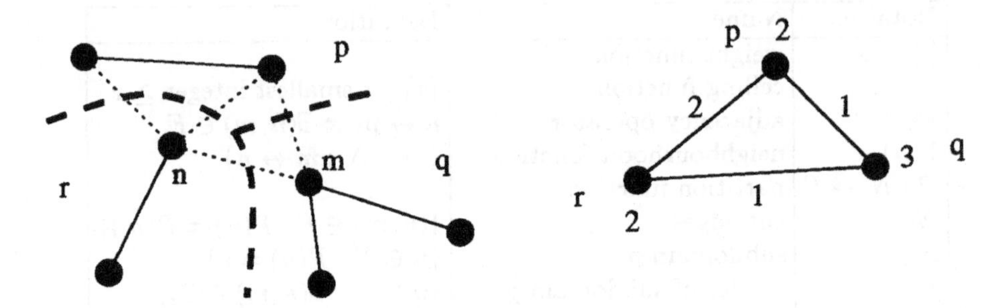

Fig. 1. An example graph and the resulting processor graph (node & edge weights as shown)

optimisation algorithm we localise the communication with respect to the partition $\mathcal{P}$ by only requiring subdomains to communicate with their neighbouring subdomains. Thus a processor p will only need to communicate to q if there is an edge $(n, m) \in E_c$ with $\mathcal{P}(n) = p$ and $\mathcal{P}(m) = q$.

In this way the partition $\mathcal{P}$ induces a processor graph on G which we shall refer to as $\mathcal{P}(G) = \mathcal{P}(G)(P, C)$, an undirected graph of P processors (nodes) and C connections (edges). There is an edge $(p, q) \in C$ if $S_p \leftrightarrow S_q$. In addition, the weight on each processing node $p \in P$ is given by $|p| = |S_p|$ and the weight of a connection is given by $|(p, q)| = \sum_{\{(n,m) \in E : n \in S_p, m \in S_q\}} |(n, m)|$. Thus $|P| = \sum_{p \in P} |S_p| = |N|$ and $|C| = |E_c|$. Figure 1 gives an example of a partitioned graph (with unit node and edge weights) and the resulting processor graph (with node and edge weights as shown).

Note that this processor graph may bear no direct relationship to the physical processor topology of the parallel machine, although the algorithm should be more efficient if this is the case. For the purposes of this paper, however, we shall assume that the processors are uniformly connected – that is, point-to-point communication carries a uniform latency no matter which processors are communicating. This is a realistic assumption for certain machines with small or moderate numbers of processors but for other situations it may be possible to optimise the partition for the machine topology, [22].

3 THE INITIAL PARTITION

The aim of the initial partitioning code is to divide up the graph as rapidly as possible in order that it can be distributed and the partition optimised in parallel. Whilst a loop partition (e.g. the first N/P nodes to processor 1, etc.) is probably the fastest method for doing this, intuition suggests and experience shows that it is worth a little extra effort in order to reduce the amount of optimisation that must be carried out.

The code therefore utilises a version of the Greedy algorithm [4]. This is clearly seen to be the fastest ***graph-based*** method as it only visits each graph edge once. The variant employed here differs from that proposed by Farhat only in that it works solely with a graph rather than the nodes and elements of a finite element mesh.

For experimental purposes, loop and random partitioning codes have also been implemented. Although the loop partition can be reasonably successful if the data has some special structure, in general the resulting partitions needed far more optimisation than those for the Greedy algorithm to achieve the same quality.

4 THE PARALLEL OPTIMISATION METHOD

Once the graph is partitioned, optimisation can take place to improve the quality of the partition. The method described here uses a combination of three heuristics to both achieve load-balance and to minimise the interprocessor communication. Initially the subdomain heuristic attempts to 'improve' the 'shape' of the subdomains. However, this heuristic cannot guarantee load-balance and so a second heuristic is applied to share the workload equally between all processors. Finally a localised version of the Kernighan-Lin algorithm, [10], is applied to minimise the communication cost.

4.1 Migration and the gain and preference functions

Throughout each of the three phases, it is assumed that the final partition will not deviate too far from the initial one. Thus, in general, only border nodes are allowed to migrate to neighbouring subdomains.

For these purposes, at each iteration we calculate the gain and preference values for every border node. Loosely the gain $g(n,q)$ of a node $n \in B_p$ can be calculated for each adjacent subdomain, S_q, and expresses some 'estimate' of how much the partition would be 'improved' were n to migrate to S_q. The preference $f(n)$ is then just the ***adjacent*** subdomain which maximises the gain of a node – i.e. $f(n) = q$ where q attains $\max_{r \in P} g(n,r)$. Most nodes in B_p will only be adjacent to one subdomain, q say, and in this case the preference is simply q. Where 3 or more subdomains meet, the preferences need to be explicitly calculated and if more than one subdomain attains the maximum gain, a pseudo-random choice is made, §4.4.

Note that the definition of the gain differs for each phase of the algorithm. For example, when applying load-balancing and local refinement, §4.3 & §4.4, the gain $g(n,q)$ just expresses the reduction in the cut-edge weight. Thus in figure 1 (assuming unit edge weights) there would be one less edge cut if node n migrated to subdomain p and so $g(n,p) = 1$. Similarly $g(n,q) = 0$ and hence $f(n) = p$. For node m, $g(m,p) = -1$ and $g(m,r) = -1$ and so a pseudo-random choice must be made for $f(m)$. Note that here we do not allow the preference of a node to be its current processor since we may have to migrate nodes even if the net gain is negative.

The only time that internal nodes migrate is in the first phase, §4.2, when the heuristic can detect small subsets of the subdomain which are disconnected from the main part. In this case the gain of internal and border nodes in the disconnected subset is just some arbitrarily large number; the preference is the nearest adjacent subdomain (in a graph sense).

4.2 The subdomain heuristic

Motivation

One of the problems of partition optimisation is that of local minima traps. Algorithms such as the Kernighan-Lin heuristic, [10], which can be very successful at minimising the number of cut edges at a local level, do not fully address the global problem.

A partial solution to this problem has been the introduction of non deterministic heuristics such as simulated annealing, e.g. [11, 24]. They address local minima traps by allowing transfers of graph nodes between subdomains *even if* the communication cost is increased, the transfer being accepted according to some probability based on a cooling schedule. These algorithms can be very effective, especially if the initial partition is good but they can suffer from high run times.

The new heuristic introduced here has been designed according to two constraints. We want the algorithm to

(a) address the optimisation at a subdomain level to try and attain a *global* minimum;

(b) carry out the optimisation *locally* to give an efficient parallel algorithm.

At first these two constraints may appear to frustrate each other. However, a useful analogy with a simple schoolroom experiment can be drawn. The experiment consists of bubbling a gas through a pipette into a detergent solution and the result is that the bubbles make a regular (hexagonal) pattern on the surface of the liquid. They achieve their regularity without any global 'knowledge' of the shape of the container or even of any bubbles not in immediate contact; it is simply achieved by each bubble minimising its own surface tension.

Translating this analogy to a partitioned graph we see that each subdomain must try to minimise its own surface area. In the physical 2D or 3D world the object with the smallest surface to volume ratio is the circle or sphere. Thus the idea behind the domain-level heuristic is to determine the centre of each subdomain (in some graph sense) and to then measure the distance from the centre to the edges and attempt to minimise this by migrating nodes which are furthest from the centre.

Implementation

Determining the 'centre' of the subdomain is relatively easy and can be achieved by moving in level sets inwards from the subdomain border until all the nodes in the subdomain have been visited. More precisely, for each subdomain S_p, define a series of (disjoint) level sets, $L_0 = B_p$; $L_1 = \{n \in S_p - L_0 : n \leftrightarrow L_0\}$; $\ldots L_i = \{n \in S_p - \bigcup_{j=0}^{i-1} L_j : n \leftrightarrow L_{i-1}\}$; etc. ...The final non-empty set, L_c, of this series defines the centre of the subdomain. Note that if the graph is connected (assumed) the level sets will completely cover the subdomain, i.e. $S_p = \bigcup_0^c L_i$, but that the centre may not be a connected set of nodes.

The reverse of this process can then be used to determine the (radial) distance of nodes from the centre. Thus $M_0 = L_c$; $M_1 = \{n \in S_p - M_0 : n \leftrightarrow M_0\}$;

$\ldots M_i = \{n \in S_p - \bigcup_{j=0}^{i-1} M_j : n \leftrightarrow M_{i-1}\}$; etc. ...It is easy to see (for example, figure 2) that the mapping to level sets is not necessarily the same as that given by the inwards search for the centre. Also, if S_p is not connected, then the level sets may not cover the subdomain, i.e. $\bigcup M_i \neq S_p$.

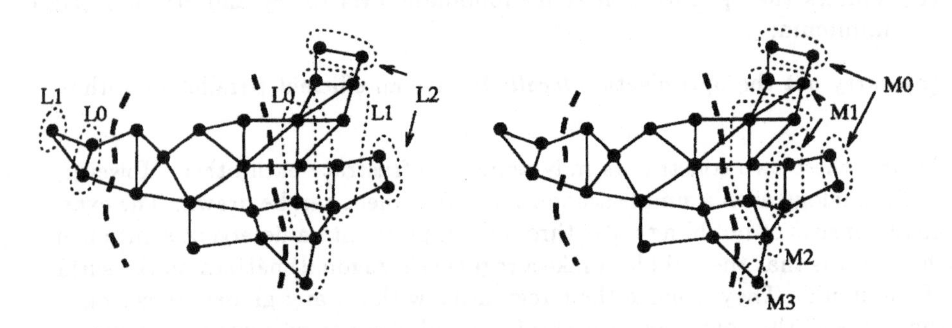

Fig. 2. An example of the level sets moving in from the boundary, L0, L1 & L2, and moving out from the centre, M0, M1, M2 & M3

Having derived these sets each node n can be marked by its radial distance, $c - i$ for $n \in M_i$. Nodes in S_p which are not connected to L_c are not marked. This is useful for migrating small disconnected parts of the subdomain to more appropriate processors, although it should be noted that centre is essentially determined by the diameter of each connected subset and not its size. Thus it is possible that a subset of the subdomain may contain more than half the weight of the subdomain but still be unmarked. Figure 2 shows a simple example of these level sets for a disconnected subdomain.

Whilst it is fairly easy to determined the radial distance for each node, the gain function must be derived empirically. Although we use separate load-balancing and local refinement heuristics, §4.3 & §4.4, it is inefficient to achieve some sort of steady-state with the subdomain heuristic only to destroy it later. For this reason we encode imbalance and communication minimisation information in the gain at this stage. We first define an imbalance adjustment for every domain by

$$\delta_p := 20 \frac{(W - |S_p|)}{|B_p|} \times \frac{|N|}{|E|}$$

and add it to the radial distance to give an adjusted radial distance, $\phi(n) = c - i + \delta_p$ for $n \in M_i \subset S_p$. This imbalance adjustment has been arrived at by experimentation and loosely expresses the amount of weight a subdomain can hope to gain or lose according to the size of its border and the average connectivity of the graph. We then define the gain on a node $n \in B_p$ by

$$g(n,p) = \phi(n) + \sum_{m \in S_p \cap \Gamma(n)} \phi(m) + \theta|(n,m)|$$

$$g(n,q) = \sum_{m \in S_q \cap \Gamma(n)} \phi(m) + \theta|(n,m)| \qquad p \neq q.$$

Again the preference $f(n)$ is just the ***adjacent*** subdomain which maximises the gain of a node and for the subdomain heuristic we migrate all nodes with positive gain.

Here θ expresses the relative influence between the radial distance, which is used to 'improve' domain shape, and the edge weight, which is used to minimise $|E_c|$. We have found a very successful policy is to gradually increment θ throughout the run. This has a dual effect, strongly reminiscent of the cooling schedule in simulated annealing, [11], which both slows down the migration towards a steady-state and which transfers the emphasis from global to local optima. For the experiments described in §6.2 we started each run with $\theta = 1$ and incremented it by 1 after every 4 iterations (except after using graph coarsening and having interpolated to the full graph – configuration 'jostle/reduction' – when we incremented θ by 1 every iteration).

4.3 Load-balancing

The load-balancing problem, i.e. how to distribute N tasks over a network of P processors so that none have more than $\lceil N/P \rceil$, is a very important area for research in its own right with a vast range of applications. The topic is comprehensively introduced in [16] and some common strategies described. In particular, much work has been carried out on parallel or distributed algorithms recently, e.g. [3], and an important feature of the problem is whether the information on the current load-state is gathered globally or locally – the former enables an algorithm to converge faster but may carry a high communication overhead.

Here we use a localised iterative algorithm for distributed load-balancing devised by Song, [17]. It builds on a proposition proved by Bertsekas and Tsitsiklis, [2, pp. 521], which states that a transfer policy, devised under certain easily satisfied conditions, will converge as $t \to \infty$. In common with the rest of our optimisation, the localised nature means that it only requires information from, and migrates nodes to, its neighbours in the processor graph. The description of the algorithm assumes that the workload consists of independent tasks of equal size and is proven to iterate to convergence with a maximum load-difference of $d/2$ where d is the diameter of the processor graph.

Translating this to the mesh partitioning problem, we use Song's algorithm to determine the *number* of nodes to migrate from a given processor to each of its neighbours and then use the gain and preference functions, §4.1, to determine *which* nodes to move. Although the algorithm does not guarantee perfect load-balance consider the example Tri60K in §6.2 of 60,005 nodes on 64 processors. The maximum diameter possible (which arises if the processor graph is a 1D linear array or chain) is 63. The optimal load is 938 and so the worst case deviation from this figure is $8/938 = 0.85\%$. Of course the finer the granularity the more exaggerated this effect can be, but experience suggests the imbalance is usually far from the worst possible and usually insignificant.

Conceptually this algorithm is ideally suited to our application; all information about the load-state is gathered locally and all transfers are between neighbouring processors. Unfortunately, however, it can suffer from very slow convergence, typically $O(P)$ and in particular the rate of convergence is directly proportional to the load-imbalance. For example, in one randomly generated processor graph of 233 nodes the algorithm took 100 iterations to go from 106.1% to 4.0% load-imbalance and then a further 200 iterations to converge while reducing the load-imbalance to 2.1%. We consider this to be too high an overhead, particularly as most of the final iterations only involved transfers of one or two nodes at a time.

Fortunately this effect can be easily countered by running the algorithm to completion based on the current load-state but without actually carrying out the transfers. The accumulated results can then be aggregated and used as a migration schedule. The only qualification is that if the processor graph loses a connection scheduled for migrating nodes, the algorithm must be rerun and the schedule revised.

Currently we use this technique for sequential partitioning and we also propose to use it in the parallel implementation, although we are investigating other alternatives such as diffusion type schemes, e.g. [3]. The problem of calculating the migration-schedule in parallel can be solved by broadcasting the processor graph around the network and replicating the calculation on every processor. This is not ideal because of the global communication involved, but each processor need only broadcast its current load plus a sparse vector listing its neighbours (or half of its neighbours since the connections are symmetric).

The gain and preference functions for load-balancing are as described in §4.1. The border nodes are sorted by gain and the load specified by the schedule migrated to preferred subdomains. Of course a subdomain may not have enough nodes along its boundary to satisfy the amount of migration chosen by the load-balancing algorithm, but this will not break the conditions of the proposition provided at least one node is transferred to the most lightly loaded neighbour at each iteration. Also the fact that, due to node migration, the processor graph may change from iteration to iteration does not seem to significantly affect the convergence.

4.4 Local partition refinement

Having achieved optimal (or near optimal) load-balance it may still be possible to move nodes around the processor network to further minimise the number of cut edges whilst retaining the load-balance. An algorithm which comes immediately to mind for this purpose is the Kernighan-Lin (KL) heuristic, which maintains load-balance by employing pairwise exchanges of nodes. Unfortunately it has $O(n^2 \log n)$ complexity but a linear-time variant which delivers similar results has been proposed by Fiduccia & Mattheyses (FM), [6]. The FM algorithm achieves this reduction partially by calculating swaps one node at a time rather than in pairs. Both algorithms have the possibility of escaping local minima traps by continuing to compute swaps even if the swap gain is negative in the hope that some later total gain may be positive.

The algorithm we describe here is largely inspired by the FM algorithm but with several simplifications to enable an efficient parallel algorithm. To motivate these simplifications we make the following observations:

- In our applications it is very unlikely that an internal node will have a higher gain than a border node – thus we only consider transferring border nodes
- It is also unlikely that an overall gain will accrue by transferring a node to a subdomain which it is not adjacent to – thus nodes only transfer to neighbouring subdomains.

One immediate advantage of these modifications is that by only considering border nodes the problem has sublinear complexity. On the debit side, it effectively destroys the possibility inherent in the KL/FM algorithms of climbing out of local minima.

For either algorithm, the swapping of nodes between two sub-domains is an inherently non-parallel operation and hence there are some difficulties in arriving at efficient parallel versions, [15]. In a naive parallel implementation one could imagine a pair of processors alternately sending each other one node at a time – effective but horrendously inefficient. A second difficulty arises from the fact that the algorithms treat bisections and so processors must exchange in pairs. Previously authors, [8, 18], have implemented this by employing pairwise exchanges (e.g. for 4 processors, 1 pairs with 2 & 3 pairs with 4; then 1 with 3 & 2 with 4 and finally 1 with 4 & 2 with 3).

We overcome both of these problems by replicating the transfer decision process on both sides of each sub-domain interface. Thus a processor p examines both its border and halo nodes to decide ***not only*** which of its nodes should migrate to neighbour q, ***but also*** which nodes should transfer from q to p. It does not use the information about transfers of nodes that it does not own, but the duplication of this transfer scheme allows any pair of processors to arrive ***independently*** at the same optimisation. Although this may be computationally inefficient (because the workload is doubled) it is almost certainly much more efficient than a 'hand-shaking' communication for each pair of nodes. It also avoids the need for pairwise exchanges between neighbouring processors. The algorithm is described more fully in [20]

5 CLUSTERING: REDUCING THE PROBLEM SIZE

For coarse granularity partitions it is inefficient to apply the optimisation techniques to every graph node as most will be internal to the subdomains. A simple technique to speed up the load-balancing process, therefore, is to group nodes together to form ***clusters***, use the clusters to define a new graph, recursively iterate this procedure until the graph size falls below some threshold and then apply the partitioning algorithm to this reduced size graph. This is quite a common technique and has been used by several authors in various ways – for example, in a multilevel way analogous to multigrid techniques, [1], and in an adaptive way analogous to dynamic refinement techniques, [23].

The technique used here for graph reduction is a variant of the Greedy algorithm [4] (as mentioned in §3, this is about the fastest graph-based algorithm), although various algorithms have been successfully used, [1, 8, 9]. It is used recursively to cluster nodes into small groups, each of which defines a node of the reduced graph. It is, of course, important that the groups of nodes are connected (in order to retain the features of the original graph) and so we relax the condition that each cluster should contain the same node weight in favour of guaranteeing that the nodes of each cluster form a connected set. The node and edge weights for the reduced graph derive simply from the sum of node weights over the cluster and sum of edge weights from the cluster to other clusters.

Two parameters arise in this method – what should be the maximum cluster weight (or size) at each iteration and when should the reduction process be terminated. In the experiments carried out for this paper, we increased the cluster weight limit by a factor of l each iteration and stopped the reduction when the number of nodes in the reduced graph fell below a certain threshold m. For the purposes of the code module we allow the user to set these values but typically (and in the experiments) we set $l = 4$ and $m = 100 \times P$. Relating the threshold to P in this way means that the reduced graph does not become meaninglessly small for large P.

Note that this algorithm is highly parallel for a distributed graph. For each iteration, every processor would cluster only the set of nodes that it owns, exchange information on the mapping of border nodes to clusters with its neighbours and form its own part of the reduced graph. A global sum of the total number of nodes allows the termination of the process to be synchronised. Alternatively each processor could terminate locally once its own number of nodes has fallen below a certain threshold. However, if the mesh is badly load-balanced, this may lead to wide variation in the amount of reduction on each processor. With a final step to exchange border node weights once the iterations have terminated, the reduced graph can be created entirely in parallel.

6 EXPERIMENTAL RESULTS

The following experiments were mostly carried out on a Silicon Graphics Indigo 2 with a 150 MHz CPU and 64 Mbytes of memory. The final set of results for the largest mesh came from a Sun SPARC station with a 50 MHz CPU and 224

Mbytes of memory; typically this processor is about 3 times slower than the Silicon Graphics machine but the memory requirements forced its usage. The code has also been compiled and run on an IBM RS6000 with similar timing ratios. Further results can be found in [20, 21].

6.1 Metrics

We use four metrics to measure the performance of the algorithms. Three of these are concerned with the partition quality – the total weight of cut edges, $|E_c|$, the maximum percentage load-imbalance, δ, and the average processor degree, D_a. The fourth is $t(s)$ the execution time in seconds of each algorithm.

Unfortunately, there are no *ideal* metrics for assessing partition quality as the parallel efficiency of the problem from which the mesh arises will depend on many things – typically the machine (size, architecture, latency, bandwidth and flop rate), the solution algorithm (explicit, implicit with direct linear solution, implicit with iterative linear solution) and the problem itself (size, no. of iterations) all play a part. However, $|E_c|$ gives a rough indication of the volume of communication traffic and the average degree of the processor graph, D_a, gives an indication of the number of messages each processor must send. From a load-balancing point of view the worst percentage imbalance is derived from $\delta := 100 \times (\max_{p \in P} |S_p| - W)/W\%$, since the speed of the calculation is determined by the processor with the largest workload, not by underloaded processors.

6.2 The results

We include results for runs on 5 different graphs. The first two, the Hammond & Barth5 meshes are available by anonymous ftp (from `riacs.edu/pub/grids`) and have been previously been used for benchmarking partitioning algorithms, [1, 8]. The Hammond mesh is a relatively small ($N = 4,720$, $E = 13,722$) two-dimensional finite-element mesh around a 3-element airfoil and Barth5 is a similar but larger ($N = 15,606$, $E = 45,878$) mesh around a 4-element airfoil. The other meshes are homegrown, [12]: Tri60K is a two-dimensional finite-volume mesh ($N = 60,005$, $E = 89,440$) arising from a casting simulation, [14]. Tet100K and Tet1M are three-dimensional finite-volume meshes ($N = 103,081$, $E = 200,976$; $N = 1,119,663$, $E = 2,212,012$ respectively) in the shape of a Y standing on a baseplate.

We have compared the method with two of the most popular partitioning algorithms, Greedy and Multilevel Recursive Spectral Bisection (MRSB). The Greedy algorithm, [4], is actually performed as part of the jostle code and is fast but not particularly good at minimising cut edges. MRSB, on the other hand, is a highly sophisticated method, good at minimising $|E_c|$ but suffering from relatively high runtimes, [1]. The MRSB code was made available to us by one of its authors, Horst Simon, and run unchanged with contraction thresholds of 100 and 500.

For each mesh the jostle code has been run in three different configurations, each easily specified to the code in a defaults file. There are two sets of results using the graph coarsening, §5; 'jostle/fast reduction' uses the full algorithm as described on the reduced graph and then only uses the load-balancing and local-refinement on the full graph; 'jostle/reduction' additionally uses the subdomain heuristic, §4.2, although with a cut off after 5 iterations. The other configuration, 'jostle', does not use any graph coarsening.

The results for both the Hammond and Barth5 meshes, tables 2 & 3, are somewhat surprising in that the value of $|E_c|$ is actually lower for the runs with graph coarsening. It is the authors' experience that the results for the other meshes are more typical, with the solution quality reflecting the amount of work that is put in to achieve it. Throughout it can be seen that jostle achieves approximately the same partition quality as MRSB though much more rapidly. The imbalances are generally small (smaller than one would expect for simulated annealing, say) and the higher figures for the Hammond and Barth5 meshes reflect the finer granularity. For example the 2.70% imbalance in the Hammond mesh arises from the most heavily loaded processor having 152 nodes as opposed to an optimum of 148.

In the final set of results, table 6, we offer an example of how fast the code can be on a workstation even for a huge mesh.

7 CONCLUSION

The work described above has outlined a new method for partitioning graphs with a specific focus on its application to the mapping of unstructured meshes onto parallel computers. In this context the graph-partitioning task can be very efficiently addressed through a two-stage procedure – one to yield a legal initial partition and the second to improve its quality with respect to interprocessor communication and load-balance. The method is further enhanced through the use of a clustering technique. The resulting software tool, JOSTLE, has been designed for implementation in parallel for both static and dynamically refined meshes. For the experiments reported in this paper on static meshes of up to one million nodes, the JOSTLE procedures are an order of magnitude faster than existing techniques, such as Multilevel Recursive Spectral Bisection, with equivalent quality.

Future work on this technique will include further study into distributed load-balancing algorithms, a parallel implementation and work on the mapping of meshes to machine topologies.

Acknowledgements

We would like to thank Horst Simon for the copy of his Multilevel Recursive Spectral Bisection code. We would also like to acknowledge Peter Lawrence and Kevin McManus for supplying the Tri60K, Tet100K and Tet1M meshes.

REFERENCES

[1] S. T. Barnard and H. D. Simon. A Fast Multilevel Implementation of Recursive Spectral Bisection for Partitioning Unstructured Problems. *Concurrency: Practice & Experience*, 6(2):101–117, 1994.

[2] D. P. Bertsekas and J. N. Tsitsiklis. *Parallel and Distributed Computation: Numerical Methods.* Prentice Hall, Englewood Cliffs, NJ, 1989.

[3] G. Cybenko. Dynamic load balancing for distributed memory multiprocessors. *J. Par. Dist. Comput.*, 7:279–301, 1989.

[4] C. Farhat. A Simple and Efficient Automatic FEM Domain Decomposer. *Comp. & Struct.*, 28:579–602, 1988.

[5] C. Farhat and H. D. Simon. TOP/DOMDEC – a Software Tool for Mesh Partitioning and Parallel Processing. Tech. Rep. RNR-93-011, NASA Ames, Moffat Field, CA., 1993.

[6] C. M. Fiduccia and R. M. Mattheyses. A Linear Time Heuristic for Improving Network Partitions. In *Proc. 19th IEEE Design Automation Conf.*, pages 175–181, IEEE, 1982.

[7] B. Hendrickson and R. Leland. Multidimensional Spectral Load Balancing. Tech. Rep. SAND 93-0074, Sandia National Labs, Albuquerque, NM., 1992.

[8] B. Hendrickson and R. Leland. A Multilevel Algorithm for Partitioning Graphs. Tech. Rep. SAND 93-1301, Sandia National Labs, Albuquerque, NM., 1993.

[9] B. W. Jones. *Mapping Unstructured Mesh Codes onto Local Memory Parallel Architectures.* PhD thesis, School of Maths., University of Greenwich, London SE18 6PF, UK, 1994.

[10] B. W. Kernighan and S. Lin. An Efficient Heuristic for Partitioning Graphs. *Bell Systems Tech. J.*, 49:291–308, 1970.

[11] S. Kirkpatrick, C. D. Gelatt, and M. P. Vecchi. Optimization by simulated annealing. *Science*, 220:671–680, 1983.

[12] P. Lawrence. *Mesh Generation by Domain Bisection.* PhD thesis, School of Maths., University of Greenwich, London SE18 6PF, UK, 1994.

[13] R. Lohner, R. Ramamurti, and D. Martin. A Parallelizable Load Balancing Algorithm. AIAA-93-0061, 1993.

[14] K. McManus, M. Cross, and S. Johnson. Integrated Flow and Stress using an Unstructured Mesh on Distributed Memory Parallel Systems. (submitted for the Proceedings, Parallel Computational Fluid Dynamics '94), University of Greenwich, London, SE18 6PF, UK, 1994.

[15] J. Savage and M. Wloka. Parallelism in Graph Partitioning. *J. Par. Dist. Comput.*, 13:257–272, 1991.

[16] N. G. Shivaratri, P. Krueger, and M. Singhal. Load distributing for locally distributed systems. *IEEE Comput.*, 25(12):33–44, 1992.

[17] J. Song. A partially asynchronous and iterative algorithm for distributed load balancing. *Parallel Comput.*, 20:853–868, 1994.

[18] P. R. Suaris and G. Kedem. An Algorithm for Quadrisection and Its Application to Standard Cell Placement. *IEEE Trans. Circuits and Systems*, 35(3):294–303, 1988.

[19] D. Vanderstraeten and R. Keunings. Optimized Partitioning of Unstructured Finite Element Meshes. CESAME Tech. Rep. 93.32 (accepted by *Int. J. Num. Meth. Engng.*), 1993.

[20] C. Walshaw and M. Cross. A Parallelisable Algorithm for Optimising Unstructured Mesh Partitions. (submitted for publication), 1994.

[21] C. Walshaw, M. Cross, S. Johnson, and M. Everett. JOSTLE: Partitioning of Unstructured Meshes for Massively Parallel Machines. (submitted for the Proceedings, Parallel CFD'94), 1994.

[22] C. Walshaw, M. Cross, S. Johnson, and M. Everett. Mapping Unstructured Meshes to Parallel Machine Topologies. (in preparation), 1994.

[23] C. H. Walshaw and M. Berzins. Dynamic Load-Balancing For PDE Solvers On Adaptive Unstructured Meshes. (accepted by *Concurrency: Practice & Experience*), 1993.

[24] R. D. Williams. Performance of dynamic load balancing algorithms for unstructured mesh calculations. *Concurrency: Practice & Experience*, 3:457–481, 1991.

Method	$\lvert E_c \rvert$	$t(s)$	$\delta\%$	D_a
greedy	1456	0.05	0.00	5.25
jostle/fast rdtn	1066	0.48	2.70	4.25
jostle/reduction	1074	0.69	1.35	4.31
jostle	1118	1.05	0.68	4.44
MRSB/100	1155	2.69	0.00	4.25
MRSB/500	1154	4.05	0.00	4.00

Table 2. Results for the Hammond mesh: $P = 32$, $N = 4,720$, $E = 13,722$

Method	$\lvert E_c \rvert$	$t(s)$	$\delta\%$	D_a
greedy	4046	0.29	0.00	5.34
jostle/fast rdtn	2940	2.04	1.23	4.59
jostle/reduction	2943	2.97	2.46	4.59
jostle	2968	4.71	1.64	4.66
MRSB/100	3122	11.59	0.00	4.62
MRSB/500	3097	17.90	0.00	4.47

Table 3. Results for the Barth5 mesh: $P = 64$, $N = 15,606$, $E = 45,878$

Method	$\lvert E_c \rvert$	$t(s)$	$\delta\%$	D_a
greedy	3834	1.00	0.00	5.88
jostle/fast rdtn	2731	4.81	0.64	4.72
jostle/reduction	2500	7.41	0.32	4.72
jostle	2541	14.79	0.64	4.78
MRSB/100	2435	35.50	0.00	4.25
MRSB/500	2495	41.25	0.00	4.31

Table 4. Results for Tri60K mesh: $P = 64$, $N = 60,005$, $E = 89,440$

Method	$\lvert E_c\rvert$	$t(s)$	$\delta\%$	D_a
greedy	26274	2.06	0.00	13.64
jostle/fast rdtn	17047	16.95	0.62	9.77
jostle/reduction	16275	24.80	0.74	9.30
jostle	15855	56.30	0.37	9.27
MRSB/100	17751	147.81	0.00	8.86
MRSB/500	17027	146.88	0.00	8.70

Table 5. Results for Tet100K mesh: $P = 128$, $N = 103,081$, $E = 200,976$

P	$\lvert E_c\rvert$	$t(s)$	$\delta\%$	D_a
10	53632	163.61	0.00	4.60
20	60660	148.79	0.29	6.50
50	71603	150.69	0.52	7.84
100	119632	229.41	0.30	12.12
200	123583	210.01	0.14	11.60

Table 6. Results for Tet1M mesh: method is 'jostle/fast rdtn', $N = 1,119,663$, $E = 2,212,012$

3

PLUMP: PARALLEL LIBRARY FOR UNSTRUCTURED MESH PROBLEMS

Ivan Beg*, Wu Ling, Andreas Müller, Piotr Przybyszewski***, Roland Rühl and William Sawyer**

Centro Svizzero di Calcolo Scientifico (CSCS-ETHZ), La Galleria, Via Cantonale, CH-6928 Manno, Switzerland

** University of Toronto, Ontario, Canada*
*** Stanford University, California, USA*
**** Technical University of Gdansk, Poland*

ABSTRACT

The Joint CSCS-ETH/NEC Collaboration in Parallel Processing is creating a tool environment for porting large and complex codes to massively parallel computers. Along with a parallel debugger and a performance monitor, this environment provides a Parallelization Support Tool (PST) to supplement the data-parallel programming language High Performance Fortran (HPF). Whereas HPF has only facilities for regular data decompositions, PST supports user-defined mappings of the global name space to individual processors, allowing for the parallelization of unstructured problems.

Since the additional directives of PST alone do not remove all of the complexity of programming parallel unstructured mesh applications, a Parallel Library for Unstructured Mesh Problems (PLUMP) is currently being developed at CSCS to support the local refinement and dynamic repartitioning of meshes distributed over a processor array. The constituent routines simplify the manipulation of the underlying dynamic data structures. The use of PLUMP in conjunction with PST can facilitate the design and implementation of a class of specific, but industrially important, applications.

In this paper we first specify the functionality of PLUMP, and then discuss its use in parallelizing two unstructured problems with dynamic aspects, namely a *tight binding molecular dynamics* code and a *finite-element package*. Finally we indicate some possible future additions to the library and discuss other future applications.

A. Ferreira and J. D. P. Rolim (eds.), Parallel Algorithms for Irregular Problems: State of the Art, 47–71.

1 INTRODUCTION

As part of *the Joint CSCS-ETH/NEC Collaboration in Parallel Processing* [1], we are currently developing a Parallel Library for Unstructured Mesh Problems (PLUMP). PLUMP is implemented in High Performance Fortran (HPF) [2] augmented by language extensions supporting user-defined mappings and allowing for the parallelization of unstructured problems on distributed memory parallel processors (DMPPs). These extensions are built into a language processor (Parallelization Support Tool, PST) which is included in the *Annai* [3] tool environment currently being implemented in the collaboration at CSCS.

Much work has been done on support for unstructured mesh problems. The currently implemented libraries, e.g., PARTI [4] and OP [5], supply routines based on message-passing primitives which support layout and reference of distributed unstructured data. The main function of this library is to provide run-time support for a global name space for unstructured problems, using basic languages such as C or Fortran. In contrast to PLUMP these languages do not provide a higher level of abstraction for specific problem classes, and they do not fit into the framework of HPF.

Another approach has been to propose language extensions to support unstructured problems, e.g., HPF+ [6], value-based distributions in Fortran D [7], and user-defined mappings in Oxygen [8], and PST [9]. These extensions may be appropriate for future revisions of HPF, but their use still leaves much of the responsibility to the developer to layout, update and access the distributed global name space—a task which is not necessarily appreciated by application developers.

PLUMP is meant to support the important class of applications in which the basic data structure can be described as a weakly interconnected graph with relatively few interactions between nodes and with dynamic changes in time [1]. The dynamic nature of the graph might reflect the refinement of an underlying finite-element mesh, or the evolution of a time dependent n-body problem, where the neighborhood around any body is changing with time. There are numerous other industrial applications within this class.

[1] The graph representation of such a problem can be misleading in that common applications deal with *several* graphs, e.g., the graph of elements in a finite element mesh, and the graph of element vertices. These underlying graphs will be related in a very specific manner, and thus it is usually possible to describe the problem through one characteristic graph, possibly supplemented with additional information.

PLUMP manages a connected graph. In order to allow the model to describe as many problems as possible, we define a graph *element* as the description of the interactions of two or more *nodes*, or equivalently *vertices*. The nodes are the fundamental unit of the problem. In the simplest case, the graph element defines the relationship between two nodes and is therefore a so-called *edge*. More generally, an element defines the r nodes belonging to it (consider a triangular finite element describing the relationship between its three vertices), and therefore an element consists of a $r \times r$ matrix of interactions, hereafter denoted as A^e.

Let m be the number of coupled physical quantities associated with each node; therefore the dimensions of A^e will increase accordingly to $mr \times mr$.

> **Example:** Assume each graph element is a prism (or *block*) which has eight vertices (necessarily shared by other elements). Furthermore there are five physical quantities associated with each node. The resulting dense matrix A^e has the dimensions 40×40.

PLUMP permits the developer to parallelize the operations on the graph without actually manipulating distributed data. It supplies commonly used routines which access the data in an efficient manner.

In what follows below, we first summarize the PST features important for understanding the design concepts behind PLUMP. These concepts of PLUMP are then discussed in Sect. 3 together with the type of applications supported by the library. The detailed library specification in Sect. 4 is followed in Sect. 5 by descriptions of two applications—a finite-element package and a molecular dynamics code—and their implementation using PLUMP. In Sect. 6 possible extensions to PLUMP are discussed and future work is proposed.

2 HPF/PST SUPPORT FOR UNSTRUCTURED PROBLEMS

PST allows the user to compile standard HPF programs extended with PST directives in specific `EXTRINSIC` routines. The directives allow for the use of a set of new parallelization directives but also provide a different programming paradigm than that of HPF. HPF provides a single-threaded program image to the user, and the compiler enforces identical state of non-distributed variables on all processors. This model makes the underlying architecture look like either a sequential von Neumann machine, or, when the data distributions are added, like a SIMD machine with a single instruction stream. The PST paradigm is more akin to paradigms provided on shared memory MIMD machines: by default a PST routine allows a different state on all processors.

In the above mentioned extrinsic routines, data can be irregularly distributed using one of the following distribution specifications (see also [9]):

- Tables containing the mapping of global array indices to processor numbers.
- Integer-valued mapping functions.
- Block-general mapping arrays. PLUMP makes extensive use of block-general distributions which we therefore explain in more detail below.

PST adds run-time global name space support to the HPF system in which it is embedded. The HPF system we use—as most other HPF compilers—only supports a global name space statically through compile-time analysis. Since the run-time analysis causes significant overhead, the user can explicitly specify in the routine head whether communications are required (so-called `PUBLIC` routines), or whether the routine can run in parallel but without communication (so-called `LOCAL` routines). Once communication patterns are generated at run-time for the access of non-local array elements, these patterns can be reused, if they do not change from one routine invocation to the next, for instance, in iterative algorithms. The user can then add the `SAVECOM` directive to the routine header. HPF only includes few extensions for the parallelization of loops. As soon as data are distributed depending on run-time information (e.g., mapping arrays), loops can not be efficiently parallelized using compile-time techniques employed in HPF compilers (e.g., the "owner-computes rule"). PST allows to *align* loops with distributed data arrays. The following example loop

```
!PST$ ALIGN 10 WITH X(I)
      do 10 i=1,n
         :
 10   continue
```

is translated into a parallel loop such that accesses to `X(I)` inside the loop are performed without communication.

A PST *block-general distribution* allows the partition of an array into contiguous blocks of array elements. In contrast to the HPF `BLOCK` distribution, the block boundaries are given through a mapping array and may *change* from processor to processor. In contrast to block-general distributions supported by other systems [6], PST permits *gaps* in the array. That is, a global array is partitioned into local parts and mapped to the processors, but the processors do not access all elements of their respective array segment. Accesses are only allowed up to a given upper bound, which is also included in the mapping array defined by the programmer when declaring a block-general distribution.

Since the underlying graph of the problem is constantly being modified, i.e., elements are being added and removed, the local data segment needs to offer enough space to support several refinements, without (excessive) communication. At some point, the local data segment will become full, or the load will become too unbalanced for efficient execution, and the data will need to be redistributed.

Two features of PST support the mesh refinement: data remapping and data redistribution. The data redistribution corresponds to the HPF `REDISTRIBUTE` primitive extended to irregular distributions. That is, the distribution of an array can be changed by an executable statement to any other distribution allowed by the extended language. Data remapping is unique to PST. Any distributed array can be remapped if the user defines a permutation array of global indices. Consider for instance the following five statements:

```
      integer p(m)
      double precision a(m,n)
!PST$ DISTRIBUTE (BLOCK) ::  p
!PST$ DISTRIBUTE (BLOCK,*) ::  a
!PST$ REMAP_1D (a,p,1)
```

The array `a` is remapped along its first dimension using a user-defined permutation array `p`. Effectively `a` is mapped to a new array `a'` with `a'(i) = a(p(i))`. Note that `p` is distributed identically to the dimension of `a` which is to be permuted.

Another PST feature crucial for the support of PLUMP is the support of dynamic allocation which, in principle, is supported by Fortran 90 but currently not available in HPF systems.

PST has inherited most of the features originally included in Oxygen [8]. We included dynamic data distributions because of our experience in a recent project [10]. Data remapping, block-general distributions with gaps, and dynamic allocation have been added to support PLUMP and its target applications.

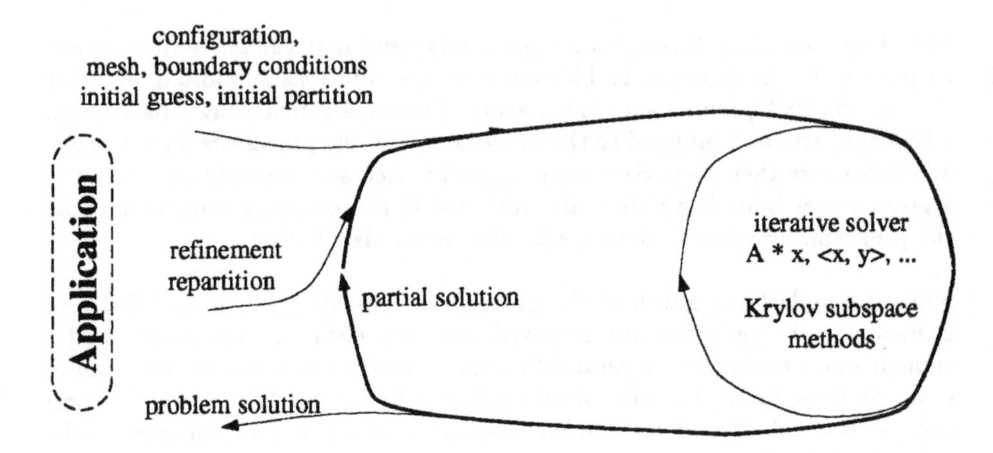

Fig. 1. Finite-element based solution process of discretized system of partial differential equations: based on an initial finite-element mesh, and an initial guess, a first solution is computed through an iterative algorithm (for instance Krylov subspace solvers). The resolution of the solution is increased by refining the discretization which adds elements to the finite-element mesh. On distributed memory parallel processors, the mesh refinement often requires redistribution of the main data structures for better load balance.

3 DESIGN OF LIBRARY

PLUMP makes use of the facilities of PST for a specific class of applications, i.e., computational solutions to problems where the basic data structure can be described as a weakly interconnected graph. Important examples of this class of applications are finite-element based solutions of partial differential equations.

Fig. 1 depicts the typical solution process of such a PLUMP target application. We note that the application can be split in three parts: the high level application part which defines a mesh and performs mesh refinement (and—on DMPPs—partitioning) based on approximate solutions. Secondly, at the low level basic linear algebra operations like vector dot products or matrix-vector products are required. These operations are coded in a language which we believe is close to what will be standard on commercially available DMPPs in few years, namely HPF extended with similar language constructs to the ones detailed in Sect. 2. In addition, the operations should be applied to a data structure both suited for numerical efficiency of these operations and for convenient use when defining and refining the mesh at the high level. Therefore, thirdly, the definition and management of this data structure is done in an efficient interface between the high level mesh refinement and the low level routines, namely within PLUMP.

The distribution of the graph nodes is by ***domain decomposition***, i.e., a cluster of neighboring nodes remains on one processor. We can assume that the nodes are distributed over the processors initially by some external partitioning algorithm which performs load balancing and minimizes the graph edges between partitions (see for instance [11]).

The distribution may be irregular due to the lack of uniformity in the graph. In addition, in order to support mesh refinement and other dynamic aspects, elements can be added or removed from the graph, requiring the nodes to be modified or removed altogether. It may be necessary after numerous updates to repartition the graph in order to keep the load roughly balanced, i.e., to call an external partitioning algorithm once more.

Internally, the partitioned graph is stored as a distributed sparse matrix containing the connectivity and nodal values. The actual storage is hidden from the developer using PLUMP, and is designed to exploit our HPF extensions for data (re)distribution and manipulation.

4 PLUMP SPECIFICATION

4.1 Overview

Ultimately PLUMP will turn into a general library to support a wide range of irregular problems. Initially it is only a research tool also used to evaluate the possibilities of the *Annai* tool environment. More general functionality will be added at a later time.

As already explained in Sect. 3, PLUMP provides an interface between an application at the high-level, and basic numeric routines at the low-level. The functionality of PLUMP is outlined in Fig. 2. The initial version of PLUMP contains the following functionality:

- A routine to allocate and initialize internal data structures for a given maximum size. The routine returns a handle to be used for matrix operations in the client library.

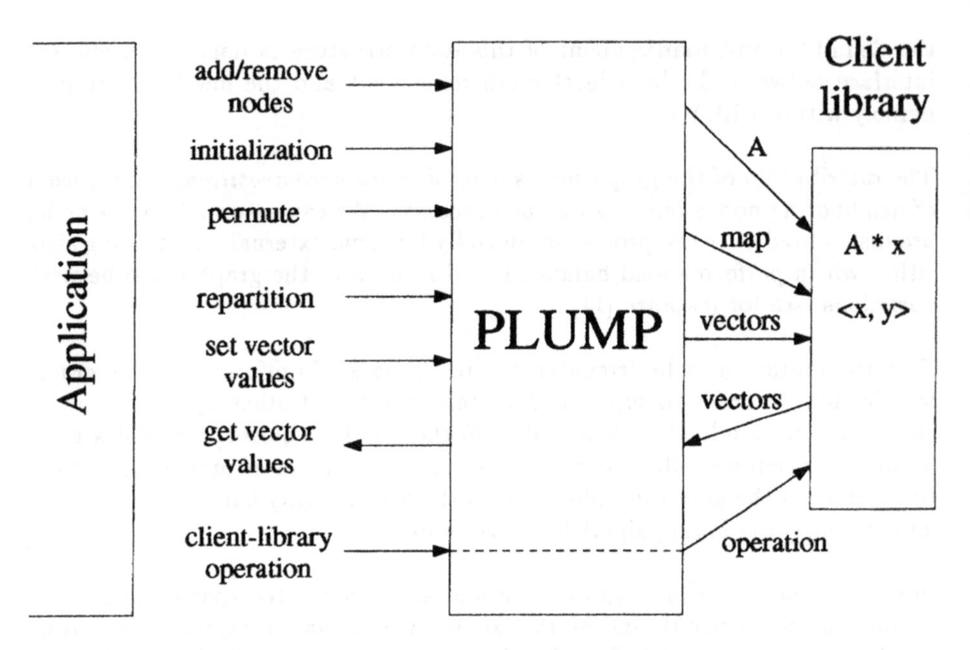

Fig. 2. PLUMP manages an internal representation of the unstructured mesh required by the application. This data structure is constructed with element addition and removal requests, and also allows for permutation (and—on DMPPs—repartition) of the nodes. Vectors of values can be allocated and initialized on request and used in basic computational operations in the numerical client-library; vectors of values, representing for instance approximate solutions to the problem, can be requested from and be returned by PLUMP.

- Routines to allocate and/or initialize vectors. Also these routines return handles which can be passed to the client library.
- A routine to define a partition (or repartition) of the problem, using as input a predetermined partitioning from an external algorithm.
- Routines to add or remove one or several elements.
- A routine to reorder the global matrix (and also vector indices) as defined by a permutation vector returned from reordering algorithms such as RCM (Reverse Cuthill-McKee) [12]. Note that the actual ordering of nodes in PLUMP is hidden from the user.

- Other routines to interface the basic linear algebra operations of the client-library to the application; this refers for instance to the computation of preconditioners, necessary for the solution of ill-conditioned problems.
- A mechanism to request actions on PLUMP handles from the client library. Practically this can be implemented using function pointers.

The client-library contains major linear algebra operations on the data structures constructed by PLUMP, for instance matrix-vector products, and vector-vector operations.

4.2 Internal Data Formats

As described in Sect. 1, we can assume that each graph element consists of $r \times r$ interactions between *nodes*. Within PLUMP the data are held in an internal format which must offer the following qualities:

- A high degree of transparency to the user.
- Flexibility to add and remove elements dynamically, without compromising efficiency.
- Expedient access to individual elements, in order to make the library routine as fast as possible.

Clearly these goals are mutually influential and their individual importance depends on the nature of the application. Thus there can be no one ideal data structure. We have evaluated the following three data formats, each having its own advantages, in order to support as wide a spectrum of applications as possible. Although the internal data formats are not of interest to the user, we mention them here briefly to highlight the strengths and weaknesses of each.

1. **Extended Compressed Sparse Row:** The ExtCSR data storage format is a modification of compressed sparse row (CSR) [13] format. The underlying matrix is explicitly assembled into this format from the values of each element, i.e., coinciding nodal values are added together. Like CSR format, ExtCSR only stores the non-zero matrix entries, along with their row indices. However, the ExtCSR format contains an additional data

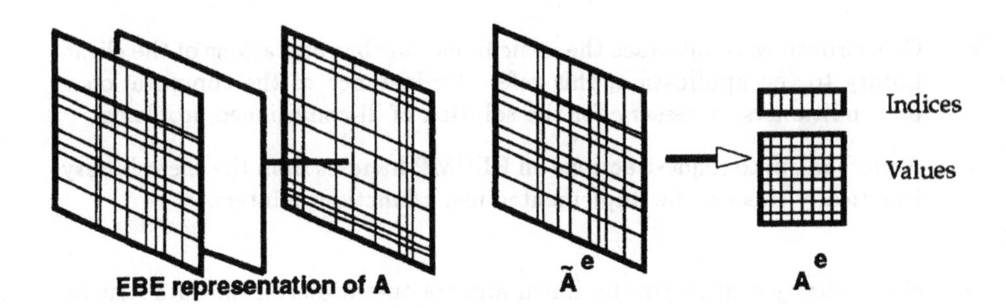

Fig. 3. The Element-by-Element (EBE) storage format

structure (link-array)—an integer array as long as those for real values of matrix entries and their column indices. Each entry in the link-array is an integer pointer to the next entry of a row. The last entry in each row has value 0. Adding and removing elements in the ExtCSR format involves altering some indices in the link-array.

The ExtCSR format is very general: it supports an arbitrary number of row entries in the matrix. Since the matrix is explicitly assembled, the construction of preconditioners is expedited. On the other hand, the matrix-vector multiplication is less efficient than others, due to the level of indirection inherent to the CSR format. Also, all the element values must be explicitly supplied again upon deletion, as these must be subtracted out of the assembled matrix.

2. **Element-by-element:** The element-by-element (EBE) [14] storage stores element data in separate *slices* A^e, i.e., one slice per element, as illustrated in Fig. 3. Thus the underlying matrix is never explicitly assembled. This storage ignores any possible overlaps of elements, thus resulting in redundancy.

   The EBE format strength is in efficient element insertion and deletion, since elements are separate entities. In spite of the implicit redundancy, the matrix-vector multiplication is efficient since it entails the repeated multiplication of dense element matrices A^e with a subset of the vector x,

$$
\begin{aligned}
y = Ax &= \sum_{e \in \text{ elements}} I^e A^e I^{eT} x \\
A^e &\in \Re^{r \times r} \\
I^e &\in \Re^{n \times r} \quad \text{Node indices} \\
\hat{x} = I^{eT} x &\Rightarrow \hat{x} = \text{gather}(x) \\
y = I^e \hat{y} &\Rightarrow y = \text{scatter}(\hat{y})
\end{aligned}
$$

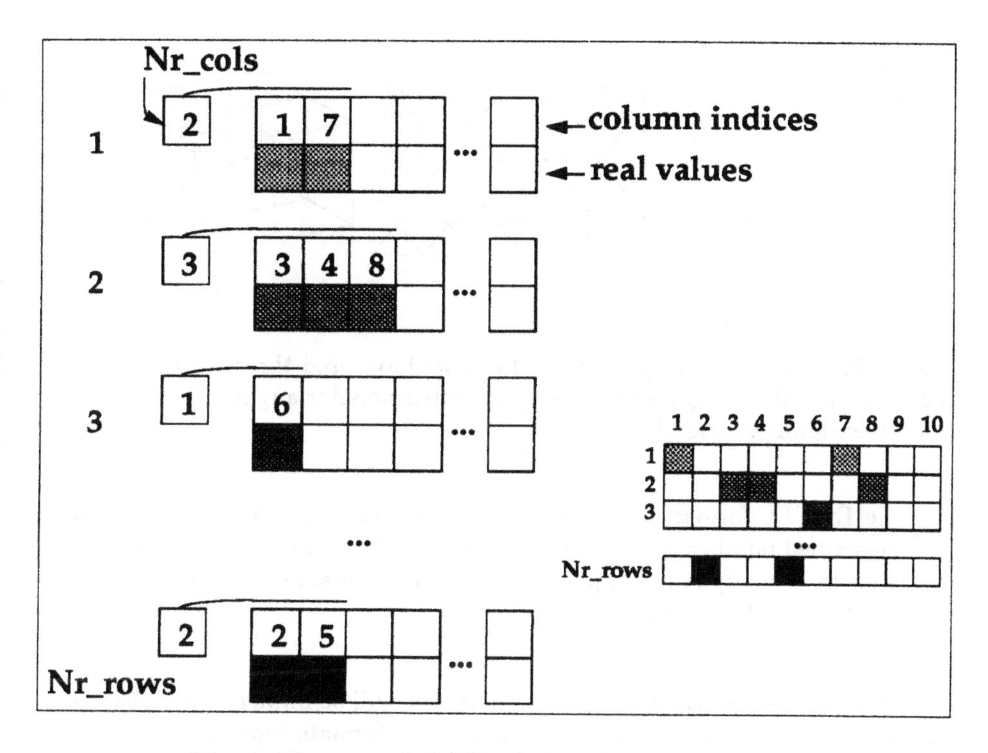

Fig. 4. The extended Ellpack-Itpack storage format

whereby an optimized Basic Linear Algebra Subroutine (*BLAS*) can be used. However, a simple implementation of the EBE format assumes $r = const$ within any given application. Furthermore, the construction of preconditioners is difficult, since the matrix is never explicitly assembled.

3. **Extended Ellpack-Itpack:** The underlying matrix A is explicitly constructed in Ellpack-Itpack (ELL) format [13] extended by *gaps*—rows filled with zeros—to allow for subsequent mesh refinement and element insertion. The extended Ellpack-Itpack (ExtELL) format is depicted in Fig. 4. The *gap*'s size has to be specified during initialization.

   Like Ellpack-Itpack storage, ExtELL format stores a fixed number of (possibly zero) matrix elements per row. Matrix-vector multiplication is efficient, since an optimized scalar product (e.g., *BLAS* `DDOT`) of a row of the matrix with a the "gathered" vector $\hat{x}$ can be performed. Since the matrix is explicitly assembled, node redundancy is avoided, and the construction of preconditioners is simplified. On the other hand, element insertion and deletion routines are more complex than for the EBE format. In addition,

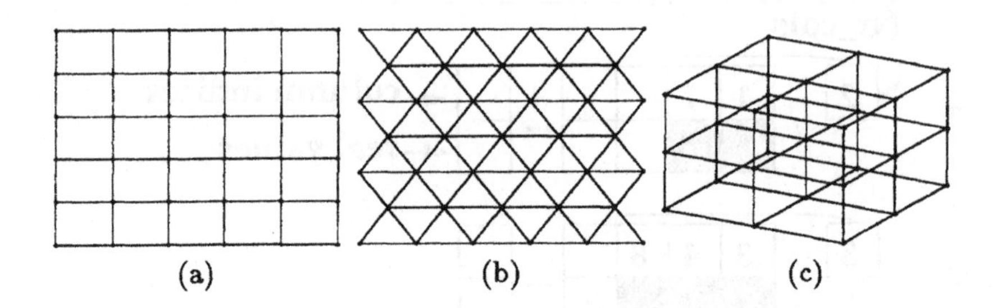

Fig. 5. For our redundancy considerations we have used three regular meshes: a 2-dimensional rectangular mesh (a), a 2-dimensional triangular mesh (b), and a 3-dimensional rectangular mesh.

the ExtELL format is also not entirely general, since it will only allow a prescribed number of matrix row elements. This is rarely a limitation, since even in most dynamic applications a realistic maximum of row elements can be predetermined.

In PLUMP all of the above data storage formats will be implemented. The performance of operations on matrices in these three formats depends on the application (e.g., on the mesh used in calculations, on the need for preconditioners, or on the frequency of element insertion or deletion) and on the hardware used. For instance, on vector processors with optimized *BLAS* routines, EBE and ExtELL formats provide much faster matrix-vector multiplication routines then ExtCSR for most of the meshes. The user should specify which method to use knowing the requirements of the application and the available hardware.

Redundancy Considerations for the EBE Data Storage Format In the EBE format, contributions between elements are stored in separate *slices*. Matrix entries are a superposition of values stored in these slices, therefore there is some redundancy in the memory used and in the calculations using EBE format. The redundancy factor depends on the mesh. Three examples of regular meshes are shown in Fig. 5. The redundancy factors are:

2-dimensional rectangular mesh:	$16/9 \approx 1.78$
2-dimensional triangular mesh:	$18/7 \approx 2.57$
3-dimensional rectangular mesh:	$64/27 \approx 2.37$

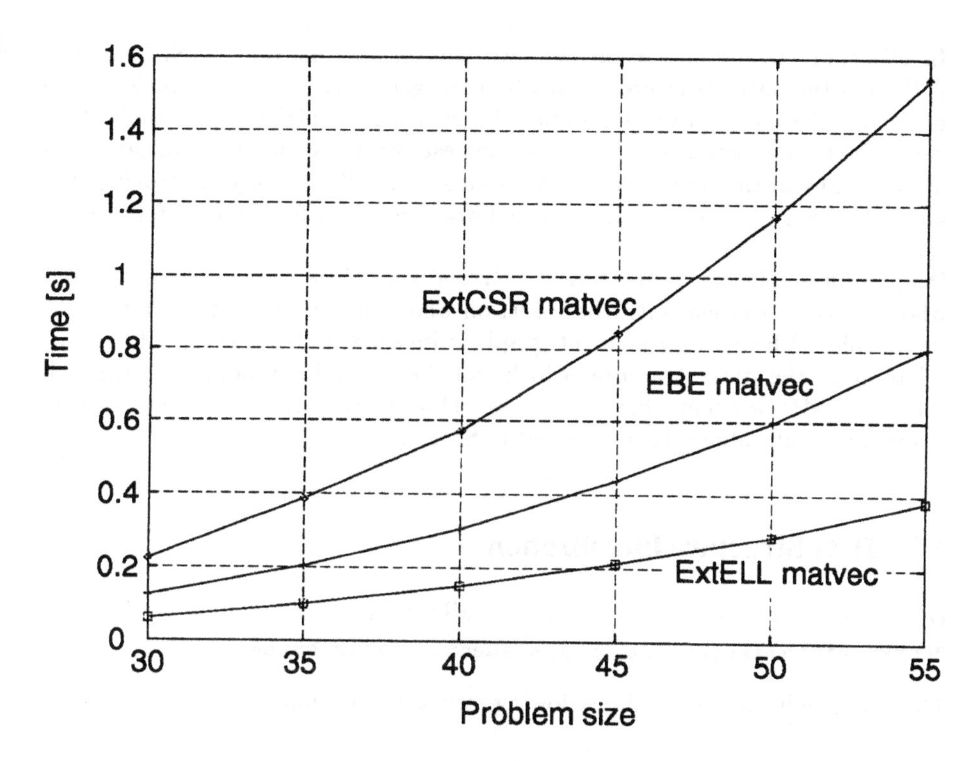

Fig. 6. Performance of matrix-vector products on the NEC SX-3 (1 processor, with optimized *BLAS*) applied to three different data matrix representations: ExtCSR, EBE, and ExtELL. The ExtELL format performs best. The underlying mesh is a 3-dimensional rectangular grid.

On the other hand, the matrix-vector multiplication using EBE format, in particular with large EBE slices and optimized level-2 BLAS routines, may outperform a matrix-vector product implemented using the ExtCSR format. In Fig. 6 we show execution times of matrix-vector multiplications using either of the three data structures. The size of the EBE slices is 8×8. EBE is about twice as fast as ExtCSR, in spite of the redundancy factor.

The matrix-vector multiplication routine with ExtELL format is always faster than one with EBE format. The speedup is approximately equal to the redundancy factor, because the ExtELL routine can also take advantage of (level-1) *BLAS*.

In the short frame of this article, we concentrate on the ExtELL format. Although the data structure cannot be used generally, it is most efficient with problems where the maximal number of non-zero row entries is known. In these cases the format usually provides the fastest matrix-vector multiplication routines. Thus in the remainder of this section we will give only examples of the above PLUMP components and client-library routines for ExtELL format.

Data Hiding The data structure mentioned should be hidden from the user, who is only concerned with the insertion and deletion of elements into/from the graph. When a new element graph is initialized, PLUMP should return a handle to the data structure, which can then only be manipulated through other PLUMP or client-library routines. This strategy implies the availability of dynamic data allocation explained in Sect. 2.

4.3 Data Structure Initialization

During initialization, memory should be allocated:

```
PLUMP_initialize(handle, max_elements, {attributes})
```

The user needs to specify the following structure attributes.

- The format of data storage.
- The maximal size of the vectors and the matrix.
- The following parameters depending on the data storage format:

ExtCSR	maximal number of elements
EBE	maximal number of elements, the size of the slices
ExtELL	the size of the row gaps

After giving initial values to the data structures used, a handle identifying the matrix data structure is returned. This handle is used later for instance in client-library subroutine calls.

4.4 Element Insertion, Removal and Refinement

The routines realizing element insertion require minimal communication, i.e., only for boundary elements. Two basic routines are supported:

```
PLUMP_insert_element(handle, element#, #_neighbors, {neighbors},
                     {values}, size)
PLUMP_remove_element(handle, element#, #_neighbors, {neighbors},
                     {values}, size)
```

An element in these cases is understood as an array of contributions of a group of nodes. If such arrays are stored in the EBE format, the array of real values of the element is not needed in the `PLUMP_remove_element` routine. This can be one reason for choosing the EBE format, since for the other formats PLUMP needs to recalculate the values of the element or get it from an additional data structure before removing the element from the global matrix.

4.5 Data Remapping

The user is mainly interested in a general renumbering of nodes, independent of the current partition of the data structures internal to PLUMP. In general such a renumbering or permutation on DMPPs requires communication of all processors with all processors. One of its major uses is in the context of repartitioning for load balancing after introduction of sufficiently many additional elements.

Since the application programmer can allocate a matrix and vectors separately, the library must provide separate routines for the permutation of vectors and the matrix, to be called for each data structure individually. In Fig. 7 we depict a PLUMP-internal matrix remapping routine based on a given block-general distribution.

Fig. 8 shows how a vector is permuted using PST's `REMAP` directive. The algorithm used for matrix remapping in the above PST code segment is depicted in Fig. 9. The user does not actually see the above routine but uses it through an additional interface routine which translates data-structure handles into the internal format.

4.6 Data Redistribution

Data redistribution is performed using the PST `REDISTRIBUTE` primitive, which is supported by the run-time system. In Fig. 10 we depict the PLUMP-internal routine for redistributing a vector. The matrix routine is very similar.

```
      subroutine PLUMP_internal_remap_matrix(yr_rows,
     &      max_nr_cols, ja, nr_cols, a, permutation,
     &      matrix_map, vector_map)
!PST$ SUB_SPEC PUBLIC
      integer nr_rows, max_nr_cols
      integer nr_cols(nr_rows)
      integer ja(max_nr_cols, nr_rows)
      double precision a(max_nr_cols, nr_rows)
      integer permutation(nr_rows)
      integer vector_map(*), matrix_map(*)
!PST$ DISTRIBUTE BLOCK_GENERAL(matrix_map) ::  a, ja
!PST$ DISTRIBUTE BLOCK_GENERAL(vector_map) ::  nr_cols,
     &      permutation
!PST$ REMAP_1D(nr_cols, permutation, 1)
!PST$ REMAP_1D(a, permutation, 1)
!PST$ REMAP_1D(ja, permutation, 1)
!PST$ ALIGN 10 WITH nr_cols(i)
      do 10 i = nr_rows
        do 11 j = 1, nr_cols(i)
          ja(j, i) = permutation(ja(j, i))
 11     continue
 10   continue
      end
```

Fig. 7. Matrix remapping in PLUMP.

4.7 Client-Library Operations

We consider here two example library components, a common vector-vector operation (**DAXPY**, see also Fig. 11), and a matrix-vector product (**DGEMV**, see also Fig. 12). Most of the vector-vector operations require either no communication (e.g., the **DAXPY** shown below) or global reduction operations (e.g., **DDOT**) which can be easily implemented by adding MPI global reduction operations to the PST code.

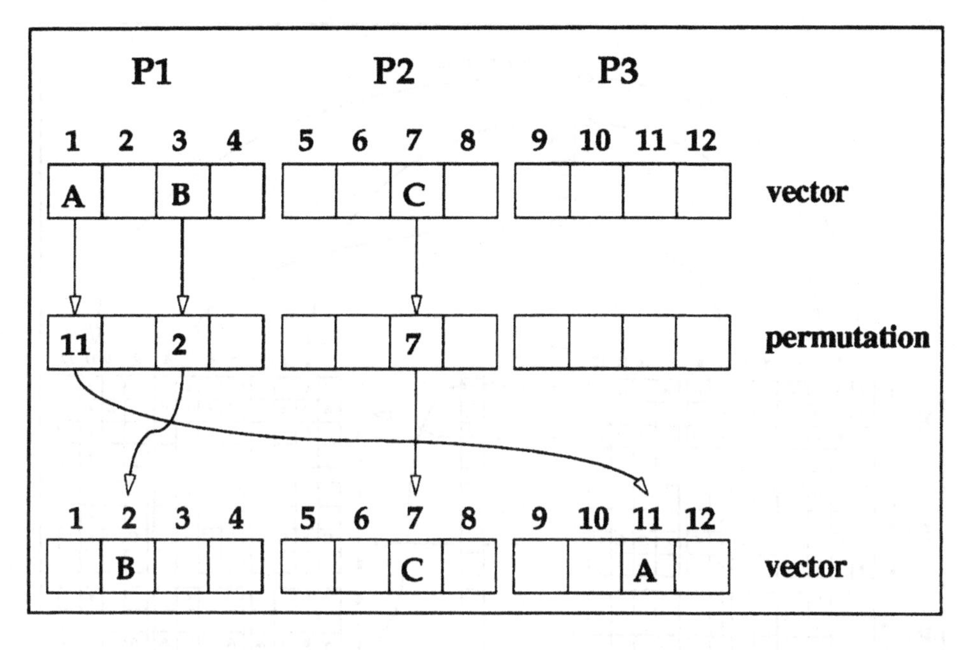

Fig. 8. Vector remapping using PST's **REMAP** primitive.

A matrix-vector product requires communication and therefore makes use of PST's run-time supported global name space.

5 EXAMPLE APPLICATIONS

5.1 SOLIDIS

SOLIDIS is a finite-element package [15] for the simulation of semiconductor micro-sensors. For three-dimensional problems the finite element is a prism or *block* with eight nodes. The graph elements therefore correspond to these blocks represented by 8×8 matrices A^e.

SOLIDIS supports dynamic grid refinement whereby one block can be split into eight sub-blocks, or eight contiguous blocks merged into one. Therefore graph elements are added and removed continually.

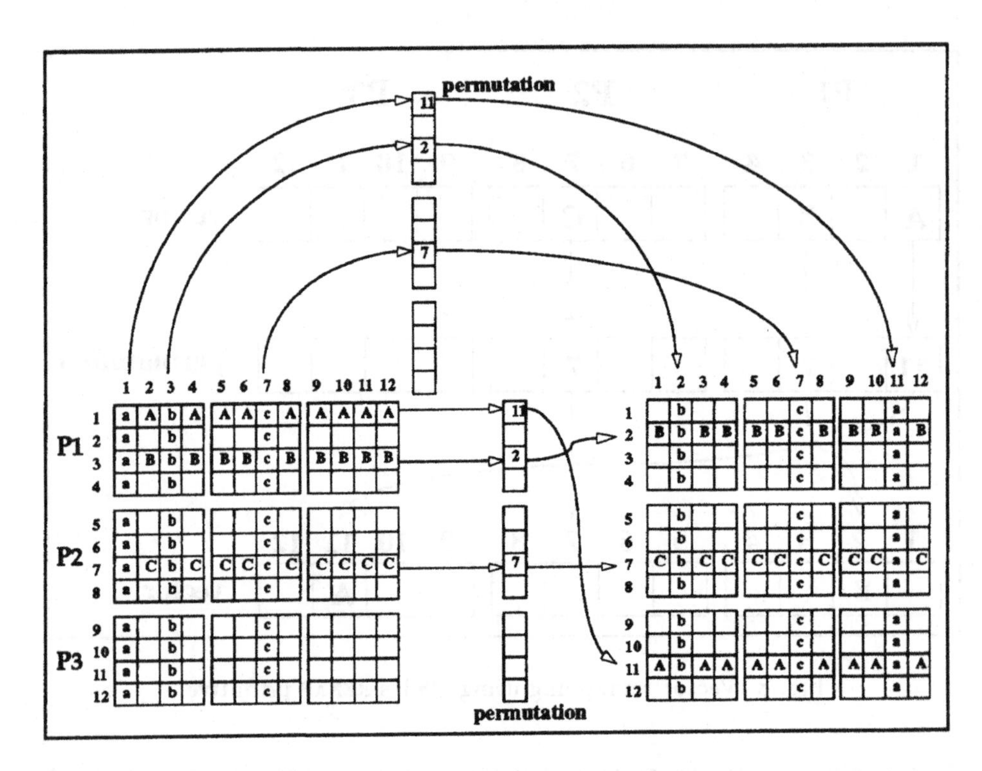

Fig. 9. Matrix remapping using PLUMP_remap_matrix as the Fortran subroutine.

```
      subroutine PLUMP_internal_redistribute_vector(nr_rows, x,
     &        old_vector_map, new_vector_map)
!PST$ SUB_SPEC LOCAL
      integer nr_rows
      double precision  x(nr_rows)
      integer permutation(nr_rows)
      integer old_vector_map(*), new_vector_map(*)
!PST$ DISTRIBUTE BLOCK_GENERAL(old_vector_map) ::  x
!PST$ REDISTRIBUTE(x, new_vector_map)
      end
```

Fig. 10. Vector redistribution in PLUMP.

```
      subroutine Client_Lib_daxpy(nr_rows, x, y, alpha,
     &          vector_map)
!PST$ SUB_SPEC LOCAL
      integer i, j
      integer nr_rows
      double precision x(nr_rows), y(nr_rows), alpha
      integer vector_map(*)
!PST$ DISTRIBUTE BLOCK_GENERAL(vector_map) ::  x, y
!PST$ ALIGN 10 WITH Y(I)
      do 10 i = 1, nr_rows
         y(i) = y(i) + alpha * x(i)
 20   continue
      end
```

Fig. 11. DAXPY in PLUMP.

In the classical version of SOLIDIS, all element matrices are assembled into a global matrix, A, which is then used in a solver to find the solution of $Ax = b$. After mesh refinement, the number of elements changes and the global matrix, A, is assembled again.

PLUMP provides a convenient way to solve finite-element problems without being involved in the matrix assembly. It allows insertion and deletion of elements without reassembling the global matrix, and supports the permutation of nodes when they are renumbered. Significant reassembly execution time overhead is avoided when solving large problems.

With PLUMP and the client-library, it is possible to implement the required Krylov subspace solver by taking advantage of the matrix-vector multiplication (**DGEMV**) and vector-vector routines (such as **DAXPY** and **DDOT**). The underlying linear problem can be ill-conditioned due to the exponentially varying spatial scales, and it is therefore necessary to use the Element-by-Element preconditioners available in PLUMP for some simulations.

SOLIDIS supplies a predetermined initial partition of *macro-elements*, which are accessed via global element numbers. SOLIDIS provides refinement information, and each processor can then refine elements locally. When a large load imbalance is determined, i.e., when one partition is refined considerably more

```
      subroutine Client_Lib_dgemv(nr_rows,
     &        max_nr_cols, nr_cols, ja, a, x, y,
     &        matrix_map, vector_map)
!PST$ SUB_SPEC PUBLIC SAVECOM
      integer i, j
      integer nr_rows, max_nr_cols
      integer nr_cols(nr_rows)
      integer ja(max_nr_cols, nr_rows)
      double precision a(max_nr_cols, nr_rows)
      double precision  x(nr_rows), y(nr_rows)
      integer vector_map(*), matrix_map(*)
!PST$ DISTRIBUTE BLOCK_GENERAL(matrix_map) ::  a, ja
!PST$ DISTRIBUTE BLOCK_GENERAL(vector_map) ::  x, y, nr_cols
!PST$ ALIGN 10 WITH Y(I)
      do 10 i = 1, nr_rows
        y(i) = 0.
        do 20 j = 1, nr_cols(i)
          y(i) = y(i) + x(ja(j, i)) * a(j, i)
 20     continue
 10   continue
      end
```

Fig. 12. DGEMV in PLUMP.

than others, we redistribute and renumber the mesh. These operations are supported by PLUMP.

By definition, element-by-element storage wastes space due to storage if element overlaps, but we anticipate that the PLUMP version will faster than version that reassembles the global matrix. We currently are implementing the PLUMP version of the code and will make a critical comparison to the matrix assembly version. Tests have shown that more than 90% of the execution time is spent in the linear solver BiCGSTAB [16], and most of that time within the matrix-vector multiplication, we anticipate the parallel performance of the algorithm to compare to that of the `Client_Lib_dgemv` routine in PLUMP's client-library.

5.2 TBMD

TBMD is a tight binding molecular dynamics code [17] which simulates the dynamic behavior of n atoms in either a cubic cell, or a thin surface. The former has periodic boundary conditions in all three dimensions, the latter only in two, with Dirichlet boundary conditions on the exposed surface. In this model, silicon atoms (elements in the graph) interact through four different orbitals with atoms in its immediate neighborhood. Therefore for each of the few atoms in its proximity, an atom generates a 4×4 matrix A^e.

$2n$ eigenvalues and eigenvectors of the resulting $4n \times 4n$ matrix are sought in order to update atomic positions. Due to the dynamic nature of the simulation—atomic interactions begin or end at a given cutoff distance—elements are inserted or deleted continually, with statistically equal frequency.

In order to find the position updates after any one time step, the eigenvalues and eigenvectors of the (symmetric) $4n \times 4n$ matrix of orbital interactions are sought. First, the Lanczos tridiagonalization of the matrix is performed—a procedure which requires only the matrix-vector multiplication Ax and vector-vector operations, both supplied by the client-library. Then the eigenvalues of the resulting tridiagonal problem are found using a parallel divide-and-conquer algorithm suggested by Cuppen [18], using the implementation of Dongarra and Sorensen [19]. The resulting eigenvectors need to be distributed or "striped" over the processors in order to save space.

Initially TBMD starts with a regular lattice of atoms (such as the face-centered cubic structure of silicon), which is easily partitioned on the parallel machine. With increasing temperature the atoms move about; any one atom loses interactions with another atom and gains new interactions, thus elements—or more appropriately 4×4 matrices A^e of inter-orbital interactions between two atoms—are added or removed using the corresponding PLUMP routines. With time, the atoms need to be renumbered globally to insure locality, and then the data are redistributed accordingly. We make use of the corresponding routines in PLUMP for this purpose. We expect the frequency of renumberings and redistributions (depending on the simulation temperature) not to be necessary more often than every 10 time-steps.

The TBMD is currently being parallelized with PLUMP and we anticipate performance results by late 1994.

6 CONCLUSIONS AND FUTURE WORK

In the preceding sections we have presented PLUMP and discussed two applications which can be parallelized with ease by exploiting its functionality. We have illustrated the potential usefulness of PLUMP for applications which can be described by a dynamic element graph.

This document serves as the specification for PLUMP, which is still in its implementation phase and will be integrated into the above-mentioned applications in Autumn 1994 as part of the CSCS Summer Student Internship Program (SSIP94). Performance results will be available by Dec. 1994.

The library currently has only the limited functionality necessary to test the HPF/PST compiler and to support the above-mentioned applications. In the future we intend to make extensions to PLUMP in order to provide support for a wider range of applications. Here are some of the extensions we are considering:

- **Modifications to internal data formats:** The efficiency of the operations on the internal data format depends strongly on the requirements of the application. For example, some internal formats result in faster matrix-vector multiplication at the cost of a slower insertion and removal routines. We intend to add to and modify the existing formats in order to offer routines which are optimal for certain computational aspects required by the application and to ensure *efficiency-preserving portability*, i.e., high performance on a variety of DMPPs. The user can choose the most efficient format for the application by passing a parameter to the initialization routine.
- **Updates to reflect changes in HPF/PST:** It is likely that numerous changes will be made to the HPF/PST compiler which will make it necessary to alter PLUMP. It is possible that PST facilities may be improved to better support PLUMP. Thus these performance enhancements will be continually added to the library.
- **Automated Redistribution:** Currently the PLUMP redistribution routines rely on a element to processor remapping provided by an external partitioning algorithm. We plan to integrate such an algorithm into PLUMP, in order to provide the user with transparent load balancing.

Concurrently we plan to acquire further applications with dynamically changing graph structure, and parallelize them with PLUMP accordingly. This will make

PLUMP a general-purpose library which will allow comfortable parallelization of many industrially important unstructured problems.

Acknowledgements

We would like to thank Jan Korvink (ETH Zürich) and Luciano Colombo (Università di Milano) for their valuable technical assistance and for making their codes, SOLIDIS and TBMD respectively, available to us for parallelization. We also acknowledge many useful guiding comments and careful proofreading of Karsten Decker.

REFERENCES

[1] C. Clémençon, K. M. Decker, A. Endo, J. Fritscher, N. Masuda, A. Müller, R. Rühl, W. Sawyer, E. de Sturler, B. J. N. Wylie, and F. Zimmermann, "Application-Driven Development of an Integrated Tool Environment for Distributed Memory Parallel Processors," in *Proceedings of the First International Workshop on Parallel Processing (Bangalore, India, Dec. 27–30)*, Dec. 1994.

[2] C. Koelbel, D. Loveman, R. Schreiber, G. Steele, Jr, and M. Zosel, *The High Performance Fortran Handbook.* The MIT Press, 1994.

[3] C. Clémençon and A. Endo and J. Fritscher and A. Müller and R. Rühl and B. J. N. Wylie, "The "Annai" Environment for Portable Distributed Parallel Programming," in *28th Hawaii International Conference on System Sciences*, (Maui International Resort, Wailea, Maui, Hawaii), IEEE, May 1995.

[4] H. Berryman, J. Saltz, and J. Scroggs, "Execution time support for adaptive scientific algorithms on distributed memory machines," *Concurrency: Practice and Experience*, vol. 3, June 1991.

[5] D. A. Burgess, P. I. Crumpton, and M. B. Giles, "A parallel framework for unstructured grid solvers," in *Proceedings of the IFIP WG10.3 Working Conference on Programming for Massively Parallel Distributed Systems, April 25–29, 1994*, (California), April 1993.

[6] B. Chapman, P. Mehrotra, and H. Zima, "Extending HPF for advanced data parallel applications," Tech. Rep. TR 94-7, Institute for Software Technology and Parallel Systems, University of Vienna, Austria, 1994.

[7] R. v. Hanxleden, K. Kennedy, and J. Saltz, "Value-based distributions in fortran d." Draft, 1994.

[8] R. Rühl, *A Parallelizing Compiler for Distributed-Memory Parallel Processors.* PhD thesis, ETH-Zürich, 1992. Published by Hartung-Gorre Verlag, Konstanz, Germany.

[9] A. Müller and R. Rühl, "Extending High Performance Fortran for the Support of Unstructured Computations," Technical Report SeRD-CSCS-TR-94-08, Centro Svizzero di Calcolo Scientifico, CH-6928 Manno, Switzerland, Nov. 1994.

[10] C. Pommerell and R. Rühl, "Migration of Vectorized Iterative Solvers to Distributed Memory Architectures," in *Colorado Conference on Iterative Methods (Breckenridge, Colorado, April 1994)*, 1994. Preliminary proceedings, accepted for publication in SIAM J. Sci. Comput.

[11] A. Pothen, H. D. Simon, and K.-P. Liou, "Partitioning sparse matrices with eigenvectors of graphs," *SIAM Journal on Matrix Analysis and Applications*, vol. 11, pp. 430–452, 1990.

[12] E. H. Cuthill and J. McKee , "Reducing the bandwidth of sparse symmetric matrices," in *Proceedings of the 24th Nat. Conf. Assoc. Comp. Mach.*, pp. 157–172, ACM, 1969.

[13] Y. Saad, "SPARSKIT: A basic tool kit for sparse matrix computation," Tech. Rep. CSRD Report no. 1029, University of Illinois, CSRD, August 1990.

[14] T. J. R. Hughes, J. Winget, I. Levit, and T. E. Tezduyar, "New alternating direction procedures in finite element analysis based upon EBE approximate factorization," in *Computer Methods for Nonlinear Solids and Mechanics*, pp. 75–109, New York: ASME, 1983.

[15] J. Korvink, *An Implementation of the Adaptive Finite Element Method for Semiconductor Sensor Simulation.* PhD thesis, ETH-Zürich, Nov. 1993. Verlag der Fachvereine Zürich, Bericht Nr. 8.

[16] H. A. van der Vorst, "Bi-CGSTAB: A fast and smoothly converging variant of Bi-CG for the solution of nonsymmetric linear systems," *SIAM J. Sci. Stat. Comput.*, vol. 13, pp. 631–644, Mar. 1992.

[17] L. Colombo, "Tight-binding molecular dynamics: Present status and perspectives," in *Proceedings of PC'94*, CSCS, Aug. 1994.

[18] J. J. M. Cuppen, "A divide and conquer method for the symmetric tridiagonal eigenproblem," *Numerische Mathematik*, vol. 36, pp. 177–195, 1981.

[19] J. J. Dongarra and D. C. Sorensen, "A fully parallel algorithm for the symmetric eigenvalue problem," *Scientific and Statistical Computing*, vol. 8, pp. 139–154, Mar. 1987.

4

PARALLELIZING VISION COMPUTATIONS ON CM-5: ALGORITHMS AND EXPERIENCES

Viktor K. Prasanna and Cho-Li Wang

Department of EE-Systems

University of Southern California

Los Angeles, CA 90089-2562, USA

ABSTRACT

This chapter summarizes our work in using Connection Machine CM-5 for vision. We define a realistic model of CM-5 in which explicit cost is associated with data routing and cooperative operations. Using this model, we develop scalable parallel algorithms for representative problems in vision computations at all three levels: low-level, intermediate-level and high-level.

1 INTRODUCTION

Vision problems are generally classified into three levels: low level, intermediate level and high level. From a computational perspective, the low-level processing is mostly iconic and the data communication is local to each pixel. At the higher levels, symbolic techniques are used. The operations performed on each data item can be nonlocal and the communication is irregular. This combination of computational needs provides a significant challenge in designing fast parallel algorithms as well as in implementing the algorithms on available parallel machines to obtain large speed-ups.

This paper summarizes our work in using CM-5 for vision. The Connection Machine CM-5, a *synchronized* MIMD machine [7], is operated in SPMD (Single-Program Multiple-Data) mode, half way between the highly synchronized SIMD

A. Ferreira and J. D. P. Rolim (eds.), Parallel Algorithms for Irregular Problems: State of the Art, 73–95.

model and the message passing MIMD model. This provides the desired capabilities to solve vision tasks at different levels. For scalability analysis of our algorithms and accurate prediction of the performance of the vision tasks, we model the CM-5 as a set of SISD nodes connected by a high-bandwidth low-latency data network and a control network. In this model, the communication startup time, data transmission rate, and synchronization overheads are considered.

The computational requirements of the vision tasks depend on the data structures and algorithms used to implement specific tasks and vary from one task to another. In this paper, we identify a set of generic problems in vision and propose scalable solutions based on the computation and communication characteristics of CM-5. We develop data partitioning and mapping techniques to reduce the communication time and balance the computation among processors. Theses techniques can further be adapted for solving various vision problems on the other message passing parallel machines such as IBM SP-2 and Intel Paragon.

To illustrate our algorithms and implementations we have considered parallelizing tasks arising in low-level and intermediate-level processing in a building detection system and a high-level task for object recognition on CM-5. First, we discuss a low-level vision task, *Contour Detection*. This involves detection of edges using convolution, removal of the "false" edges using a thinning operation, removal of the "weak" edges using a thresholding operation, and linking of the edge pixels to form contours. Second, we discuss two intermediate-level vision tasks : *Linear Approximation* and *Perceptual Grouping*. Linear approximation is a data reduction technique used to extract linear features from contours. The input to the task is linked contour pixels embedded in the image. The output is a collection of line segments representing the contours. Perceptual grouping is performed to impose structural organization onto sensory data. The task groups the primitive features detected by low-level processing recursively to form structural hypotheses. Both linear approximation and perceptual grouping are directed towards closing the gap between what is produced by low-level processing (such as edge detectors) and what is desired as input (perfect contours, no noise, no fragmentation, etc.) to high-level analysis. In the last part of our work, we consider a high-level vision task: *object recognition using geometric hashing*. In general, recognition systems work by hypothesizing matches between scene features and model features, predicting new matches, and verifying or changing the hypotheses through a search. In geometric hashing [12], a hashing technique is employed to speed-up the matching between models and input scene points.

Vision tasks at the three levels exhibit different types of computation and communication features. In contour detection, the operations can be performed in a synchronized fashion with data communication between the pixels that is regular and local to each pixel. In linear approximation, a contour can cross over several sub-image blocks. The communication can extend over large windows. In perceptual grouping, the data communication patterns are generally global and irregular as a token may perform a search crossing over several sub-image blocks and the distribution of tokens depends on the input image. In geometric hashing, the task is communication intensive. Large amount of communication is performed to access the hash bins and to accumulate votes for the matching results. In the worst case, the communication can cause congestion in the network and at the processors.

Based on the proposed scalable algorithms, we have implemented the low- and intermediate-level tasks on CM-5 using C language and CMMD message passing libraries. Our implementations show that, given a 1024×1024 grey level image as input, the extraction of linear features, which includes contour detection and linear approximation, can be performed in less than 3.6 seconds on a partition of CM-5 having 32 PNs. The serial implementation in LISP can take a few hours to extract features in a 1024×1024 image on a state-of-the-art Sparc System 10. In perceptual grouping, using 8K line segments (detected from a 1024×1024 image) as input to perform the grouping process, our parallel implementation takes nearly 0.742 seconds for grouping line segments into longer lines and nearly 0.731 seconds for junction detection on a partition of CM-5 having 32 PNs. For the high-level processing, the code was written in CM Fortran data parallel language and linked with Connection Machine Scientific Software Library (CMSSL). An earlier implementation of the geometric hashing [12] results in 1.52 *sec* for a probe of the recognition phase on an 8K processor CM-2 on an input scene consisting of 200 feature points. We show that using the same hash function and the same number of hash bins as in [12], a recognition step on a scene consisting of 1024 feature points can be performed in less than 200 *msec* on a partition of CM-5 having 32 PNs.

The organization of this paper is as follows. Steps of the vision tasks considered in this paper are outlined in Section 2. Section 3 discusses a model of CM-5. Section 4 describes the key ideas of our algorithms and asymptotic analysis of these algorithms. Experimental results are shown and analyzed in Section 5. Finally, conclusions are presented in Section 6.

2 REPRESENTATIVE VISION TASKS

To illustrate our algorithms and implementations we have considered parallelizing tasks arising in low-level and intermediate-level tasks in a building detection system [4] and a high-level vision task: ***object recognition using geometric hashing*** [12]. In this section, we briefly discuss the main steps performed in these tasks.

2.1 Overview of Low-Level Processing

In this work, we consider a contour detection task in the building detection system. The objective of the task is to extract contour pixels from an input image. Input to the task is a 2-D image array of pixels (grey levels) and the output is a same sized array with linked contour pixels embedded in the array. The main processing steps are [9]:

1. *Edge detection* - convolve the input image with masks corresponding to ideal step edges in a selected number of directions.
2. *Thinning and thresholding* - compare each edge with its neighbors (in the direction orthogonal to the edge's orientation) and retain the edge if it is of greater magnitude and the magnitude is also greater than a fixed threshold.
3. *Edge linking* - compare each edge with its neighbors in the direction of the edge's orientation and form a link to the neighbors if they are of similar orientation.

The contour-detection task consists of edge detection, thinning, and linking steps. These are *window operations*, in which the output at a pixel is based on the value of the input pixel and the value of its neighboring pixels. The neighborhood is defined by the window size. Given an $n \times n$ image and the largest window of size $m \times m$, extraction of contours, can be performed in $O(m^2n^2)$ time on a serial machine.

2.2 Intermediate-Level Analysis

We discuss two intermediate-level vision tasks : Linear Approximation and Perceptual Grouping steps in the building detection system.

A. Linear Approximation

Linear approximation is a data reduction technique used for representing contours (intensity data) by piecewise line segments. Several heuristics are available for approximating a contour by a set of line segments. A general discussion and evaluation of such techniques can be found in [3]. In this paper, we employed a *strip-based* algorithm [13]. However, the algorithm can be easily modified to adapt other heuristics.

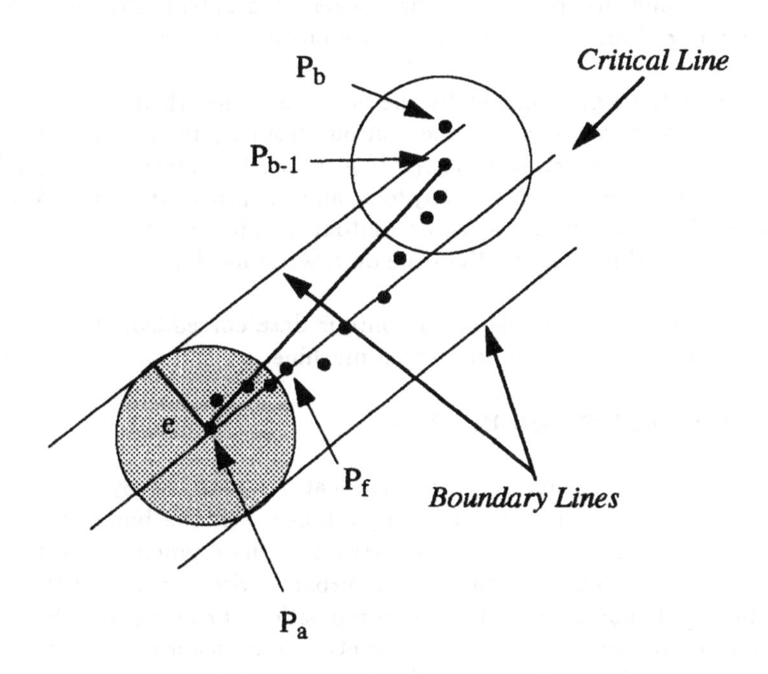

Fig. 1. Strip-based linear approximation.

The input to this algorithm is a set of (open) contours produced by the contour detection task. The contour data is embedded in the 2-D image array with a link specifying the successor as shown in Figure 1. The length of a contour denotes the number of contour pixels in the linked list. It is assumed that the maximum length of a contour in an $n \times n$ image is $O(n)$. Note that in the linking step, edges are linked based on orientation; sharp corners can be detected and the contour can be disconnected at those pixels.

The algorithm proceeds as follows [13]: given a starting pixel p_a of a contour and an error bound e for controlling the quality of the approximation, it selects a pixel p_f on the contour which is at a distance $> e$ from p_a. If it can not find such a pixel, the algorithm stops and forms a line from p_a to the last pixel of the contour. Otherwise, draw a line using p_a and p_f. This line is referred to as the *critical line*. Next form two lines parallel to the critical line at a distance e. These lines are referred to as the ***boundary lines*** as shown in Figure 1. Beginning with pixel p_{f+1}, examine the remaining pixels on the contour until a pixel p_b is found, where p_b is the first pixel lying outside the region formed by the boundary lines or p_b is the last pixel on the contour.

If p_b is outside the region formed by the boundary lines, then the line segment $\overline{p_a p_{b-1}}$ is the approximation to the contour from p_a to p_{b-1} and the same procedure is applied starting at p_{b-1}. If p_b is the last pixel, then $\overline{p_a p_b}$ is the approximation to the contour from p_a to p_b and the procedure stops. Assuming that the total number of pixels on the contour as l, it is easy to verify that the strip-based algorithm runs in $O(l)$ time on a serial machine.

Given an $n \times n$ image with detected contour data embedded, the strip-based algorithm runs in $O(n^2)$ time on a serial machine.

B. Perceptual Grouping

Perceptual grouping groups the primitive features detected by low-level processing recursively to form structural hypotheses. In the building detection system, the input to the task is the extracted line segments detected from aerial image of buildings in urban and suburban environments and the output is the buildings hypotheses modeled as composition of rectangular blocks. The grouping process makes explicit the geometric relationships among the image data and provides *focus of attention* for detection of the building structures in the image. The main grouping steps are [4]:

1. **Line Grouping** - groups line segments which are closely bunched, overlapped, and parallel to each other to form a *line* (a linear structure at a higher granularity level). For each line segment, a search is performed within the region on both sides of the line segment to find other line segments which are parallel to it. The detected segments are grouped to form a *line*.

2. **Junction Grouping** - groups two close right-angled *lines* to form a *Junction*. For each *line*, a search is performed within the region on both sides

of the *Line* and near its end-points to find *lines* which may jointly form right-angled corner(s).

3. **Parallel Grouping** - groups two *lines* which are parallel to each other and have high percentage of overlap. For each *line*, a search is performed on a sorted *line* list (based on the slope of the *line*) to form a *parallel* by grouping with *line* having a difference of slope within given threshold value and satisfying certain constraints with respect to overlap.

4. **U-contour Grouping** - forms a *U-contour* if any *parallel* having its two lines aligned at one end. A search is performed within the window near the aligned end of each *parallel* to group with *lines* possibly connecting the end-points at the aligned end.

5. **Rectangle Grouping** - forms a rectangle if any two *U-contours* share the same *parallel*.

Let *token* denote any of a line segment, a *line*, a pair of parallel lines, a *U-contour*, or a rectangle. The sequential algorithm [4], as described above, employs a search-based approach. The token data is first stored in the image plane before using it in Line, Junction, U-contour Grouping steps. It is also stored as a sorted list in the Parallel Grouping step. Each token performs a search within a *window* in the image (or in the sorted list) to group the tokens in the window satisfying certain geometric constraints. Let S denote the set of input tokens to a grouping step and $W(S)$ denote the total area of the search windows generated by the grouping step. A grouping step can be performed in $O(W(S))$ time on a serial machine. Note that the input set S to a grouping step is usually the output of an earlier grouping step.

2.3 High-Level Task - Geometric Hashing

Geometric hashing offers a different and more parallelizable paradigm for object recognition. The algorithm consists of two procedures, *preprocessing* and *recognition*. These are shown in Figures 2 and 3 respectively. Additional details can be found in [12].

Preprocessing: The preprocessing procedure is executed off-line and only once. In this procedure, the model features are encoded and are stored in a hash table data structure. However, the information is stored in a highly redundant multiple-viewpoint way. Assume each model in the database has n feature points. In Step a in Figure 2, for each ordered pair of feature points in

Preprocessing Procedure :
For each model i such that $1 \leq i \leq M$ **do**
Extract n feature points from the model;
For $j = 1$ to n **do**
For $k = 1$ to n **do**
a. Compute the coordinates of all other features points in the model by taking this pair as basis.
b. Using the hash function on the transformed coordinates, access a bin in the hash table. Add the (*model, basis*) pair (*i.e.*, (i, jk)) to the bin.
end

Fig. 2. Outline of steps in preprocessing.

the model chosen as a basis, the coordinates of all other points in the model are computed in the orthogonal coordinate frame defined by the basis pair. In Step *b*, (*model, basis*) pairs are entered into the hash table bins. The complexity of this preprocessing procedure is $O(n^3)$ for each model, hence $O(Mn^3)$ for M models.

Recognition Procedure:
1. Extract the set of feature points S from the input.
2. *Selection:* Select a pair of feature points as basis.
3. *Probe:*
   a. Compute the coordinates of all other features points in the scene relative to the selected basis.
   b. Using the given hash function on the transformed coordinates access the hash table obtained in the Preprocessing phase.
   c. Vote for the entries in the hash table.
   d. Select the (*model, basis*) pair receiving maximum number of votes as the matched model in the scene.
4. *Verification:* Verify the candidate model edges against the scene edges.
5. If a model wins the verification process, remove the corresponding feature points from the scene.
6. Repeat steps 2, 3, 4, and 5 (until some specified condition).

end

Fig. 3. Outline of steps in recognition using geometric hashing.

Recognition: In the recognition procedure, a scene S of feature points is given as input. An arbitrary ordered pair of feature points in the scene is chosen. Taking this pair as a basis, the coordinates of the remaining feature points are computed. Using the hash function on the transformed coordinates, access a bin in the hash table (constructed in the preprocessing phase), and for every recorded (*model, basis*) pair in the bin, collect a vote for that pair. The pair winning maximum number of votes is taken as the matching candidate. The execution of the recognition phase corresponding to one basis pair is termed as a *probe.* If no (*model, basis*) pair scores high enough, another basis from the scene is chosen and a probe is performed.

The time taken to perform a probe depends on the hash function employed to run. Assuming a scene S results in $V(S)$ votes, a probe of the recognition phase can be implemented in $O(|V(S)|)$ time on a serial machine.

3 A MODEL OF CM-5

A Connection Machine Model CM-5 system contains between 32 and 16,348 processing nodes (PNs). Each node is a 32 MHz Sparc processor with up-to 32 Mbytes of local memory. The peak performance of a node having 4 vector units is 128 MFLOPS. The PNs are interconnected by three networks: a data network, a control network, and a diagnostic network. The data network provides point-to-point data communication between any two PNs. Communication can be performed concurrently between pairs of PNs and in both directions. The data network is a 4-ary *fat tree.* The bandwidth continues to scale linearly up to 16,384 PNs. The control network provides cooperative operations, including broadcast, synchronization, and scans (parallel prefix and suffix). The control network is a complete binary tree with all the PNs as leaves. Each partition consists of a control processor, a collection of processing nodes, and dedicated portions of data and control networks. Throughout this chapter, size of the machine refers to the number of PNs in a partition.

For our analysis, we will model the CM-5 as a set of high performance SISD machines interacting through the data and control networks. We assume *cooperative message passing* [14]: the sending and receiving PNs must be synchronized before sending the message. Thus, the *startup* cost including software overhead and synchronization overhead is associated with each message. In our analysis, we consider machine sizes that are not "large." The hardware latency (network interface overhead and network latency) in data network is small and is *hidden* by the software overheads.

Let T_d and T_c denote the startup times (seconds/message) for sending a message using the data network and control network respectively. Let τ_d denote the transmission rate (seconds per unit of data) for data communication using the data network and τ_c denote the transmission rate for broadcasting data using control network. We make the following assumptions for our analysis:

1. In a unit of time, a PN can perform an arithmetic/logic operation on local data.
2. Sending a message containing m units of data from a PN to another PN or exchanging a message of size m between a pair of PNs takes $T_d + m\tau_d$ time using the data network.
3. Suppose each PN has m units of data to be routed to a single destination using the data network and the set of all destinations is a permutation, then the data can be routed in $T_d + m\tau_d$ time.
4. Broadcasting a message containing m units of data from a PN to all PNs can be performed in $T_c + m\tau_c$ time using the control network, where T_c is the startup time for broadcasting and τ_c is the transmission rate (seconds per unit of data) for broadcasting using control network.
5. Combining (max, sum, prefix sum, etc.) a unit of data from each PN can be performed in τ_g time using the control network.

It has been measured [6], the startup time T_d is around 60-90 μsec and T_c is around 2 μsec. These times are measured by sending a 0 byte message between two PNs using the data network or broadcasting a 0 byte message from a PN to all the PNs using the control network. For regular data communication pattern, τ_d is at the range of 0.100 to 0.123 μsec/byte. This model favors communicating long messages to communicating a large number of short messages. The transmission rate τ_c has been observed to be 1.25 μsec/byte. The times for performing a combining operation (τ_g) are 6 to 12 μsec for integer value and 39 to 56 μsec for double precision. In our analysis, a unit of data is defined as a fixed size data structure to contain image data such as a contour pixel, a label, a Line token etc. The size of the data structure can be varied due to different types of image data. Using this model, we can quantify the communication times and predict the running times of our implementations.

4 SCALABLE PARALLEL ALGORITHMS

A parallel algorithm is considered scalable if the execution time of the algorithm on a machine with P processors varies as $\frac{1}{P}$ [5]. In this section, we present scalable algorithms for the low-level, intermediate-level, and high-level vision tasks described in Section 2.

4.1 Low-Level Processing - Scalable Contour Detection Algorithm

The contour-detection task consists of edge detection, thinning, and linking steps which are *window operations.* The operations can be performed in a synchronized fashion with data communication between the pixels that is regular and local to each pixel. To exploit this type of parallelism on CM-5, we partition the image array into P blocks, where P is the number of processors available. Each block is of size $\frac{n}{\sqrt{P}} \times \frac{n}{\sqrt{P}}$. The blocks are numbered in a shuffled row-major order. The block i is mapped to PN_i as shown in Figure 4. This mapping reduces the communication towards the root of the *fat* tree.

An important step of the implementation is the *image boundary padding* which involves communication between PNs. In this step, the boundary data between neighboring image blocks is exchanged and stored in a local buffer. This step is required because the window operations performed on the boundary pixels of an image block may need pixel data stored in the neighboring PNs. The communication time depends on the size of the window.

The processing steps for the contour detection are outlined in Figure 5. Given an $n \times n$ image and P PNs, we assume that each PN contains an image block of size $\frac{n^2}{P}$ and the largest window is of size $m \times m$. The serial complexity of contour detection is $O(n^2m^2)$, assuming a constant number of window operations are performed.

Theorem 1 *Given an image of size $n \times n$, the contour detection procedure can be performed in $O(n^2m^2/P)$ computation time and $12T_d + (12\lfloor\frac{m}{2}\rfloor\frac{n}{\sqrt{P}} + 24\lfloor\frac{m}{2}\rfloor)\tau_d$ communication time using a partition of CM-5 having P PNs.*

Note that four boundary data exchange steps are performed in each PN to update padded data in eight neighboring sub-image blocks. The communication

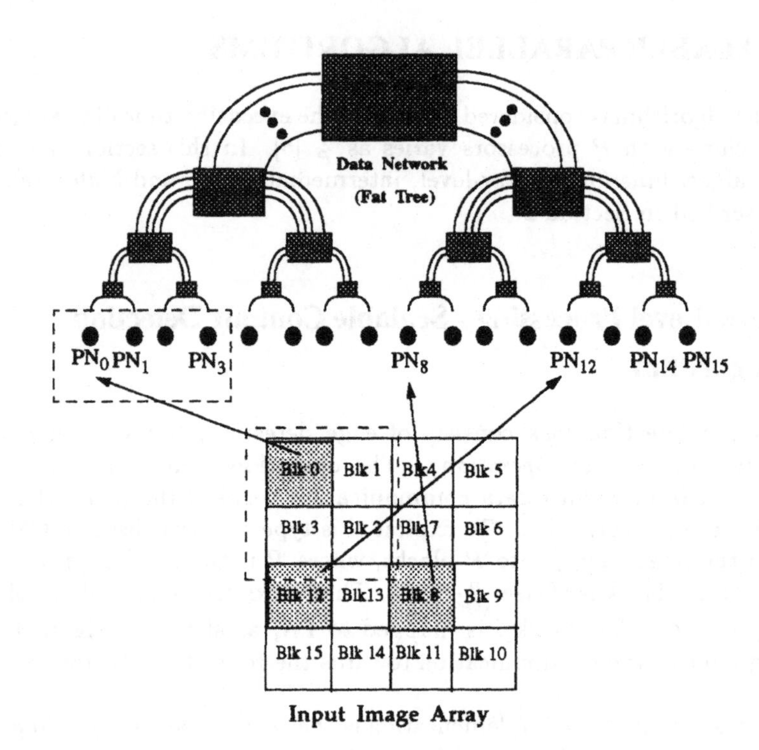

Fig. 4. Partitioning and mapping of the input image to CM-5 PN array

Procedure : Contour-Detection
Input: $n \times n$ image and each PN has image block of size $\frac{n}{\sqrt{P}} \times \frac{n}{\sqrt{P}}$
 Step 1: Perform *Boundary Data Padding*
 Step 2: Perform *Edge Detection*
 Step 3: Perform *Boundary Data Padding*
 Step 4: Perform *Thinning and Thresholding*
 Step 5: Perform *Boundary Data Padding*
 Step 6: Perform *Edge Linking*
end

Fig. 5. A skeleton of contour detection procedure.

time for each Boundary Data Padding step is $4(T_d + \lfloor \frac{m}{2} \rfloor (\frac{n}{\sqrt{P}} + +2\lfloor \frac{m}{2} \rfloor)\tau_d)$.

The total computation time for Steps 2, 4, and 6 is $O(\frac{m^2n^2}{P})$. This assumes $\lfloor\frac{m}{2}\rfloor < \frac{n}{\sqrt{P}}$. In practice, $n \geq 256, m \leq 10$, and $P \leq 1K$.

4.2 Intermediate-Level Processing

Intermediate-level tasks operate on collection of image data extracted in low-level processing. In parallel implementations of intermediate-level tasks, the processors that hold these image data must coordinate their activity because the output depends on image data stored in different processors.

A. Scalable Linear Approximation Algorithm

Most of the known heuristics for linear approximation are inherently sequential in nature. One of the solutions to speed up the process is to let each PN concurrently approximate the contours whose starting pixels are located in its image block; if the contour crosses over the boundary, the approximation is continued by the PN containing the the next part of the contour. The process continues until all contours are approximated. We refer to this as Algorithm A. However, in this approach, some PNs can become bottlenecks, if many contours pass through them and the approximation processes for each of these contours arrive at these PNs at the same time.

We propose a scalable algorithm for performing linear approximation. Our approach is to assign the same label to each pixel of a contour and group all the pixels having the same label to a single PN to perform the linear approximation. Using this data allocation strategy, the contour redistribution is performed once to localize the contours into the PNs and the communication needed during the execution of the approximation procedure is minimal. An advantage of this approach is that any linear approximation heuristic (that has been proposed for serial machines) can be adapted without having to pay a communication penalty after the contours have been localized. We refer to the second approach as Algorithm B.

We define *workload* on a PN as the total number of contour pixels to be processed by the PN. We employ a *load balancing* strategy that relocates the contour pixels such that all the connected contour pixels are moved to the same PN (or at most two PNs) and each PN has no more than $O(\frac{n^2}{P})$ *workload*.

Theorem 2 *Given an image of size $n \times n$, the Linear Approximation step can be performed in $O(n^2/P)$ computation time and $(20P + 3\log P)T_d + (24n +$*

Procedure : Linear-Approximation
Input: PN_i has contour data extracted from the ith image block, $0 \leq i \leq P-1$.
Step 1: Perform *Linear Approximation*
Step 2: Perform *Connected Component Labeling*
Step 3: Group contour data having the same label to a PN
Step 4: Perform *Linear Approximation*
end

Fig. 6. A skeleton of linear approximation procedure.

$\frac{80n^2}{P})\tau_d$ *communication time on a partition of CM-5 having P PNs, where* $n \geq \frac{1}{2}P^{\frac{3}{2}}$.

The processing steps for the linear approximation on CM-5 are outlined in Figure 6. Given an $n \times n$ image and P PNs, we assume that each PN contains an image block of size $\frac{n^2}{P}$ and all the contour pixels in the image block have been detected.

In Step 1, any contour that does not cross an image boundary can be approximated within a PN. This approximation can be carried out for all the contours local to the PNs in $O(\frac{n^2}{P})$ time. In the reminder of this discussion, we only consider contours that cross image boundaries. In Step 2, a divide-and-conquer strategy can be used to complete the component labeling in $O(\frac{n^2}{P}+n\sqrt{P})$ computation time and $24n\tau_d+3\log PT_d$ communication time. In Step 3, Leighton's *column sort* [8] can be used to move contours. The sorting algorithm consists of a local sort followed by a data communication. These two steps are repeated four times. This algorithm is constrained by $r \geq 2(P-1)^2$, where r is the number of data in each PN and P is the number of PNs. We use a modified *columnsort* to sort the data in which a local sort step in *columnsort* is performed by a group of PNs by using *columnsort*. We partition the processor array into $P^{\frac{3}{5}}$ disjoint groups, each group having $P^{\frac{2}{5}}$ PNs. The local sort is performed on $r \times P^{\frac{2}{5}}$ elements in a group of $P^{\frac{2}{5}}$ PNs. The sorted elements are then shuffled among the groups. We repeat the local sort and the shuffle step four times. In this algorithm, each PN performs 16 sorting steps and there are 20 data communication steps. However, the constraint on r and P can be relaxed to $r \geq 2P^{\frac{4}{5}}$. We refer to this algorithm as 2-Level columnsort. Note that we only sort the labels and the number of labels in a PN is $O(\frac{n}{\sqrt{P}})$. Contour data having the same label are moved along with the label. Thus, the computation time for sorting is $O(\frac{n\log n}{\sqrt{P}})$. If $n \geq \frac{1}{2}P^{\frac{3}{2}}$, the total communication time in step 3 is $\frac{80n^2}{P}\tau_d + 20PT_d$. At the end of the sorting step, the contour pixels are redistributed such that each PN has at most $\frac{n^2}{P}$ contour pixels and each contour is in at most two adjacent PNs and given two PNs at most one contour extends over them. The strip-based heuristic for approximating the contours can be applied within each PN. Step 4 can be completed in $O(\frac{n^2}{P})$ time. On a partition of CM-5 having P PNs, the computation time is $O(n^2/P)$ and the communication time is $(20P + 3\log P)T_d + (24n + \frac{80n^2}{P})\tau_d$, where $n \geq \frac{1}{2}P^{\frac{3}{2}}$.

B. Scalable Perceptual Grouping Algorithm

The perceptual grouping is parallel and distributed in nature as the grouping process can be performed independently at each token and the geometric structure may occur over arbitrary spatial extents. However, it is possible that some tokens produce a large number of grouping operations and result in large number of new tokens, while others don't, leading to an uneven distribution of load on the processors. This situation in the worst case can lead to severe congestion at the processors.

Let *token* denote any of a line segment, a pair of parallel lines, a junction, or a rectangle. The *search space* in this application is a $n \times n$ grid structure corresponding to the input image or a hash table having fixed number of bins (e.g. hash bins numbered from 0 to 359) corresponding to the range of the slope of line segments. Each token performs search within a *window* on the grid structure or over a fixed number of hash bins on the hash table to group with tokens satisfying some geometric constraints. We only analyze the Junction Grouping step. The same approach can be applied to all the other grouping steps. In Junction Grouping, we have line segment data (endpoint location, length, or slope of the line etc.) associated with two endpoints of the line on the grid structure. We assume that a constant number of token data is stored at a grid point. In fact, using the thinning technique described in [9], at most three line segments overlap on a pixel to form a "fork" or "join" branch. The input to the process is a set S of tokens distributed in an arbitrary fashion in the PNs. The serial time complexity is $O(W(S))$, where $W(S)$ is the total search area generated by the input tokens.

Theorem 3 *Given a set S of tokens as input to a grouping step, the grouping step can be performed in $O(W(S)/P)$ computation time and $|S|\tau_c + PT_c$ communication time using a partition of CM-5 having P PNs.*

Procedure : Perceptual-Grouping
Input: set S of tokens
 Step 1: Broadcast tokens to all PNs
 Step 2: Perform *Load balancing*
 Step 3: Perform grouping operations
end

Fig. 7. A skeleton of parallel algorithm for perceptual grouping.

The steps of our perceptual grouping algorithm are outlined in Figure 7. The algorithm does not assume any distribution of the tokens among the PNs. The distribution depends on the input image, the algorithms employed in the earlier processing steps. The idea is to let each PN receive all the token data by broadcasting. We then balance the workload among the PNs and perform the grouping operations without further data communication.

We assume that $W(S) \geq P|S|$. For example, we assume partition size $P \leq 256$. The constraint $W(S) \geq P|S|$ is satisfied as the average window size $W(S)/|S|$ is usually more than a few hundreds in performing a grouping step. In Step 1, each PN broadcasts its token data to all the PNs using the control network. The P broadcast steps take $|S|\tau_c + PT_c$ time. In Step 2 we do load balancing. Each PN first constructs the search space by storing the $|S|$ token data on the grid structure. This can be performed in $O(|S|)$ computation time using a well-known technique to store a sparse table without initializing all the entries in the grid structure (see, for example, [1]). We then compute the workload $W(S)$ by summing up the size of search window of each token. We partition the set S such that only portion of S is assigned to a PN and the workload of each PN is nearly $W(S)/P$. The load balancing step can be performed in $O(|S|)$ time. In Step 3, each PN performs the search operations on the grid structure for the tokens assigned to it. This step takes $O(W(S)/P)$ computation time. Thus, the total computation time is $O(W(S)/P)$ and the total communication time is $\tau_c|S| + PT_c$. A slight variation of the above grouping algorithm can be employed in each of the grouping steps defined in Section 2.

4.3 High-Level Processing - Geometric Hashing

In this section, we present scalable algorithm for performing recognition phase in geometric hashing. In the recognition phase, possible occurrence of the models in the scene is checked. The models are available in the hash table created during the preprocessing phase. An arbitrary ordered pair of feature points in the scene is chosen and broadcast to all the PNs. Taking this pair as a basis, a probe of the model database is performed.

In our algorithm, each bin is partitioned across the PNs. Thus, if a bin has k entries then each PN receives $\lceil \frac{k}{P} \rceil$ entries. We first outline the procedures employed in our algorithm.

1. The *Compute_Keys* procedure computes the transformed coordinates of the scene points and quantizes them according to a hash function.
2. The *Access_Bin* procedure distributes the encoded scene points from each PN to all the other PNs. Thus, all PNs have the complete set of encoded scene points.
3. The *Generate_Vote* procedure accesses the hash bin entries locally and copies the (*model*, *basis*) pairs into a list.
4. The *Compute_Winner* procedure determines the (*model*, *basis*) pair receiving the maximum number of votes (*winner*) by sorting the (*model*, *basis*) list produced by *Generate_Vote* procedure.

Based on the procedures described above, a parallel procedure to perform a probe is outlined in Figure 8.

Procedure : Geometric-Hashing
Input: set S of scene points
Step 1: Broadcast a basis to all PNs.
Step 2: Perform *Comp_Keys* procedure.
Step 3: Perform *Access_Bin* procedure.
Step 4: Perform *Generate_Vote* procedure.
Step 5: Perform *Compute_Winner* procedure.
end

Fig. 8. A parallel procedure to process a probe in the recognition phase.

In the following analysis, we ignore the initialization costs, such as loading the scene points to the processor array and loading the hash table into the processor array. We assume that each PN has $\frac{|S|}{P}$ scene points initially. The hash table is given as a fixed data and the input is the set of feature points S. The analysis is performed in terms of the total number of votes cast $|V(S)|$.

Theorem 4 *Given a model of CM-5 having P PNs in which each PN has* $\frac{|S|}{P}$ *scene points and a copy of hash table, a probe can be performed in* $O(\frac{|V(S)|}{P})$ *time in computation,* $(20P+1)T_d+(\frac{20|V(S)|}{P}+1)\tau_d+(P+1)T_c+(|S|+1)\tau_c+\tau_g$ *time in communication, where* $V(S)$ *denotes the set of all votes cast by points in* S.

Proof: The steps of the probe procedure are shown in Figure 8. Step 1 takes $T_c + \tau_c$ time to broadcast a selected basis to all the PNs using control network. Step 2 consists of local computation only which can be completed in $O(\frac{|S|}{P})$ time. Now copy $\frac{|S|}{P}$ encoded scene points from each node to all PNs so that each PN has a copy of $|S|$ encoded scene points. Step 3 can be completed in $P(T_c + \frac{|S|}{P}\tau_c)$ time using the control network. In this algorithm, we uniformly distribute the entries of each bin to all PNs such that each PN generates the same number of votes for any given S in Step 4. This results in an even distribution of workload over the PN array. Thus, step 4 which performs a table lookup can be completed in $O(\frac{|V(S)|}{P})$ computation time. The last step which computes the winner can be performed by sorting the $|V(S)|$ votes using $(model, pair)$ as the key. The 2-Level columnsort described in Section 4.2 can be employed in $Compute\_Winner$ step. As the range of these keys is 0 to Mn^2, a radix sort can be used in a local sort step in a PN. Thus, the sort of $|V(S)|$ votes can be performed in $O(\frac{|V(S)|}{P})$ computation time and $20PT_d + \frac{20|V(S)|}{P}\tau_d$ communication time. At the end of the sort, the votes are distributed over the P PNs such that votes corresponding to a particular $(model, basis)$ pair are in contiguous PNs. By counting the votes within PNs and within adjacent PNs, the $(model, basis)$ pair receiving the maximum number of votes can be computed in $O(\frac{|V(S)|}{P})$ computation time and $T_d + \tau_d + \tau_g$ communication time. The total execution time per probe is $O(\frac{|V(S)|}{P})$ computation time and $(20P + 1)T_d + (\frac{20|V(S)|}{P} + 1)\tau_d + (P+1)T_c + (|S|+1)\tau_c + \tau_g$ time in communication. □

5 IMPLEMENTATION DETAILS AND EXPERIMENTAL RESULTS

We implemented our solutions on CM-5 partitions of 32, 64, 256, and 512 PNs. Our goal is to provide fast execution of the vision tasks on available partition sizes of machines for problem sizes of interest to the vision community. In our implementations, parallel I/O was not employed. All the reported times do not consider I/O time for initial loading of image data.

Total Execution Time (in *seconds*) on CM-5					
Machine Size	Image Size				
(No. of PNs)	128X128	256X256	512X512	1KX1K	2KX2K
32	0.097	0.347	1.001	3.526	14.958
64	0.065	0.239	0.537	1.853	7.602
256	0.053	0.170	0.086	0.599	2.010
512	0.052	0.179	0.123	0.374	1.118

Table 1. Execution times for extracting linear features on various sizes of image, using various partitions of CM-5.

5.1 Contour Detection and Linear Approximation

At the time of this writing, the scalable linear approximation algorithm discussed in Section 4.2 was not implemented. We implemented Algorithm A which is also discussed in Section 4.2 This algorithm consists of two main steps: (1) Perform local approximation (2) Exchange boundary information. These two steps are repeated until there are no more contours to process. Indeed, to reduce frequent communications, Step (1) was performed until all the PNs had finished their local work. The rationale for choosing this approach is as follows. Note that the convolution step takes up a major part of the total execution time in the serial implementation (see Table 2). Thus, a reasonably good non-optimal parallel algorithm for implementation of a linear approximation task probably will not lead to severe degradation of the overall performance. The communication time of our scalable algorithm is $(20P+3\log P)T_d+(24n+\frac{80n^2}{P})\tau_d$. The serial time for linear approximation on a $2K \times 2K$ image (the largest image we processed) is only 54 seconds. If $P = 512$ and we let $T_d = 60$ μsec and $\tau_d = 0$ μsec, the communication time is ≈ 600 *m*sec. The speed-up achievable is less than 100. For the values of P (≤ 512) and n (≤ 2048) considered here, the alternate approach seems to lead to a reasonably fast implementation.

In Table 1, we show the total execution time for extracting linear features on various images using various partition sizes of CM-5. The code was written using C and CMMD 3.2 message passing library [14]. Vector units were not employed. For the sake of comparison, the execution times of individual steps in contour detection and linear approximation procedures on Sun Sparc 400 are shown in Table 2. The largest image we have processed is of size $2K \times 2K$. The total execution time for processing such an image (including contour detection and linear approximation) is 1.118 seconds. The same image processed by a

Execution Time (in *seconds*) on Sun Sparc 400					
Processing	Image Sizes				
Steps	128X128	256X256	512X512	1KX1K	2KX2K
Convolution	1.23	4.90	19.73	80	314
Thinning	0.18	0.72	2.95	7.87	47
Linking	0.26	1.18	5.05	14.65	91
Approximation	0.18	0.70	2.95	8.41	54
Total Time	1.86	7.50	30.68	112	506

Table 2. Execution times on various image sizes using Sun Sparc 400 operating at 33MHz.

Execution Time (in *msec*)				
Step	Line Grouping		Junction Grouping	
No. of Tokens	8943		3736	
Partition size	32	64	32	64
Broadcast Tokens	333	351	145	148
Load Balancing	70	70	51	51
Grouping	339	179	535	326
Total time	742	600	731	525

Table 3. Execution times (in *msec*) for the Line Grouping and the Junction Grouping steps on various partitions of CM-5. Total number of junctions detected was around 3000.

Sun Sparc 400 station operating at 32 MHz using a optimized code written in C takes more than 8 minutes.

5.2 Perceptual Grouping

We have implemented the Line Grouping and Junction Grouping steps. The code was written using C and CMMD 3.2 message passing library. Vector units were not employed. The input was 8943 line segments detected from a $1K \times 1K$ image. The serial implementation written in C results in 14.166 seconds for Line Grouping and 8.680 seconds for Junction Grouping using Sun Sparc 400. In Table 3, we show the execution times of Line Grouping and

Junction Grouping steps. Given a $1K \times 1K$ input image, processing steps including contour detection, linear approximation, Line grouping, and Junction Grouping can be performed in 5.0 seconds using a partition of CM-5 having 32 PNs.

5.3 Geometric Hashing

We have used a synthesized model database containing 1024 models. Each model consists of 16 randomly generated models in 2 dimensions. This results in a hash table having 4M entries. We implemented our algorithm on CM-5

Execution Time (in *msec*)					
Number of PNs	Number of votes generated in a *probe* step				
	32K	64K	128K	256K	512K
32	29	40	63	97	185
64	23	30	37	66	111
256	16	21	23	31	46
512	17	19	20	25	34

Table 4. Execution times of a *probe* step on various partitions of **CM-5** on a scene consisting of **1024** points.

using partitions of 32, 64, 256, and 512 PNs. The code was written in CM Fortran version 2.1.1 and linked with Connection Machine Scientific Software Library (CMSSL) version 3.2. The performance-enhanced *CM_sort*() library function was used to sort the votes in *Compute_Winner* step. In addition, the vector units were used.

Our experimental results for the case of 1024 scene points are shown in Table 4. An earlier implementation [12] results in 1.52 *sec* for a probe of the recognition phase on an 8K processor CM-2 on an input scene consisting of 200 feature points. In this experiment, the model database had 1024 models, each model was represented using 16 feature points and approximately $100K$ votes generated. This performance is obtained under the following assumptions: the number of the processors $\approx$ number of hash bins, a distribution of hash table entries and the resulting votes such that no congestion occurs at a PN during hash bin access or during voting. Using the same hash function and the same number of hash bins (8K bins) as in [12], our implementations show that a probe on a scene consisting of 1024 feature points which generates approximately 512K votes can be performed in 185 *msec* on a 32 processor CM-5. The execution time is reduced to 34 *msec* using 512 processor CM-5. In the implementation in [12], certain inputs result in hash bin access congestion wherein many processors access the data in a single PN. Such congestion does not occur in our implementation. Our algorithm evenly distributes the generated votes. The performance of our algorithm depends on the total number of votes only.

6 CONCLUDING REMARKS

We have presented scalable algorithms for low-level, intermediate-level, and high-level vision tasks on a model of CM-5. Our implementations exploit the features of CM-5. The experimental results are very encouraging and brings a promising future in using parallel machines to realize an integrated vision systems to support interactive execution mode. Indeed, vision computations have significantly different characteristics compared with other grand challenge problems in scientific and numerical computations. We believe modeling features of parallel machines, designing data partitioning techniques, and mapping the computations to balance the load using explicit message passing is a feasible approach to solve vision problems on message passing parallel machines.

REFERENCES

[1] A. Aho, J. Hopcroft, and J. Ullman, *Data Structures and Algorithms*, Addision-Wesley, 1983.

[2] O. Bourdon and G. Medioni, “Object Recognition Using Geometric Hashing on the Connection Machine,” *International Conference on Pattern Recognition*, pages 596–600, 1988.

[3] J. G. Dunham, “Optimum Uniform Piecewise Linear Approximation of Planar Curves,” *IEEE Transactions on Pattern Analysis and Machine Intelligence*, Vol. 8, No. 1, pages 67-75, 1986.

[4] A. Huertas, C. Lin, and R. Nevatia, “Detection of Buildings from Monocular Views of Aerial Scenes Using Perceptual Grouping and Shadows,” *Image Understanding Workshop*, pages 253-260, 1993.

[5] V. Kumar, A. Grama, A. Gupta, and G. Karypis, *Introduction to Parallel Computing: Design and Analysis of Parallel Algorithms*, Benjamin/Cummings, 1994.

[6] T. Kwan, B. Totty, and D. Reed, "Communication and Computation Performance of the CM-5," *Proc. of Supercomputing '93*, pages 192-201, 1993.

[7] C. Leiserson et.al, "The Network Architecture of Connection Machine CM-5," Technical Report, Thinking Machines Corporation, 1992.

[8] F. Leighton, "Tight Bounds on the Complexity of Parallel Sorting," *IEEE Transactions on Computers*, Vol. 34, No. 4, pages 344–354, 1985.

[9] R. Nevatia and K. Babu, "Linear Feature Extraction and Description," *Computer Graphics and Image processing*, Vol. 13, pages 257-269, 1980.

[10] V. Prasanna and C. Wang, "Scalable Parallel Implementations of Perceptual Grouping on Connection Machine CM-5," in *International Conference on Pattern Recognition*, 1994.

[11] V. Prasanna and C. Wang, "Image Feature Extraction on Connection Machine CM-5," in *Image Understanding Workshop*, pages 595-602, 1994.

[12] I. Rogoutsos and R. Hummel, "Massively Parallel Model Matching: Geometric Hashing on the Connection Machine," *IEEE Computer*, pages 33–42, 1992.

[13] J. Roberge, "A Data Reduction Algorithm for Planar Curves," *Computer Vision, Graphics, and Image Processing*, Vol. 29, pages 168-195, 1985.

[14] Thinking Machines Corporation, CMMD Reference Guide Version 3.0, 1992.

[15] C. Wang, V. K. Prasanna, H. Kim, and A. Khokhar, "Scalable Data Parallel Implementations of Object Recognition using Geometric Hashing," Journal of Parallel and Distributed Computing, pages 96-109, March 1994.

PART II

DISCRETE OPTIMIZATION

5

SCALABLE PARALLEL ALGORITHMS FOR UNSTRUCTURED PROBLEMS

Vipin Kumar, Ananth Grama, Anshul Gupta, and George Karypis

Computer Science Department, University of Minnesota, Minneapolis, Minnesota 55455

ABSTRACT

In this paper we summarize our work on development of parallel algorithms for searching large unstructured trees, and for finding solution of large sparse systems of linear equations. Search of large unstructured trees is at the core of many important algorithms for solving discrete optimization problems. Solution of large sparse systems of equations is required for solving many important scientific computing problems. For both of these domains, we show that highly scalable parallel algorithms can be developed, and these algorithms can obtain high speedup on a large number of processors.

1 INTRODUCTION

The driving force behind high performance computing is the need to solve large scale application problems. To effectively perform such tasks using parallel computers, scalable and portable algorithms must be developed. In this paper, we describe scalable parallel algorithms used in a variety of unstructured domains. Issues of load balancing and reducing processor idling become particularly important in such domains. Furthermore, combined with the unstructured nature of communication patterns, parallel algorithms in these domains tend to be significantly more involved.

A. Ferreira and J. D. P. Rolim (eds.), Parallel Algorithms for Irregular Problems: State of the Art, 99–113.

2 PARALLEL ALGORITHMS FOR DISCRETE OPTIMIZATION

Discrete optimization problems (DOPs) arise in various applications such as planning, scheduling, computer aided design, robotics, game playing and constraint directed reasoning. Often, a DOP is formulated in terms of finding a (least cost) solution path in a graph from an initial node to a goal node and solved by graph/tree search methods such as branch-and-bound and dynamic programming.

Given that DOP is an NP-hard problem, one may argue that there is no point in applying parallel processing to these problems, as the worst-case run time can never be reduced to a polynomial unless we have an exponential number of processors. However, the average time complexity of heuristic search algorithms for many problems is polynomial. Also, there are some heuristic search algorithms which find suboptimal solutions in polynomial time (e.g., for certain problems, approximate branch-and-bound algorithms are known to run in polynomial time). In these cases, parallel processing can significantly increase the size of solvable problems. Some applications using search algorithms (e.g. robot motion planning, task scheduling) require real time solutions. For these applications, parallel processing is perhaps the only way to obtain acceptable performance. For some problems, optimal solutions are highly desirable, and can be obtained for moderate sized instances in a reasonable amount of time using parallel search techniques (e.g. VLSI floor-plan optimization [1]).

In this section, we provide an overview of our research on parallel algorithms for solving discrete optimization problems.

2.1 Parallel Depth First Search

Depth-first search (DFS), also referred to as Backtracking, is a general technique for solving a variety of discrete optimization problems [17, 3, 1]. Since many of the problems solved by DFS are computationally intensive, there has been a great interest in developing parallel versions of DFS. For most applications, state space trees searched by DFS tend to be highly irregular, and any static allocation of subtrees to processors is bound to result in significant load imbalance among processors.

In [11, 19, 24, 31], we presented and analyzed a number of parallel formulations of DFS which retain the storage efficiency of DFS, and can be implemented on any MIMD multiprocessor. In these parallel formulations, the search space is dynamically distributed among a number of processors. These formulations differ in the methods used for distributing work among processors. To study the effectiveness of these formulations, we have applied these to a number of problems. Parallel formulations of IDA* for solving the 15 puzzle problem yielded near linear speedup on Sequent Balance up to 30 processors and on the Intel Hypercube^{TM} and BBN Butterfly^{TM} up to 128 processors [23, 24, 31]. Parallel formulation of PODEM, which is the best known sequential algorithm for solving the test pattern generation problem, provided linear speedup on a 128 processor Symult^{TM} [3]. Parallel depth-first branch-and-bound for floorplan optimization for VLSI circuits yielded linear speedups on a 1024 processor Ncube^{TM}, a 128 processor Symult^{TM} and a network of 16 SUN workstations [1]. Linear speedups were also obtained for parallel DFS for the tautology verification problem for up to 1024 processors on the Ncube/10^{TM} and the Ncube/2^{TM}[2, 11, 19]. [1] In [14], we have presented new methods for load balancing of unstructured tree computations on large-scale SIMD machines such as CM-2 ^{TM}[2]. The analysis and experiments show that our new load balancing methods provide good speedups for parallel DFS on SIMD architectures. In particular, their scalability is no worse than that of the best load balancing schemes on MIMD architectures.

Scalability Analysis

From experimental results for a particular architecture and a range of processors alone, it is difficult to ascertain the relative merits of different parallel algorithm-architecture combinations. This is because the performance of different schemes may be altered in different ways by changes in hardware characteristics (such as interconnection network, CPU speed, speed of communication channels etc.), number of processors, and the size of the problem instance being solved [21]. Hence any conclusions drawn from experimental results on a specific parallel computer and problem instance are rendered invalid by changes in any one of the above parameters. Scalability analysis of a parallel algorithm and architecture combination has been shown to be useful in extrapolating these conclusions [20, 24].

[1] Our work on tautology verification, test pattern generation and floorplan optimization received honorable mention for the Gordon Bell Award for outstanding research in practical parallel computing. This is significant considering that this award has historically been given to researchers working on numerical problems.

[2] CM-2 is a registered trademark of the Thinking Machines Corporation.

We have developed a scalability metric, called *isoefficiency* , which relates the problem size to the number of processors necessary for an increase in speedup in proportion to the number of processors used [24]. In general, for a fixed problem size W, increasing the number of processors P causes a decrease in efficiency because parallel processing overhead will increase while the sum of time spent by all processors in meaningful computation will remain the same. Parallel systems that can maintain a fixed efficiency level while increasing both W and P are defined as *scalable* [20]. If W needs to grow as $f(P)$ to maintain an efficiency E, then $f(P)$ is defined to be the **isoefficiency function** for efficiency E and the plot of $f(P)$ with respect to P is defined to be the isoefficiency curve for efficiency E. An important feature of isoefficiency analysis is that it succinctly captures the effects of characteristics of the parallel algorithm as well as the parallel architecture on which it is implemented, in a single expression. The isoefficiency metric has been found to be very useful in characterizing scalability of a number of algorithms [25].

Scalability Analysis of Parallel Depth-First Search

The primary reason for loss of efficiency in our parallel formulations of DFS is the communication overhead incurred by different processors in finding work and in contention over shared resources. By modeling these, we were able to determine the isoefficiency functions (and thus scalability) of a variety of work-distribution schemes[11, 19, 24] for the ring, cube, mesh, network of workstations, and shared memory architectures. Some of the existing parallel formulations were found to have poor scalability. This motivated the design of substantially improved schemes for various architectures. We established lower bounds on the scalability of any possible parallel formulation for various architectures. For each of these architectures, we determined near optimal load balancing schemes. The performance characteristics of various schemes have been proved and experimentally verified in [11, 19, 24, 31]. Schemes with better isoefficiency functions performed better than those with poorer isoefficiency functions for a wide range of problem sizes and processors. Table 1 presents isoefficiency functions of various parallel formulations on various architectures. For a more detailed discussion and a description of the schemes ARR, NN, GRR, GRR-M, and RP, see [11, 19].

From our scalability analysis of a number of architecture-algorithm combinations, we have been able to gain valuable insights into the relative performance of parallel formulations for a given architecture. For instance, in [1] an implementation of parallel depth first branch and bound for VLSI floorplan optimization is presented, and speedups obtained on a network of 16 workstations.

	ARR	NN	GRR	GRR-M	RP
SM	$O(P^2 \log P)$	$O(P^2 \log P)$	$O(P^2 \log P)$	$O(P \log P)$	$O(P \log^2 P)$
Cube	$O(P^2 \log^2 P)$	$\Theta(P^{\log_2 \frac{1+\frac{1}{\alpha}}{2}})$	$O(P^2 \log P)$	$O(P \log^2 P)$	$O(P \log^3 P)$
Ring	$O(P^3 \log P)$	$\Theta(K^P)$	$O(P^2 \log P)$		$O(P^2 \log^2 P)$
Mesh	$O(P^{2.5} \log P)$	$\Theta(K^{\sqrt{P}})$	$O(P^2 \log P)$	$O(P^{1.5} \log P)$	$O(P^{1.5} \log^2 P)$

Table 1. Scalability results of various parallel DFS formulations for different architectures.

The essential part of this branch-and-bound algorithm is the GRR load balancing technique. Our scalability analysis can be used to investigate the viability of using a much larger number of workstations for solving this problem. Note from Table 1 that GRR has an overall isoefficiency of $O(P^2 \log P)$ for this platform. Hence, if we had 1024 workstations on the network, we can obtain the same efficiency on a problem instance which is 10240 ($= \frac{1024^2 \log 1024}{16^2 \log 16}$) times bigger compared to a problem instance being run on 16 processors. This result is of significance, since it indicates that it is indeed possible to obtain good efficiencies with large number of workstations. Scalability analysis also sheds light on the degree of scalability of such a system with respect to other parallel architectures such as hypercube and mesh multicomputers. For instance, the best applicable technique implemented on a hypercube has an isoefficiency function of $O(P \log^2 P)$. With this isoefficiency, we would be able to get identical efficiencies as those obtained on 16 processors by increasing the problem size 400 fold (which is $\frac{1024 \log^2 1024}{16 \log^2 16}$). We can thus see that it is possible to obtain good efficiencies even with smaller problems on the hypercube. We can thus conclude from isoefficiency functions that the hypercube offers a much more scalable platform compared to the network of workstations for this problem.

These parallel formulations were extensively tested on a 1024 node nCUBE2 parallel computer. Table 2 shows average speedup obtained in parallel algorithm for solving instances of the satisfiability problem using different load balancing techniques.

2.2 Parallel Best First Search

The A* algorithm is a well known search algorithm that can use problem-specific heuristic information to prune search space. As discussed in [17], A* is essentially a "best-first" branch-and-bound (B&B) algorithm. A number

Scheme→ P↓	ARR	GRR-M	RP	NN	GRR
8	7.506	7.170	7.524	7.493	7.384
16	14.936	14.356	15.000	14.945	14.734
32	29.664	28.283	29.814	29.680	29.291
64	57.721	56.310	58.857	58.535	57.729
128	103.738	107.814	114.645	114.571	110.754
256	178.92	197.011	218.255	217.127	184.828
512	259.372	361.130	397.585	397.633	155.051
1024	284.425	644.383	660.582	671.202	

Table 2. Average speedups for various parallel formulations.

of researchers have investigated parallel formulations of A*/B&B algorithms [16, 22]. An important component of A*/B&B algorithms is the priority queue which is used to maintain the "frontier" (*i.e.*, unexpanded) nodes of the search graph in a heuristic order. In the sequential A*/B&B algorithm, in each cycle a most promising node from the priority queue is removed and expanded, and the newly generated nodes are added to the priority queue.

In most parallel formulations of A*, different processors concurrently expand different frontier nodes. Conceptually, these formulations can be viewed to differ in the data structures used to implement the priority queue of the A* algorithm. Some formulations are suited only for shared-memory architectures, whereas others are suited for distributed-memory architectures as well. The effectiveness of different parallel formulations is also strongly dependent upon the characteristics of the problem being solved. We have investigated a number of parallel formulations of A*[22]. Some of these formulations are new, and the others are very similar to the ones developed by other researchers. We have tested the performance of these formulations on the 15-puzzle, the traveling salesman problem(TSP), and the vertex cover problem (VCP) on the BBN Butterfly^{TM} multiprocessor. The results for the 15-puzzle and VCP are very similar, but very different from results obtained for the TSP. The reason is that the TSP and VCP generate search spaces that are qualitatively different from each other, even though both problems are NP-hard problems. We have also performed a preliminary analysis of the relationship between the characteristics of the search spaces and their suitability to various parallel formulations[22].

2.3 Speedup Anomalies in Parallel Search

In parallel DFS, the speedup can differ greatly from one execution to another, as the actual parts of the search space examined by different processors are determined dynamically, and can be different for different executions. Hence, for some execution sequences the parallel version may find a solution by visiting fewer nodes than the sequential version thereby giving superlinear speedup, whereas for others it may find a solution only after visiting more nodes resulting in sublinear speedup. It may appear that on the average the speedup would be either linear or sublinear. This phenomenon of speedup greater than P on P processors in isolated executions of parallel DFS has been reported by many researchers. for a variety of problems and is referred to by the term speedup anomaly.

In [32], we present analytical models and experimental results on the average case behavior of parallel backtracking. We consider two types of backtrack search algorithms: (i) simple backtracking (which does not use any heuristic information); (ii) heuristic backtracking (which uses heuristics to order and prune search). We present analytical models to compare the average number of nodes visited in sequential and parallel search for each case. For simple backtracking, we show that the average speedup obtained is (i) linear when distribution of solutions is uniform and (ii) superlinear when distribution of solutions is non-uniform. For heuristic backtracking, the average speedup obtained is at least linear (i.e., either linear or superlinear), and the speedup obtained on a subset of instances (called difficult instances) is superlinear. We also present experimental results over many synthetic and practical problems on various parallel machines, that validate our theoretical analysis.

3 HIGHLY PARALLEL DIRECT SOLVERS FOR SPARSE LINEAR SYSTEMS

Solving a large sparse system of linear equations is often one of the core computations in large-scale unstructured problems that arise in structural engineering, fluid dynamics, power networks, modeling, linear programming, and various other engineering and scientific applications. Sparse linear systems can be solved using either iterative or direct methods—both have their own advantages and disadvantages and the more suitable method depends on the type of the problem. Although direct methods require more memory and are often costlier than iterative methods, they are important because of their generality and robustness. For solving sparse linear systems arising in certain applications, they are the only feasible methods known.

In [12], we describe an optimally scalable parallel algorithm for factorization of sparse matrices. This algorithm incurs strictly less communication overhead than any known parallel formulation of sparse matrix factorization, and hence, can utilize a higher number of processors effectively. It is well known that dense matrix factorization can be implemented efficiently on distributed-memory parallel computers [29, 18]. However, despite inherent parallelism in sparse sparse direct methods, not much success has been achieved to date in developing their scalable parallel formulations [13, 34]. We have shown that the parallel Cholesky factorization algorithm described in [12] is as scalable as the best parallel formulation of dense matrix factorization on both mesh and hypercube architectures. We also show that our algorithm is equally scalable for sparse matrices arising from two- and three-dimensional finite element problems.

It is difficult to derive analytical expressions for the number of arithmetic operations in factorization and for the size (in terms of number of nonzero entries) of the factor for general sparse matrices. This is because the computation and fill-in during the factorization of a sparse matrix is a function of the the number and position of nonzeros in the original matrix. In the context of the important class of sparse matrices that are adjacency matrices of graphs whose n-node subgraphs have $O(\sqrt{n})$-node separators (this class includes sparse matrices arising out of all two-dimensional finite difference and finite element problems), the contribution of this work can be summarized by Figure 1. A simple fan-out algorithm [8] with column-wise partitioning of an $N \times N$ matrix of this type on p processors results in an $O(Np \log N)$ total communication volume [10] (box A). The communication volume of the column-based schemes represented in box A has been improved using smarter ways of mapping the matrix columns onto processors, such as, the subtree-to-subcube mapping [8] (box B). A number of column-based parallel factorization algorithms [27, 4, 8, 7, 13, 34] have a lower bound of $O(Np)$ on the total communication volume. Since the overall computation is only $O(N^{1.5})$ [9], the ratio of communication to computation of column-based schemes is quite high. As a result, these column-cased schemes scale very poorly as the number of processors is increased [34, 33]. A few schemes with two-dimensional partitioning of the matrix have been proposed [33], and the total communication volume in the best of these schemes [33] is $O(N\sqrt{p}\log p)$ (box C).

In summary, the simple parallel algorithm with $O(Np \log p)$ communication volume (box A) has been improved along two directions—one by improving the mapping of matrix columns onto processors (box B) and the other by

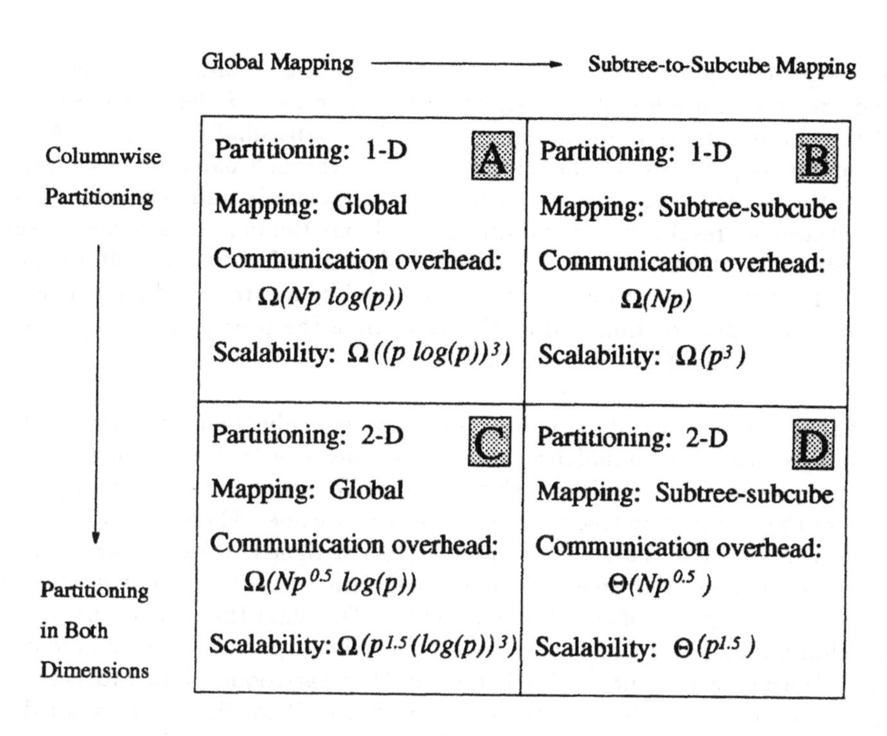

Fig. 1. An overview of the performance and scalability of parallel algorithms for factorization of sparse matrices resulting from two-dimensional N-node grid graphs. Box D represents our algorithm, which is a significant improvement over other known classes of algorithms for this problem.

The algorithm in [12] combines the benefits of improvements along both these lines. The total communication overhead of our algorithm is only $O(N\sqrt{p})$ for factoring an $N \times N$ matrix on p processors if it corresponds to a graph that satisfies the separator criterion. Our algorithm reduces the communication overhead by a factor of at least $O(\log p)$ over the best algorithm implemented to date. It is also significantly simpler in concept as well as in implementation, which helps in keeping the constant factors associated with the overhead term low. Furthermore, this reduction in communication overhead by a factor of $O(\log p)$ results in an improvement in the scalability of the algorithm by a factor of $O((\log p)^3)$; i.e., the rate at which the problem size must increase with the number of processors to maintain a constant efficiency is lower by a factor of $O((\log p)^3)$. This can make the difference between the feasibility and non-feasibility of parallel sparse factorization on highly parallel ($p \geq 256$) computers.

The algorithm presented in [12] is based on the multifrontal principle [5, 26] and the computation is guided by an elimination tree. Independent subtrees of the elimination tree are initially assigned to individual processors. As the computation proceeds up the elimination tree, through pairwise exchanges of data, the matrix is progressively split both horizontally and vertically among the processors resulting in a two-dimensional partitioning. This strategy keeps the communication low in the initial stage of parallel factorization. In the later part, the increase in communication is balanced by a corresponding increase in computation per column, and at the same time the processors do not starve due to narrowing of the elimination tree.

Although, asymptotically scalable, subtree-to-subcube mapping often suffers from significant load-imbalance. In [15], we describe a new elimination tree mapping scheme that minimizes this problem. We assign many subtrees (subforest) of the elimination tree to each processor subcube. These trees are chosen in such a way that the total amount of work assigned to each subcube is as equal as possible. The tree partitioning algorithm uses a set Q that contains the unassigned nodes of the elimination tree. The algorithm inserts the root of the elimination tree into Q, and then it recursively partitions the elimination tree. During each recursive level, the set Q is partitioned into two parts, L and R and each part is checked for acceptability. If so, then L is assigned to half of the processors, and R is assigned to the remaining half. If the partitioning is not acceptable, then one node v of Q is selected and is assigned to all the p processors. Node v is deleted from Q, and its children are inserted into Q. The algorithm then continues by repeating the whole process. Since it assigns subforests of the elimination tree to processor subcubes, we refer to it as subforest-to-subcube mapping scheme. The analysis presented in [15] shows that this new mapping scheme does not increase the overall communication overhead, and its scalability is similar to the earlier algorithm.

We implemented our new parallel sparse multifrontal algorithm on a 1024-processor Cray T3D parallel computer. Each processor on the T3D is a 150Mhz Dec Alpha chip, with peak performance of 150MFlops for 64-bit operations (double precision). However, the peak performance of most level three BLAS routines is around 50 Mflops. The processors are interconnected via a three dimensional torus network that has a peak unidirectional bandwidth of 150Bytes per second, and a very small latency. Even though the memory on T3D is physically distributed, it can be addressed globally. That is, processors can directly access (read and/or write) other processor's memory. T3D provides a library interface to this capability called SHMEM. We used SHMEM to develop a lightweight message passing system. Using this system we were able to achieve unidirectional data transfer rates up to 70Mbytes per second. For the computation performed during the dense Cholesky factorization, we used single-processor implementation of BLAS primitives. These routines are part of the standard scientific library on T3D, and they have been fine tuned for the Alpha chip. The algorithm was tested on matrices from a variety of sources. Four matrices (BCSSTK30, BCSSTK31, BCSSTK32, and BCSSTK33) come from the Boeing-Harwell matrix set. MAROS-R7 is from a linear programming problem taken from NETLIB. COPTER2 comes from a model of a helicopter rotor. CUBE35 is a $35 \times 35 \times 35$ regular three-dimensional grid. NUG15 is from a linear programming problem derived from a quadratic assignment problem obtained from AT&T. In all of our experiments, we used spectral nested dissection [30] to order the matrices.

The performance obtained by our multifrontal algorithm in some of these matrices is shown in Table 3. The operation count shows only the number of operations required to factor the nodes of the elimination tree.

				Number of Processors			
Problem	n	$\|L\|$	OPC	64	256	512	1024
2pt MAROS-R7	3136	1345241	720M	1.41	3.02	4.07	4.48
FLAP	51537	4192304	940M	1.27	2.87	3.83	4.25
BCSSTK33	8738	2295377	1000M	1.30	2.90	4.36	6.02
BCSSTK30	28924	5796797	2400M	1.48	3.59	5.56	7.54
BCSSTK31	35588	6415883	3100M	1.45	3.97	6.26	7.93
BCSSTK32	44609	8582414	4200M	1.51	4.16	6.91	8.90
COPTER2	55476	12681357	9200M	1.94	5.76	9.55	14.78
CUBE35	42875	11427033	10300M	2.26	6.46	10.33	15.70
NUG15	6330	10771554	29670M		7.54	12.53	19.92

Table 3. The performance of sparse direct factorization on Cray T3D. For each problem the table contains the number of equations n of the matrix A, the original number of nonzeros in A, the nonzeros in the Cholesky factor L, the number of operations required to factor the nodes, and the performance in gigaflops for different number of processors.

On large enough problems, one of our algorithms [15] delivers up to 20 GFLOPS for medium-size structures and linear programming problems on a Cray T3D. To the best of our knowledge, this is the first parallel implementation of sparse Cholesky factorization that has delivered speedups of this magnitude and has been able to benefit from several hundred processors. Our work so far has focussed on Cholesky factorization of symmetric positive definite matrices; however, the same ideas can be adapted for performing Gaussian elimination on diagonally dominant matrices that are almost symmetric in structure [6] and for solving sparse linear least squares problems [28].

REFERENCES

[1] S. Arvindam, Vipin Kumar, and V. Nageshwara Rao. Floorplan optimization on multiprocessors. In *Proceedings of the 1989 International Conference on Computer Design*, 1989. Also published as Technical Report ACT-OODS-241-89, Microelectornics and Computer Corporation, Austin, TX.

[2] S. Arvindam, Vipin Kumar, and V. Nageshwara Rao. Efficient parallel algorithms for search problems: Applications in VLSI CAD. In *Proceedings of the Frontiers 90 Conference on Massively Parallel Computation*, 1990.

[3] S. Arvindam, Vipin Kumar, V. Nageshwara Rao, and Vineet Singh. Automatic test pattern generation on multiprocessors. *Parallel Computing*, 17(12):1323–1342, December 1991.

[4] Cleve Ashcraft, S. C. Eisenstat, J. W.-H. Liu, and A. H. Sherman. A comparison of three column based distributed sparse factorization schemes. Technical Report YALEU/DCS/RR-810, Yale University, New Haven, CT, 1990. Also appears in *Proceedings of the Fifth SIAM Conference on Parallel Processing for Scientific Computing*, 1991.

[5] Iain S. Duff and J. K. Reid. The multifrontal solution of indefinite sparse symmetric linear equations. *ACM Transactions on Mathematical Software*, 9:302–325, 1983.

[6] Iain S. Duff and J. K. Reid. The multifrontal solution of unsymmetric sets of linear equations. *SIAM Journal on Scientific and Statistical Computing*, 5(3):633–641, 1984.

[7] G. A. Geist and E. G.-Y. Ng. Task scheduling for parallel sparse Cholesky factorization. *International Journal of Parallel Programming*, 18(4):291–314, 1989.

[8] A. George, M. T. Heath, J. W.-H. Liu, and E. G.-Y. Ng. Sparse Cholesky factorization on a local memory multiprocessor. *SIAM Journal on Scientific and Statistical Computing*, 9:327–340, 1988.

[9] A. George and J. W.-H. Liu. *Computer Solution of Large Sparse Positive Definite Systems*. Prentice-Hall, Englewood Cliffs, NJ, 1981.

[10] A. George, J. W.-H. Liu, and E. G.-Y. Ng. Communication results for parallel sparse Cholesky factorization on a hypercube. *Parallel Computing*, 10(3):287–298, May 1989.

[11] Ananth Grama, Vipin Kumar, and V. Nageshwara Rao. Experimental evaluation of load balancing techniques for the hypercube. In *Proceedings of the Parallel Computing '91 Conference*, pages 497–514, 1991.

[12] Anshul Gupta and Vipin Kumar. A scalable parallel algorithm for sparse matrix factorization. Technical Report 94-19, Department of Computer Science, University of Minnesota, Minneapolis, MN, 1994. A short version appears in *Supercomputing '94 Proceedings*. TR available in *users/kumar* at anonymous FTP site *ftp.cs.umn.edu*.

[13] M. T. Heath, E. G.-Y. Ng, and Barry W. Peyton. Parallel algorithms for sparse linear systems. *SIAM Review*, 33:420–460, 1991. Also appears in K. A. Gallivan et al. *Parallel Algorithms for Matrix Computations*. SIAM, Philadelphia, PA, 1990.

[14] George Karypis and Vipin Kumar. Unstructured Tree Search on SIMD Parallel Computers. Technical Report 92–21, Computer Science Department, University of Minnesota, 1992. Appears in IEEE Transactions on Parallel and Distributed Systems, Volume 5, Number 10, pp. 1057-1072, October 1994. A short version appears in *Supercomputing '92 Proceedings*, pages 453–462, 1992. Available via anonymous ftp from ftp.cs.umn.edu at users/kumar/lb-SIMD.ps.

[15] George Karypis and Vipin Kumar. A high performance sparse Cholesky factorization algorithm for scalable parallel computers. Technical Report TR 94-41, Department of Computer Science, University of Minnesota, Minneapolis, MN, 1994. Submitted to the *Eighth Symposium on the Frontiers of Massively Parallel Computation*, 1995.

[16] V. Kumar and L. N. Kanal. Parallel branch-and-bound formulations for and/or tree search. *IEEE Transactions Pattern Analysis and Machine Intelligence*, PAMI–6:768–778, 1984.

[17] Vipin Kumar. Depth-first search. In Stuart C. Shapiro, editor, *Encyclopaedia of Artificial Intelligence: Vol 2*, pages 1004–1005. John Wiley and Sons, New York, NY, 1987. Revised version appears in the second edition of the encyclopedia to be published in 1992.

[18] Vipin Kumar, Ananth Grama, Anshul Gupta, and George Karypis. *Introduction to Parallel Computing: Design and Analysis of Algorithms*. Benjamin/Cummings, Redwood City, CA, 1994.

[19] Vipin Kumar, Ananth Grama, and V. Nageshwara Rao. Scalable load balancing techniques for parallel computers. *Journal of Parallel and Distributed Computing*, 22(1):60–79, July 1994. Also available as Technical Report 91-55 (November 1991), Department of Computer Science, University of Minnesota, Minneapolis, MN. Available via anonymous ftp from ftp.cs.umn.edu at users/kumar/lb-MIMD.ps.

[20] Vipin Kumar and Anshul Gupta. Analyzing the scalability of parallel algorithms and architectures: A survey. In *Proceedings of the 1991 International Conference on Supercomputing*, 1991. Also appears in September 1994 issue of JPDC. Available via anonymous ftp from ftp.cs.umn.edu at users/kumar/survey-scalability.ps.

[21] Vipin Kumar and Anshul Gupta. Analyzing scalability of parallel algorithms and architectures. *Journal of Parallel and Distributed Computing (special issue on scalability)*, 22(3):379–391, September 1994. A short version of the paper appears in the Proceedings of the 1991 International Conference on Supercomputing. Available via anonymous ftp from ftp.cs.umn.edu at users/kumar/survey-scalability.ps.

[22] Vipin Kumar, K. Ramesh, and V. Nageshwara Rao. Parallel best-first search of state-space graphs: A summary of results. In *Proceedings of the 1988 National Conference on Artificial Intelligence*, pages 122–126, August 1988.

[23] Vipin Kumar and V. N. Rao. Scalable parallel formulations of depth-first search. In Vipin Kumar, P. S. Gopalakrishnan, and L. N. Kanal, editors, *Parallel Algorithms for Machine Intelligence and Vision*. Springer-Verlag, New York, NY, 1990.

[24] Vipin Kumar and V. Nageshwara Rao. Parallel depth-first search, part II: Analysis. *International Journal of Parallel Programming*, 16 (6):501–519, December 1987.

[25] Vipin Kumar and Vineet Singh. Scalability of Parallel Algorithms for the All-Pairs Shortest Path Problem: A Summary of Results. In *Proceedings of the International Conference on Parallel Processing*, 1990. An extended version appears in *Journal of Parallel and Distributed Processing*, 13:124–138, 1991. Available via anonymous ftp from ftp.cs.umn.edu at users/kumar/shortest-path.ps.

[26] J. W.-H. Liu. The multifrontal method for sparse matrix solution: Theory and practice. Technical Report CS-90-04, York University, Ontario, Canada, 1990. Also appears in *SIAM Review*, 34:82–109, 1992.

[27] Robert F. Lucas, Tom Blank, and Jerome J. Tiemann. A parallel solution method for large sparse systems of equations. *IEEE Transactions on Computer Aided Design*, CAD-6(6):981–991, November 1987.

[28] Pontus Matstoms. *The multifrontal solution of sparse linear least squares problems*. PhD thesis, Department of Mathematics, Linkoping University, S-581 83 Linkoping, Sweden, March 1992.

[29] Dianne P. O'Leary and G. W. Stewart. Assignment and scheduling in parallel matrix factorization. *Linear Algebra and its Applications*, 77:275–299, 1986.

[30] Alex Pothen, H. D. Simon, and K.-P. Liou. Partioning sparce matrices with eigenvectors of graphs. *SIAM Journal of Mathematical Analysis and Applications*, 11(3):430–452, 1990.

[31] V. Nageshwara Rao and V. Kumar. Parallel depth-first search, part I: Implementation. *International Journal of Parallel Programming*, 16 (6):479–499, December 1987.

[32] V. Nageshwara Rao and Vipin Kumar. On the efficicency of parallel backtracking. *IEEE Transactions on Parallel and Distributed Systems*, 4(4):427–437, April 1993. Also available as Technical Report TR 90-55, Department of Computer Science, University of Minnesota, Minneapolis, MN. Available via anonymous ftp from ftp.cs.umn.edu at users/kumar/suplin.ps.

[33] Edward Rothberg. Performance of panel and block approaches to sparse Cholesky factorization on the iPSC/860 and Paragon multicomputers. In *Proceedings of the 1994 Scalable High Performance Computing Conference*, May 1994.

[34] Robert Schreiber. Scalability of sparse direct solvers. Technical Report RIACS TR 92.13, NASA Ames Research Center, Moffet Field, CA, May 1992. Also appears in A. George, John R. Gilbert, and J. W.-H. Liu, editors, *Sparse Matrix Computations: Graph Theory Issues and Algorithms* (An IMA Workshop Volume). Springer-Verlag, New York, NY, 1993.

6

A PARALLEL GRASP IMPLEMENTATION FOR THE QUADRATIC ASSIGNMENT PROBLEM

P.M. Pardalos*, L.S. Pitsoulis*
and M. G. C. Resende**

** Center for Applied Optimization and*
Department of Industrial and Systems Engineering,
University of Florida, Gainesville, FL 32611 USA.
e-mail: pardalos@ufl.edu, *e-mail:* leonidas@deming.ise.ufl.edu

*** AT&T Bell Laboratories,*
Murray Hill, NJ 07974 USA.
e-mail: mgcr@research.att.com

ABSTRACT

In this paper we present a parallel implementation of a Greedy Randomized Adaptive Search Procedure (GRASP) for finding approximate solutions to the quadratic assignment problem. In particular, we discuss efficient techniques for large-scale sparse quadratic assignment problems on an MIMD parallel computer. We report computational experience on a collection of quadratic assignment problems. The code was run on a Kendall Square Research KSR-1 parallel computer, using 1, 4, 14, 24, 34, 44, 54, and 64 processors, and achieves an average speedup that is almost linear in the number of processors.

1 INTRODUCTION

Nonlinear assignment problems, such as quadratic, cubic, and N-adic assignment problems were formulated by Lawler [11]. One of the most extensively studied nonlinear assignment problems is the quadratic assignment problem (QAP). The QAP was first introduced by Koopmans and Beckmann in 1957 as a mathematical model for locating a set of indivisible economic activities

A. Ferreira and J. D. P. Rolim (eds.), Parallel Algorithms for Irregular Problems: State of the Art, 115–133.

[9]. Consider the problem of allocating a set of facilities to a set of locations to minimize the cost associated not only with the distance between locations but also to take into account the flows between facilities. Specifically, given n facilities and n locations, two $n \times n$ matrices, $F = (f_{ij})$ and $D = (d_{kl})$, where f_{ij} is the flow between the facility i and facility j, and d_{kl} is the distance between the location k and location l, and a set of integers $N = \{1, 2, \ldots, n\}$, the QAP can be stated as

$$\min_{p \in \Pi_N} \sum_{i=1}^{n} \sum_{j=1}^{n} f_{ij} d_{p(i)p(j)},$$

where Π_N is the set of all permutations of the set N. The above formulation of the QAP will be used throughout this paper. We assume that at least one of the matrices F or D is symmetric, and that the elements of both matrices are nonnegative, with all diagonal elements of both matrices equal to zero. These assumptions are standard. If only one of the input matrices is nonsymmetric, then there exists an equivalent QAP in which both matrices are symmetric. The second and third assumptions are appropriate when the problem is treated as a facility location problem.
Although extensive research has been done on the QAP for over three decades, the quadratic assignment problem, in contrast with its linear counterpart, the linear assignment problem (LAP), remains one of the hardest optimization problems. Today, exact algorithms cannot solve general problems of size $n > 20$. In fact, Sahni and Gonzalez [22] proved that the QAP is NP-complete and that finding an ϵ-approximate solution to the QAP in polynomial time would imply that $P = NP$. In addition to facility layout problems, the QAP appears in many applications, such as backboard wiring, manufacturing scheduling, airport terminal design, and process communications [19].

2 PARALLEL EXACT ALGORITHMS FOR THE QAP

Branch and bound algorithms have been applied successfully to many combinatorial optimization problems, and they appear to be the most efficient exact algorithms for solving the QAP. Although many distinct branch and bound techniques have been proposed, we summarize below a general framework for branch and bound algorithms. As an initialization procedure, one obtains a feasible solution (upper bound) to the problem by means of a heuristic. The problem is decomposed into a finite number of subproblems and a lower bound is established for each subproblem. Each subproblem is again decomposed in the same fashion. This way, a search tree is constructed. Subproblems in the search tree can be ignored if one determines that the best solution that can be derived from it exceeds its lower bound. This process is known as pruning and, by implementing it, one avoids complete enumeration of the feasible set. For the QAP, three classes of branch and bound algorithms have been developed:

- *Single assignment algorithms* (Gilmore [7], Lawler [11]),
- *Pair assignment algorithms* (Gavett and Plyter [5], Land [10], Nugent et al. [17]),
- *Relative positioning algorithm* (Mirchandani and Obata [16]).

The first parallel exact branch-and-bound algorithms for the QAP were proposed by Roucairol [21], and Pardalos and Crouse [18], but computational experiments therein were limited to problems of size $n \leq 15$. During the last few years, extensive work has been done on efficient parallelization of branch-and-bound algorithms for the QAP [6, 8, 20].

3 PARALLEL HEURISTICS FOR THE QAP

Although several parallel exact algorithms for the QAP have been implemented on different parallel computer architectures, the size of the problems that can be solved exactly has not improved much, relative to what has been done with sequential implementations. On the other hand, several heuristics for the QAP have been developed and efficiently implemented in parallel, providing good quality solutions in reasonable CPU time. Heuristics developed for the QAP fall in the following seven categories:

- *Construction methods*, first introduced by Gilmore [7], work iteratively by starting with an empty solution, and construct a complete solution by adding one element at every iteration.
- *Limited enumeration methods* produce good suboptimal solutions, obtained in the early stages of the enumerative process.
- *Improvement methods* consist of an iterative process, which starts with a feasible solution and improves it until some stopping criteria is reached.
- *Simulated annealing* is also an iterative process, where in early stages of the search it is more probable to move to a worse solution than in later stages [2].

```
procedure grasp()
1       InputInstance();
2       do stopping criterion not satisfied →
3           ConstructGreedyRandomizedSolution(solution);
4           LocalSearch(solution);
5           SaveSolution(solution,bestsolution);
6       od;
7       return(bestsolution)
end grasp;
```

Fig. 1. A generic GRASP pseudo-code

- *Genetic algorithms*, unlike most of the heuristics described above, conduct the search based on information about a population consisting of a subset of individuals (solutions) [16].
- *Tabu search* forbids certain moves (that are taboo) that would revisit solutions already examined and this way avoids local minima [23, 24].
- *Greedy randomized adaptive search procedure (GRASP)* is an iterative procedure where a solution is initially constructed in a greedy randomized adaptive fashion, and then this solution is further improved by local search [3, 13].

Next, we focus on the sparse GRASP and its parallel implementation on a Kendall Square Research (KSR-1) parallel computer.

3.1 GRASP

GRASP is an iterative procedure that combines many favorable properties of other heuristics. More specifically, each iteration of GRASP consists of two stages, a construction phase and a local search phase. At each iteration a solution is found and the best solution among all iterations is kept as the final solution. The GRASP procedure is presented in pseudo-code in Figure 1.

The stopping criterion can be, for example, a solution value with which we are satisfied, or a specified maximum number of iterations. The procedure in line 3 of the pseudo-code of Figure 1 builds a greedy randomized solution that is

```
procedure ConstructSolution(solution)
1       solution={};
2       do Solution is not complete →
3           MakeRCL(RCL);
4           s=SelectRandomElement(RCL);
5           solution=solution ∪{s};
6           AdaptGreedy(s,RCL);
7       od;
8       return(solution)
end ConstructSolution;
```

Fig. 2. GRASP construction stage pseudo-code

passed for further improvement in the local search in line 4. The procedure in line 5 saves the best overall solution.

In the construction stage, we start with an empty solution and with each iteration construct the greedy randomized solution, one element at a time. The procedure is summarized in the pseudo-code of Figure 2.

In line 1, the solution set is initialized empty. A *restricted candidate list* (RCL) is constructed in line 3. The RCL is a partial list of the best elements, from which a selection will be made. In line 4, the algorithm randomly selects an element s from the RCL, and adds it to the solution set. The greedy function is evaluated on the remaining elements (excluding s), where a new RCL will be constructed at the beginning of the loop. The greedy randomized component of the heuristic is due to the fact that a selection is made at random from a list of candidates generated in a greedy fashion. The technique is adaptive in the sense that for every element s that is added to a constructed solution, the best candidates to be considered thereafter are changed due to the insertion of element s. A more elaborate description of the construction stage, as adapted for the QAP, will be presented in the next section. After the completion of the construction phase, a constructed solution is passed to the local search phase for further improvement.

At first glance, GRASP can be viewed as a combination of the construction and local search heuristics. Starting with an empty solution, it constructs a final solution using some heuristic procedure, and then local search is applied to the constructed solution. Such a classification, however, is inaccurate since the random and the greedy components of the algorithm are not properly emphasized. The random component is the key for the algorithm to avoid the entrapment in local minima. It also makes it very difficult to construct examples designed

to make the heuristic perform poorly. For example, tabu search uses the *tabu list* to get away from local minima, while simulated annealing accepts cost-increasing moves with a positive probability. The greedy component is used in conjunction with randomness to construct each solution, and finally for moving from one solution to the other GRASP uses local search, which is also often used by tabu search and some implementations of simulated annealing.
Of major importance to an efficient implementation of GRASP is the development of appropriate data structures to assure quick data manipulation. Having developed efficient data structures, the algorithm can be easily implemented, since only the candidate list size and the maximum number of permutations need to be set. GRASP has been successfully implemented for numerous combinatorial optimization problems. See Feo and Resende [3] for a tutorial and survey of GRASP.

3.2 A GRASP for the QAP

In this section a detailed description of the GRASP for the QAP is presented. This paper builds upon the work on GRASP for dense quadratic assignment problems described in [13, 14].

Initial Construction Phase

In the QAP, the solution space is the set of all $n!$ possible permutations of the set $N = \{1, 2, \ldots, n\}$, denoted by Π_N. When applied to the QAP, the GRASP starts with an empty permutation vector and constructs a complete permutation, one element at a time. The constructed solution is passed to the local search for possible improvement. In the initial construction phase, the first two assignments are made simultaneously, by choosing from the set of assignments of low costs. The initial two assignments are made simultaneously because no cost occurs if less than two assignments are made. More specifically, when we say that facility i is assigned to location k and facility j is assigned to location l, then the corresponding cost of this assignment pair is $f_{ij}d_{kl}$.
In sorting the costs the following procedure is followed. Let α and β, $(0 < \alpha, \beta < 1)$, denote the candidate list restriction parameters. Also, recall that the two $n \times n$ input matrices are denoted by $F = (f_{ij})$ and $D = (d_{kl})$, and let the notation $\lfloor x \rfloor$ denote the largest integer smaller or equal to x. Finally, since only off-diagonal entries are of concern, for ease of notation let $m = n^2 - n$. Proceed by sorting the entries in the distance matrix D in ascending order and

the entries in the flow matrix F in descending order, i.e.

$$d_{k_1l_1} \leq d_{k_2l_2} \leq \cdots \leq d_{k_ml_m},$$

$$f_{i_1j_1} \geq f_{i_2j_2} \geq \cdots \geq f_{i_mj_m}.$$

Take the $\lfloor \beta m \rfloor$ smallest d_{kl} elements and the $\lfloor \beta m \rfloor$ largest f_{ij} elements and find the associated costs, i.e.

$$f_{i_1j_1}d_{k_1l_1}, f_{i_2j_2}d_{k_2l_2}, \ldots, f_{i_{\lfloor \beta m \rfloor}j_{\lfloor \beta m \rfloor}}d_{k_{\lfloor \beta m \rfloor}l_{\lfloor \beta m \rfloor}}.$$

These costs are sorted in ascending order, and only the $\lfloor \alpha\beta m \rfloor$ smallest elements are kept. Each element represents a cost due to some assignment pair. Since no change will be made in this set during the actual GRASP iterations, the procedure can be done in constant time without affecting the computational performance of the algorithm.
In the initial construction phase, two elements are randomly selected from the above set, corresponding to two assignment pairs. Note at this point, the significance of efficient data structures to store not only the set containing the sorted elements, but also another set which maps a specific assignment pair with the corresponding indices of the matrices F and D. A heap-sort is used to implement the above procedure. See [14] for implementation details.

Construction Phase for GRASP

The construction phase for the GRASP exploits the sparsity of either one of the matrices F and D, to reduce the computational effort required to construct a solution from an empty permutation. The construction phase is composed of a sparse procedure and a dense procedure, where both perform the same function. On sparse problems, the sparse procedure is substantially faster than its dense counterpart. We have observed empirically that the sparse procedure is about 25 to 30 percent faster than the dense procedure. To describe the sparse procedure the following notation is needed:

- Γ is the set of already made assignments, i.e.

$$\Gamma = \{(j_1, l_1), (j_2, l_2), \ldots\},$$

- Δ is the set of zero-cost assignment pairs with respect to the first two assignments, i.e.

$$\Delta = \{(i,k) \mid \sum_{(j,l)\in\Gamma} f_{ij}d_{kl} = 0, \ |\Gamma| = 2\},$$

- Δ' is the set of zero-cost assignment pairs with respect to the last element in set Γ, i.e.

$$\Delta' = \{(i,k) \mid f_{ij_{|\Gamma|}} d_{kl_{|\Gamma|}} = 0, |\Gamma| \geq 3, (j_{|\Gamma|}, l_{|\Gamma|}) \in \Gamma\}.$$

The algorithm is described in pseudo-code in Figure 3.
The two assignments made in the initial construction phase are input to the sparse GRASP construction phase. In line 1, we check if their resulting cost is zero. Since the first two assignments are randomly chosen among the smallest cost assignments, if their resulting cost is not zero, then the problem data has little or no sparsity, and we disregard the sparse procedure. Otherwise, in lines 2 and 3, the appropriate initializations for the sets Γ and Δ are made. In lines 4 through 12, the set Δ, consisting of all zero-cost assignments with respect to the two initial assignments, is constructed. From the set Δ, a random selection of the third assignment pair to be added to the set Γ is made in lines 13 through 16.
The most beneficial property of the sparsity of the data is illustrated in lines 18 through 27. Specifically, after the set Δ is constructed and one of its elements added to Γ, the algorithm repeats the same procedure not having to check all the possible assignments that result in zero-cost with respect to the assignments in set Γ, but only those associated with the last inserted element of set Γ. By doing this, another set of zero-cost assignments Δ' is constructed, and the same procedure is repeated until no further zero-cost assignments can be made, i.e. $\Delta' = \emptyset$. Therefore, the set of zero-cost assignments starts with some number of elements, and its size is decreased each time an assignment is made. In lines 30 through 33, the algorithm randomly chooses an assignment pair to be added to the set Γ.
It is clear that the performance of the above algorithm improves as the size of Δ and Δ' increases. The number of zero-cost assignment pairs depends both on the sparsity and the structure of the problem. Although we could have increased the size of Δ and Δ' by not choosing randomly the new assignment to be added in Γ, experiments have shown that this reduces drastically the solution quality of the overall heuristic.
The sparse procedure will terminate if the set $\Delta' = \emptyset$, i.e. there are no zero-cost assignment pairs left. Upon termination, the set Γ will contain m zero-cost assignment pairs, and we are left with completing the set Γ to include the remaining $n - m$ assignment pairs. This is done by passing the partial set Γ to the dense procedure described in Figure 4.
With the `inheap` procedure, elements are inserted into a heap, and every time the `outheap` procedure is called, the smallest element the heap is removed. The construction procedure starts with the m assignments made previously in the sparse procedure, and proceeds with the rest of the assignments until

```
procedure ConstructSparse((j_1, l_1), (j_2, l_2))
1        if f_{j_1 j_2} d_{l_1 l_2} ≠ 0 → return;
2        Γ = {(j_1, l_1), (j_2, l_2)};
3        Δ = ∅;
4        do i = 1, ..., n →
5            do k = 1, 2, ..., n →
6                if (i, k) ∉ Γ →
7                    if Σ_{(j,l)∈Γ} f_{ij} d_{kl} = 0 →
8                        Δ = Δ ∪ {(i, k)};
9                    fi;
10               fi;
11           od;
12       od;
13       s = |Δ|;
14       r = random[1, s];
15       (i, k) = r-th element of set Δ;
16       Γ = Γ ∪ {(i, k)};
17       do assignments = 4, ..., n →
18           Δ' = ∅
19           do i = 1, 2, ..., n →
20               do k = 1, 2, ..., n →
21                   if (i, k) ∈ Δ \ Γ →
22                       if f_{i j_{|Γ|}} d_{k l_{|Γ|}} = 0 →
23                           Δ' = Δ' ∪ {(i, k)};
24                       fi;
25                   fi;
26               od;
27           od;
28           if Δ' = ∅ → return;
29           Δ = Δ';
30           s = |Δ|;
31           r = random[1, s];
32           (i, k) = r-th element of set Δ;
33           Γ = Γ ∪ {(i, k)};
34       od;
end ConstructSparse;
```

Fig. 3. Sparse GRASP construction stage pseudo-code

```
procedure ConstructDense(α, Γ, m)
1       do assignments = m + 1, ..., n →
2           q = 0;
3           do i = 1, 2, ..., n →
4               do k = 1, 2, ..., n →
5                   if (i, k) ∉ Γ →
6                       C_ik = Σ_{(j,l)∈Γ} f_ij d_kl;
7                       inheap(C_ik, i, k);
8                       q = q + 1;
9                   fi;
10              od;
11          od;
12          r = random[1, ⌊αq⌋];
13          do q = 1, 2, ..., r →
14              outheap(C_ik, i, k);
15          od;
16          Γ = Γ ∪ {(i, k)};
17      od;
18      return(Γ)
end ConstructDense;
```

Fig. 4. Dense GRASP construction stage pseudo-code

$|\Gamma| = n$. For each assignment it produces all the possible pairs and keeps only the ones which have not been assigned yet. For these assignment pairs, it enters their corresponding cost C_{ik} into a heap, and updates a counter q, which identifies the number of elements in the heap. This procedure is illustrated in lines 3 through 11. In lines 12 through 15, the algorithm chooses randomly an assignment pair from the αq pairs, which form the restricted candidate list. In line 16, the chosen assignment pair is added into the set Γ and the algorithm proceeds with the next iteration. After the dense procedure has completed, a constructed solution is on hand, and local search is applied to improve the existing solution.

Local Search for the QAP

The local search procedure improves a current solution s by searching in the neighborhood $N(s)$ of that solution. For the QAP, a permutation corresponds to a solution, and the solution space of the problem consists of all the possible $n!$ permutations.
Let us define the difference between two permutations p and q to be $\delta(p,q) = \{i \mid p(i) \neq q(i)\}$ and their distance as $d(p,q) = |\delta(p,q)|$. There are different ways to construct $N(s)$ for various problems, but some commonly used principles are the following (Li [12]):

- A *reasonable neighborhood size* makes it possible to search the neighborhood in a reasonable number of computations.
- A *large variance in the neighborhood* implies that if a specific neighborhood consists of permutations $p_1, \ldots, p_n$, then the maximum distance of all permutations is large.
- *Connectivity in the neighborhood* implies that when moving from p_i to p_j, there exists a sequence of consecutive permutations $\{p_k\}$, with small $\delta(p_k, p_{k+1})$, $k = i, \ldots, j-1$.

For the GRASP local search, a k-exchange neighborhood is used, and in particular in this implementation $k = 2$. For a permutation $p \in \Pi_N$ the k-exchange neighborhood is defined to be

$$N_k(p) = \{q \mid d(p,q) \leq k, 2 \leq k \leq n\}.$$

The k-exchange local search starts with an initial permutation p, and then it examines all permutations generated by the exchange of k elements among

```
procedure kexchange(k, p)
1       StopFlag = false;
2       do StopFlag = false →
3           StopFlag = true;
4           do m = 1, 2, ..., n!/(k!(n-k)!) →
5               ProducePermutation(q, m, N_k(p));
6               if cost(q) < cost (p) →
7                   p = q;
8                   StopFlag = false;
9                   break;
10              fi;
11          od;
12      od;
13      return (p);
end kexchange;
```

Fig. 5. GRASP local search phase pseudo-code

each other, keeping the best permutation as the local optimum. The size of the above neighborhood is $P_k^n = \frac{n!}{k!(n-k)!}$, which implies that, for large k, the computational effort can be enormous. The k-exchange procedure for the GRASP is described in Figure 5.

The procedure ProducePermutation$(q, m, N_k(p))$ generates a permutation q by exchanging k elements, where the elements to be exchanged depend on the value of m, where $1 \leq m \leq P_k^n$. For example, if $k = 2$ and $p = (1, 2, \ldots, n)$ then:

$$
\begin{array}{rcll}
m & = & 1 & q = (2,1,3,\ldots,n-2,n-1,n) \\
m & = & 2 & q = (3,2,1,\ldots,n-2,n-1,n) \\
& \vdots & & \qquad\vdots \\
m & = & P_k^n - 1 & q = (1,2,3,\ldots,n-1,n-2,n) \\
m & = & P_k^n & q = (1,2,3,\ldots,n-2,n,n-1).
\end{array}
$$

When searching the neighborhood, the k-exchange local search procedure described above stops when an improved solution is found, and begins searching in the neighborhood of the best solution found in the same fashion. This type of search is called *first decrement* search, in contrast with *complete enumeration* search, where all the solutions in the neighborhood are examined, and the search continues, in the neighborhood of the best solution found, as before. In both cases, however, the worst case scenario is to search the complete ne-

```
procedure ParallelGRASP(n, F, D, m)
1       BestSol = ∞;
2       GenerateRandomNumberSeeds(s_1, s_2, ..., s_m);
3       do (in parallel) i = 1, ..., m →
4           GRASP(n, F, D, s_i, obj_i, perm_i);
5           if obj_i < BestSol →
6               UpdateBestSol(obj_i, BestSol, perm_i, BestPerm);
7           fi;
8       od;
9       return(BestSol, BestPerm);
end ParallelGRASP;
```

Fig. 6. Parallel GRASP for QAP pseudo-code

ighborhood, consisting of P_k^n solutions. However, since the initial permutation, produced by the GRASP construction phase, is usually of good quality, the local search usually searches only a small portion of the neighborhood.

3.3 Parallel GRASP for the QAP

The implementation of GRASP on an MIMD parallel architecture with m processors is straightforward [4]. The parallel implementation is given in Figure 6. Each processor is given its own random number sequence and input data. The cost of interaction is minimal, since the only global variable required to be shared is the value of the best solution found so far, `BestSol`. This is a very appealing characteristic of GRASP, since although over the last years there has been a considerable development in parallel processing, many optimization algorithms present major difficulties for their efficient implementation in parallel. In distributing the heuristic procedure and the input data to the m processors, care should be taken not to have an intersection between the m random number sequences. Based on these guidelines, we implemented GRASP on a KSR-1 parallel machine.
In line 2 of the pseudo-code of Figure 6, we generate m random number seeds, $s_1, s_2, \ldots, s_m$, whose corresponding random number sequences will be disjoint. These seeds, together with the instance data n, F, and D, are distributed to the m processors, and each processor independently solves the instance using the GRASP in lines 3 through 8. In line 9, the best solution, among the m, is the output.

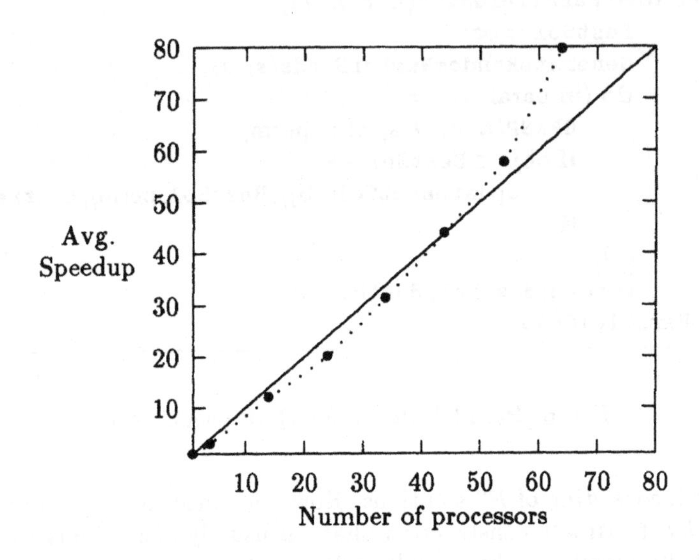

Fig. 7. Average Speedup

The above algorithm was coded in Fortran and tested on a Kendall Square Research (KSR-1) parallel computer. The KSR-1 has 128 processors with 20MHz clock and 32 MB of memory in each, resulting in a total of 4 GB. In our experiments, we were limited to a maximum of 64 processors. The 64 random number seeds were generated by a different sub-program, and the GRASP code with the instance data and the independent seeds were distributed to run independently on each processor, using the `allocate pset` command. The returned solutions from each processor were compared and the best solution was kept as the optimal.

4 COMPUTATIONAL RESULTS

In this section, computational experience with the parallel implementation of the sparse GRASP for the QAP is reported. The efficiency of the parallel GRASP implementation was tested on 20 test problems of the same size and sparsity. These problems were generated using the QAP test problem generator described

Problem name	n	Best known value	$\max(S_D, S_F)$	GRASP-sequential-sparse		
				solution	iterations	CPU time
esc32a	32	130	0.82	130	6935	1872
esc32b	32	168	0.75	168	60	18.7
esc32d	32	200	0.85	200	83	19.94
esc32h	32	438	0.90	438	162	42.86
esc128	128	64	0.98	64	22	369.2
chr22a	22	6156	0.87	6176	100000	8761.7
chr22b	22	6194	0.87	6278	100000	10340
chr25a	25	3796	0.88	3954	13660	2151
nug30	20	6124	0.20	6130	20000	7471
kra30a	20	88900	0.63	88900	12004	3809.7

Table 1. GRASP-sequential-sparse

Problem name	n	Best value	max (S_D, S_F)	GRASP-parallel-sparse			Speedup (64 proc.)
				solution	iterations	CPU time	
esc32a	32	130	0.82	130	743	323	5.8
esc32b	32	168	0.75	168	60	0.77	24.3
esc32d	32	200	0.85	200	1	0.42	31.7
esc32h	32	438	0.90	438	2	0.63	47.5
esc128	128	64	0.98	64	1	21.8	68.0
chr22a	22	6156	0.87	6156	394	39.61	9.3
chr22b	22	6194	0.87	6246	2733	285	221.2
chr25a	25	3796	0.88	3796	233	36.06	36.3
nug30	20	6124	0.20	6128	413	171	43.7
kra30a	20	88900	0.63	88900	84	29.15	130.7

Table 2. GRASP-parallel-sparse

in [15] and were of dimension $n = 20$ and of sparsity $0.3 \leq S_D, S_F \leq 0.32$, where

$$S_M = \frac{\text{number of zeros in matrix } M}{\text{number of elements in matrix } M}.$$

The best solution found by running 1000 GRASP iterations on each processor of the 64-processor machine, was used as the stopping criterion for solving the problem on 54, 44, 34, 24, 14, 4, and 1 processors. The random number seeds for each different number of processors were 1, 2, 3, 4, 5, 6, 7, and 8. In calculating the speedup, the average times for solving the 20 instances for each set of processors were used. Thus, the results reflect the actual performance of the heuristic without being significantly affected by the randomized nature of the heuristic. The average speedup is almost linear, as can be seen in Figure 7.

The parallel GRASP was also tested on selected problems from the QAPLIB [1] test problem set, that were also solved sequentially for comparison. Two stopping criteria were used: Stop if (1) best known solution is found; or (2) if 100,000 iterations do not result in finding the best known solution. In all but two problems, the parallel GRASP found the best known solution, and in the two cases that it did not, the solution was within 1% of the best known. In comparison with the sequential GRASP, the 64-processor parallel implementation was, on average, 61.9 times faster, and up to 221 times faster in one instance. Note that since for two instances (chr22a and chr22b), the serial implementation reached the iteration limit of 100,000 prior to finding the sought-after solution, the average speedup is actually greater than 61.9. Table 1 contains the results for a collection of problems from the QAPLIB for sequential implementation of GRASP, while Table 2 has the corresponding results for the parallel implementation, and the speedups of the 64-processor implementation with respect to the single processor runs.

5 CONCLUDING REMARKS

In this paper we present a parallel implementation of a GRASP for solving large sparse quadratic assignment problems.
At every iteration, the GRASP heuristic constructs an initial solution starting from an empty permutation, and local search is applied to the constructed solution for further improvement. The best solution over all iterations is kept as the output. In the construction phase of the heuristic, a complete solution is constructed by adding one element at a time in a greedy adaptive randomized fashion. In order to exploit the sparsity of the input data, an additional procedure is incorporated in the construction phase of the heuristic, which reduces the computational effort depending on the magnitude of the sparsity. Due to the nature of the heuristic, its parallel implementation on an MIMD parallel computer is very efficient. The iterations are partitioned over m-processors, with each processor having its copy of the instance data and an independent random number sequence. The heuristic was implemented on a KSR-1 parallel computer utilizing 64 processors, and the computational results indicate an almost linear average speedup. In comparison with the sequential implementation on instances from a collection of quadratic assignment test problems, the parallel implementation performed on average about 62 times faster and on one instance over 221 times faster.
The GRASP heuristic and its parallel implementation can be applied to more general nonlinear assignment problems.

Acknowledgements

The authors would like to thank the Cornell Theory Center for making available the KSR-1 parallel computer for the computational experiments.

REFERENCES

[1] R.E. Burkard, S. Karisch, and F. Rendl. QAPLIB – a quadratic assignment problem library. *European Journal of Operations Research*, 55:115–119, 1991. Updated version – Feb. 1993.

[2] R.E. Burkard and F. Rendl. A thermodynamically motivated simulation procedure for combinatorial optimization problems. *European Journal of Operations Research*, 17:169–174, 1984.

[3] T.A. Feo and M.G.C. Resende. Greedy randomized adaptive search procedures. Technical report, AT&T Bell Laboratories, Murray Hill, NJ 07974-2070, February 1994. To appear in *Journal of Global Optimization.*

[4] T.A. Feo, M.G.C. Resende, and S.H. Smith. A greedy randomized adaptive search procedure for maximum independent set. *Operations Research*, 42:860–878, 1994.

[5] J.W. Gavett and N.V. Plyter. The optimal assignment of facilities to locations by branch and bound. *Operations Research*, 14:210–232, 1966.

[6] B. Gendronu and T.G. Crainic. Parallel branch-and-bound algorithms: Survey and synthesis. Technical report, Center for Research on Transportation, University of Montréal, Montréal, Canada, 1994.

[7] P.C. Gilmore. Optimal and suboptimal algorithms for the quadratic assignment problem. *J. SIAM*, 10:305–313, 1962.

[8] A.Y. Grama, V. Kumar, and P.M. Pardalos. Parallel processing of discrete optimization problems. In *Encyclopedia of Microcomputers*, pages 129–147. Marcel Dekker, 1993.

[9] T.C. Koopmans and M.J. Beckmann. Assignment problems and the location of economic activities. *Econometrica*, 25:53–76, 1957.

[10] A. M. Land. A problem of assignment with interrelated costs. *Operations Research Quarterly*, 14:185–198, 1963.

[11] E.L. Lawler. The quadratic assignment problem. *Management Science*, 9:586–599, 1963.

[12] Y. Li. *Heuristic and exact algorithms for the quadratic assignment problem.* PhD thesis, The Pennsylvania State University, 1992.

[13] Y. Li, P.M. Pardalos, and M.G.C. Resende. A greedy randomized adaptive search procedure for the quadratic assignment problem. In P.M. Pardalos and H. Wolkowicz, editors, *Quadratic assignment and related problems*, volume 16 of *DIMACS Series on Discrete Mathematics and Theoretical Computer Science*, pages 237–261. American Mathematical Society, 1994.

[14] Y. Li, P.M. Pardalos, and M.G.C. Resende. Fortran subroutines for approximate solution of dense quadratic assignment problems using GRASP. Technical report, AT&T Bell Laboratories, 1994.

[15] Yong Li and Panos M. Pardalos. Generating quadratic assignment test problems with known optimal permutations. *Computational Optimization and Applications*, 1(2):163–184, 1992.

[16] P.B. Mirchandani and T. Obata. Locational decisions with interactions between facilities: the quadratic assignment problem a review. Working Paper Ps-79-1, Rensselaer Polytechnic Institute, Troy, New York, May 1979.

[17] C.E. Nugent, T.E. Vollmann, and J. Ruml. An experimental comparison of techniques for the assignment of facilities to locations. *Journal of Operations Research*, 16:150–173, 1969.

[18] P.M. Pardalos and J. Crouse. A parallel algorithm for the quadratic assignment problem. In *Proceedings of the Supercomputing 1989 Conference*, pages 351–360. ACM Press, 1989.

[19] P.M. Pardalos, F. Rendl, and H. Wolkowicz. The quadratic assignment problem: A survey and recent developments. In P.M. Pardalos and H. Wolkowicz, editors, *Quadratic assignment and related problems*, volume 16 of *DIMACS Series on Discrete Mathematics and Theoretical Computer Science*, pages 1–42. American Mathematical Society, 1994.

[20] P.M. Pardalos, M.G.C. Resende, and K.G. Ramakrishnan, editors. *Parallel processing of discrete optimization problems.* DIMACS Series on Discrete Mathematics and Theoretical Computer Science. American Mathematical Society, 1995.

[21] C. Roucairol. A parallel branch and bound algorithm for the quadratic assignment problem. *Discrete Applied Mathematics*, 18:211–225, 1987.

[22] S. Sahni and T. Gonzalez. P-complete approximation problems. *Journal of the Association of Computing Machinery*, 23:555–565, 1976.

[23] J. Skorin-Kapov. Tabu search applied to the quadratic assignment problem. *ORSA Journal on Computing*, 2(1):33–45, 1990.

[24] E. Taillard. Robust tabu search for the quadratic assignment problem. *Parallel Computing*, 17:443–455, 1991.

7

CONCURRENT DATA STRUCTURES FOR TREE SEARCH ALGORITHMS

Bertrand Le Cun*, Catherine Roucairol**

INRIA-Rocquencourt, Laboratoire PRiSM
Domaine de Voluceau Bat.17,
B.P. 105 F-78153 Le Chesnay Cedex, France.
Email: Bertrand.Le_Cun@inria.fr
Laboratoire PRiSM
Université de Versailles - St. Quentin-en-Y.
45, Avenue des Etats-Unis, F-78000 Versailles, France.
Email: Catherine.Roucairol@prism.uvsq.fr

ABSTRACT

Exact and approximate parallel methods for solving difficult discrete optimization problems typically work on irregular data structures. Many of these problems are solved by tree search strategies and demand suitable concurrent data structures. In this article, we discuss the data structure used for implementating parallel Best First Branch and Bound and parallel A* algorithm.

Therefore, we study first the method called *Partial Locking* which allows concurrent access on tree data structure on shared memory machines. We also present an improvement of this method which reduces the number of mutual exclusion primitives and thus reduces the implementation overhead.

Parallel Best First Branch and Bound needs suitable Parallel Priority Queues. We show the usefulness of our Partial Locking method to many of them (Skew-Heap, Splay-Trees, Funnel-Tree, Funnel-Table). We give results on our Parallel Best First Branch and Bound for the Quadratic Assignment Problem, and the Travelling Salesman problem.

We also present a new parallel implementation of the heuristic state space search A^* algorithm. We show the efficiency of a new concurrent double criteria data structure called *treap*, instead of usual priority queues (heaps), with the partial locking method. Results on the 15 puzzles are presented, which were obtained on a virtual shared memory machine: the *KSR1*.

A. Ferreira and J. D. P. Rolim (eds.), Parallel Algorithms for Irregular Problems: State of the Art, 135–155.

1 INTRODUCTION

In this paper, we analyse and experiment several concurrent data structures which are used in particular parallel Tree Search Algorithms : parallel best first branch and bound and A* algorithms. Tree Search Algorithms (TSA) are used to solve difficult discrete optimization problems. They explore a tree T of subproblems in which the original problem P occurs at the root. The children of a given subproblem are those obtained from it by breaking it up into smaller problems. The leaves of T represent solutions to P. The goal of the TSA is to find the *optimal* leaves.

Using a parallel machine seems to be a good way, of increasing the size of the solved problem, and reducing the execution time. Intuitively, the parallelization of a TSA is simple, each processor can explore a set of subtrees. But the tree is generated during the exploration, the different subtrees can not be assigned to each processor a priori. Each processor will get the unexplored nodes from a *global data structure* (GDS).

The aim of this paper is to give a solution that allows concurrent access to the GDS in the context of parallel machine with shared memory. Each processor will perform operations on the GDS which is stored in the shared memory. The main difficulty is to keep the GDS consistent, and to allow the maximum of concurency. The simplest method is that each processor locks the entire data structure whenever it makes an access. This solution is very simple but no parallelism is possible. A better method is the partial locking protocol. Each process only locks a little part of the GDS that it actually modify with mutual exclusion primitives. Each processor must follow a locking scheme to avoid deadlocks. On real parallel machines, mutual exclusion primitives have a non negligeable execution time. In the case of tree data structures if each processors follows a particular locking scheme, most of the mutual exclusion primitives could be replaced by a boolean-marked protocol. The mark algorithm is very simple, and the overhead is greatly reduced in that case.

In the next section, we describe the principle of TSA. Furthermore, we discuss our approach to the parallelization of the TSA. In Section 3, we first present the partial locking protocol and then our modified version where the overhead is reduced in the case of tree data structures. In the next two sections we show the efficiency of our partial locking protocol applied to data structures used in best first Branch and Bound and A* algorithms. Best first Branch and Bound needs suitable concurrent priority queues. We discuss several of them : Skew-Heap, Splay-Trees, and the Funnels. Then, we give some experimental results on parallel best first Branch and Bound solving the Quadratic assignment problem and the Travelling salesman problem. Parallel A* algorithm needs a particular data structure, which can support double criterion operations. A* explores a graph of solutions, so each operation have to check if a vertex of the graph is already generated or explored. We show the efficiency of the Treap data structure using our partial locking protocol for A* algorithms solving the 15-puzzle. Each experiment was performed on a KSR1 machine. Conclusions are in section 6.

2 TREE SEARCH ALGORITHMS

First we describe the principles of the *TSA*. They use in their implementation a set of nodes/subproblems obtained by decomposition of the root (the original problem). Following the given search strategy, a subproblem (i.e. an item of this set) is selected, and then is broken up into smaller problems, unless it can be proved that the resulting subproblems cannot yield an optimal solution [1] or if it can no longer be decomposed. The goal of the *TSA* is to find the ***optimal*** leaves representing an optimal solution to the problem.

Unlike the Numerical Computation which generally manipulates regular and static data structures (e.g. arrays), *TSA* uses irregular and dynamic data structures (e.g. Priority queues, hash tables, linked list) for the storage of the set, as its size is not known at the outset. It grows exponentially with the size of the problems and often remains too large to be solved successfully as the program stops due to lack of time or memory space. Parallelism is then used to speed up the construction of search trees and to provide more memory space.

2.1 Parallelization models

Research over the last decade in the parallelization of the *TSA* has shown that MIMD architecture is suitable. There exist two main high-level programming models : the shared memory model and the distributed memory model. Vendors are beginning to offer parallel machines (KSR1, CRAY T3D) with programming environments supporting both models. In these two models, each processor executes the same algorithm as the serial one. The difference is in managing the global data structure (*GDS*).

[1] its evaluation is greater than the cost of the best known solution

In the distributed model, the *GDS* could be a real distributed data structure or local data structures with load balancing. In the first case, operations are synchronous, Each processor needs the others to perform an operation. In the second case, load balancing is the issue that must be takled. As each processor can only access to its own memory, if all the data have been processed in a local memory the corresponding processor become idle while the other processors are still running.

In the shared memory model, the *GDS* is stored in the shared memory which can be accessed by each processor. The higher the access concurrency, the higher the performance. The main issue is the contention access to the data structure. Data consistency is provided by a mutual exclusion mechanism. Load balancing is easy to achieve since data structure can be accessed by any processor. The aim of this paper is to propose a solution allowing concurrent access to the Globale data structure.

3 CONCURRENT ACCESS TO GLOBAL DATA STRUCTURES

The *GDS* is stored in the shared memory. Each asynchronous processor must be able to access it. The simplest method to allow concurrent access to a data structure is for each asynchronous process to lock the entire data structure whenever it makes an access or an operation. This exclusive use of the entire data structure to perform a basic operation serializes the access to the data structure and thus, limits the speed-up obtained with the parallel algorithm.

The next section presents a better technique which is termed **Partial Locking**.

3.1 Partial Locking protocol

An operation is composed of successive elementary operations. At a given moment, a processor only needs a few nodes of the tree. Severity of contention can be reduced if each processor locks only the part of the data structure that it actually modifies. Hence, the time delay, during which the access to the structure is blocked, is decreased.

We reiterate the definition of *user view serialization* of the structure introduced by Lehman and Yao, 1981 [25], Calhoun and Ford, 1984, [4] in the context of data base management. In this approach, the processors that access the data structure with a well-ordering scheme, inspect a part of the structure that the previous processors would never change further, and leave all parts of the data structure (modified or not) in a consistent state.

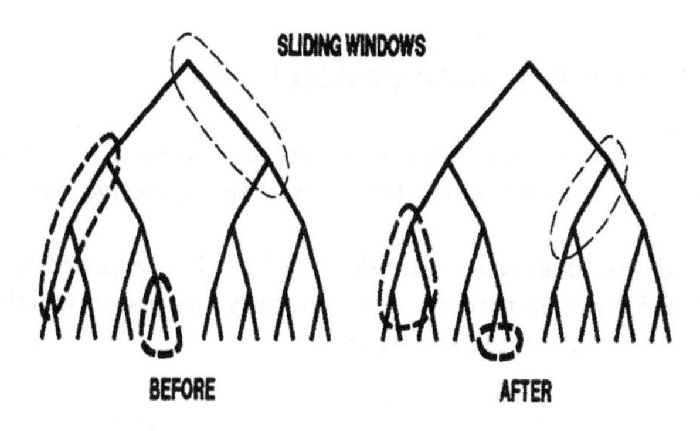

Fig. 1. Example of applying the partial locking protocol to a tree data structure.

Figure 1 shows an example of this method applied to a tree data structure. A sliding window (or a buble) of a processor executing a basic operation on a tree data structure, is defined as the set of nodes of this data structure on which the processor has exclusive access.

To forbid the creation of a cycle of processors waiting for resources (deadlocks), we have to make each processor follow a specific locking scheme.

A simple scheme of this top-down method can be described as follows. Each operating window will be moved down the path, from the root to a leaf.
Notations: an ancestor of x is defined as a node which belongs to the (unique) path between x and the root.

definition: Let S be a tree data structure in which all basic operations are executed downwards. The scheme is defined by the following rules:

Lock : a processor can request a lock on a node x only in the following cases

- **1** : if the processor has a lock on the parent of x,
- **2** : if the processor has a lock on an ancestor of x and a lock on a descendant of x,
- **3** : a processor can lock the root r of S iff the processor has no more locks in S.

Unlock : no specific rule.

The full description of the variant schemes and the proofs can be found in [6].

3.2 Partial locking Boolean protocol

If the scheme described above allows concurrent accesses, it implies an overhead due to the number of locks used : these have an expensive execution time.

With respect to the locking scheme defined in 3.1, if a processor P_i has a lock on a node A of data structure S, P_i is the only processor which can request a lock on a node B, child of A (see figure 2).

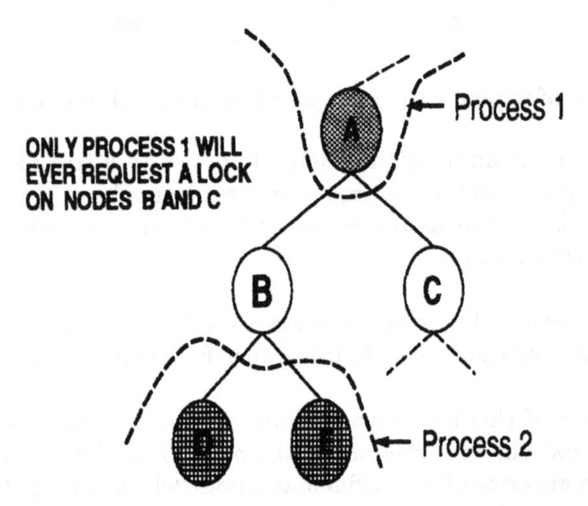

Fig. 2. Only one processor can ever request a lock on a node.

Hence, expensive mutual exclusion primitives are not needed. Locks could be replaced by a boolean-like protocol. This is just a flag, which indicates if the node is in use or free. Since the root node has no parent, several processors can try to access it. A lock will be necessary only for this node.

To illustrate this optimization, we tested a Skew-Heap using the partial locking paradigm [19] in the context of a simulation of Branch and Bound (figure 3) on the machine KSR1 with one processor. The execution time shows that the Locks protocol overhead is much greater than the boolean protocol overhead. Further, boolean protocol time is not very different than the serial time.

Serial	Locks	Boolean
22 s	57 s	28 s

Fig. 3. Times of Skew-Heap to perform 200,000 operations

3.3 Related Works

In the litterature two main kinds of techniques may be found which allow *parallel* operations on data structures for shared memory machines. They are usually called ***Parallel data structures*** and ***Concurrent data structures.***

Parallel Data structures offer several synchronous operations performed by several processors. Processors cooperate to perform their operations [10, 11, 9]. Deo and Prasad in [11] have proposed an optimal parallel heap. P processors can simultaneously perform P operations. But these P operations are synchronous. Thus, a processor that attempts to access the data structure, has to wait for the other processors. We do not want use these kinds of parallel data structures, because in our applications the time between two accesses to the data structure is never the same on two processors. Some solutions use specialized processors to handle the data structure. We prefer to employ each processor that is used to execute the tree search algorithms

Concurrent data structure algorithms allow concurrent access to a data structure by asynchronous processors. Each processor performs a serial operation which is modified in order to keep the other processors unblocked. Processors carry out their operations themselves and do not need any other processors to access to the data structure. They do not cooperate to perform their operations. These methods are very useful, if the delay between two accesses is unknown. Therefore, mutual exclusion primitives are generally used to make atomic update operations with respect to the other processors. The partial locking protocols are such techniques. Initially the idea of partial locking could be found in articles dealing with data base management. Several solutions were proposed to allow concurrent access on a Search Tree such as AVL-Tree, B-Tree and B+Tree [12, 13, 24]. Subsequently, this technique was used for different priority queue such as D-heap [2, 34], Skew-heap [19], Fibonacci Queue and priority pool [17]. Sometimes, these concurrent algorithms require specialized processors to rebuild the data structure. Our contribution is an optimized version of a partial locking method. We have reduced the overhead due to the mutual exclusion primitives. For example, Jones' Skew-heap in [19] has log n

mutual exclusion primitives calls for one operation. Our solution uses only one mutual exclusion primitive and log n boolean-like protocol calls. Figure 3 shows the importance of this optimization.

In the next two sections, we show the efficiency of our partial locking boolean protocol, applying it to data structures used in a parallel Best first Branch and Bound and in a parallel A* algorithm.

4 BEST FIRST BRANCH AND BOUND ALGORITHM

Branch and Bound algorithms are the most popular techniques for solving NP-Hard combinatorial optimization problems. Using the best first strategy, seems to be the most efficient, because the number of evaluated nodes is optimal. The data structure used in the best first branch and bound is a priority queue (PQ). From an algorithmic point of view, a Branch and Bound carries out a sequence of basic operations on the set of subproblems with their priority value. The basic operation is defined as follows :

- **DeleteMin**: gets the highest priority node. (the subproblem with the best evaluation)
- **Insert**: adds a node with a precomputed priority.
- **DeleteGreater** : removes nodes which have a lower priority than a given value (usually the upper bound).

In serial, PQ are usually represented by a heap. There exist several algorithms which manage a heap : D-heap [20], leftist-heap [20], Skew-heap [37], Binomial queue [3], Pairing heap [15, 36]. The most popular is the D-heap[2] as used in the heapsort ([38]). This is the oldest PQ implementation with $O(\log\ n)$ performance. In a heap, the priority structure is represented as a binary tree that obeys the heap invariant, which holds that each item always has a higher piority than its children. In a D-heap, the tree is embedded in an array, using the rule that location 1 is the root of the tree, and that locations 2i and 2i+1 are the children of location i. In a branch and bound, each item contains an external pointer to store the subproblem information. Bistwas and Browne [2], Rao and Kumar [34] have proposed a concurrent version of the D-heap. Both have used a partial locking protocol. Our partial locking boolean protocol cannot be used

[2] Jones in [18] called it Implicit-heap.

with the D-heap, because the operations of the algorithm require a direct access to the tree.

Serial experimental results [18, 5] show that the D-heap is not an efficient PQ implementation. Tarjan and Sleator proposed in [37] a better heap algorithm, the Skew-heap. Their algorithm is the self-adjusting version of the Leftist-heap. The basic operation used to implement DeleteMin and Insert operation is called the merge operation. Its ***amortized complexity*** is $O(\log\ n)$. It seems that the Skew-heap is one of the most efficient serial algorithms for heap implementation ([18, 5]). Jones [19] has proposed a concurrent Skew-heap using a partial locking protocol with only mutual exclusion primitives. Applying our locking boolean protocol to the Skew-heap offers better performances (see figure 3).

The complexity of the heap DeleteGreater operation is $O(n)$, because the tree must be entirely explored. Another problem is that heaps are not stable. In [27], Mans and Roucairol demonstrate that the PQ must be stable. The order of the nodes with the same priority must be fixed (LIFO or FIFO). This property avoids some anomalies of speed-up. Therefor, several PQ which are not a heap structure have been proposed. We can cite the funnels [26, 5] (table and tree) and the different Splay-trees [37, 18]

The funnel data structures were developed by Mans and Roucairol in 1990 [26] for a best-first B&B implementation. However, they can be used in many other applications. The B&B algorithms only require a small bounded interval of priority. The size of the interval is denoted by S. Initially S is equal to $ub - lb$ (ub is the cost of the best known solution and lb is the evaluation of the problem). During the execution of the algorithm, lb increases and ub decreases, then S decreases until it reaches 0. Then, there is a very simple way to achieve an efficient representation for the PQ. The idea is to use an array of FIFO of size S, where each FIFO j of the array is associated to a possible priority[3]. The serial experimental results show that the funnel table is the most efficient priority queue. Our partial locking boolean protocol cannot be used to make the access to the funnel table concurrent , because it is not a tree data structure. All details about the funnel table can be found in [26, 5]. The Funnel Tree uses a complementary binary tree with S external nodes, where S is the smallest power of 2 which is greater than $ub - lb$. Each internal node of the tree has a counter which represents the number of subproblems being contained in the FIFOs below it. This tree is used to drive the $DeleteMin$ operation to the FIFO containing the best nodes. The complexity of operation

[3] we make the assumption that priority values are integer.

is $O(\log S)$. The Funnel Tree operations are made concurrent using our partial locking boolean protocol.

The Splay-trees initially are self-adjusting binary search trees. Jones in [18] shows that Splay-trees could be used as efficient priority queues. There exist several versions of serial Splay-tree data structures depending on the Splay operation : Splay, Semi-Splay, Simple-Semi-Splay. Each of them has a Top-Down and a Bottom-Up version. These different versions of Splay-Trees support search operation on any priority. So, we create new versions of Splay-trees (called Single) where each tree node is associated to a fixed priority value. We apply our partial locking boolean protocol to the Top-Down versions of the Semi-Splay, the Simple-Semi-Splay, Single-Semi-Splay and the Single-Simple-Semi-Splay.

The funnels and splay-tree support efficient DeleteGreater operations and have the stability property.

The figure 4 recapitulates the complexity and the properties of each priority queue.

PQ	Insert	DeleteMin	DeleteGreater	Serial Perf	Stability
Funnel-Table	O(1)	O(1)	O(1)	1	Y
Splay	O(log n)	O(log n)	O(log n)	2	Y
Single-Splay	O(log S)	O(log S)	O(log S)	2	Y
Funnel-Tree	O(log S)	O(log S)	O(log S)	3	Y
Skew-Heap	O(log n)	O(log n)	O(n)	3	N
D-Heap	O(log n)	O(log n)	O(n)	4	N

Fig. 4. A summary of all priority queues

4.1 Experimental results

We have tested these data structures on a best-first B&B algorithm solving the Quadratic Assignment Problem. The Quadratic Assignment Problem (QAP) is a combinatorial optimization problem introduced by Koopmans and Beckman, in 1957, [21]. It has numerous and various applications, such as location problems, VLSI design [32], architecture design [14], etc.

The objective of QAP is to assign n units to n sites in order to minimize the quadratic cost of this assignment, which depends both on the distances between the sites and on the flows between the units. We use the algorithm of Mautor and Roucairol [28]. Their algorithm uses the depth-first strategy; we modified it to use the best-first strategy.

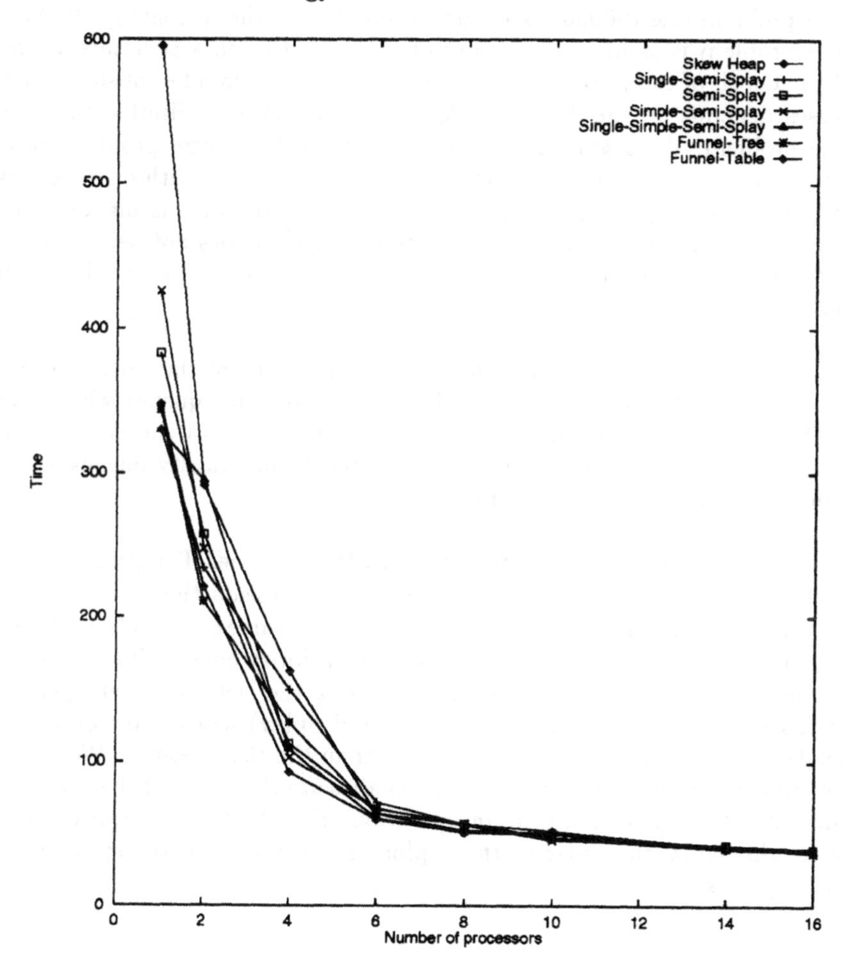

Fig. 5. Time results on KSR1 machine solving the nugent 15 problem

Figure 5 shows the results obtained on a KSR1 machine with virtual shared memory. The programs solve the nugent15 problem which was obtained from QAP Library. We can see that the Splay-Trees and the Funnels PQ are more efficient than the Skew-Heap up to 6 processors. That confirms the theoretical complexity, and that the Heap is not a good PQ representation for the Best first Branch and Bound. The single Splay-Trees seem to be more efficient than the non single. The speed-up obtained is also very good too up to these limits. With more than 6 procesors, the time no longer decreases any more and the time of the different PQ are quite the same. This phenomenon is due to the problem size (96000 nodes are evaluted with the nugent15). If the size of the problem is small, the number of nodes in the data structure is small, and the depth of the priority queue is small. Now, the number of simultaneous accesses depends on the depth of PQ. So, the speed-up is limited by the size of the problem. We began to test our program with a bigger problem (nugent 16) to verify this conclusion. However, the time obtained with 6 processors is very good according to the machine and the problem. In this instance of the problem, the Upper bound is such that the algorithm does not need to use the DeleteGreater operation. The initial upper bound is very close to the optimal value.

As the Travelling Salesman Problem is a subproblem of the QAP, we wrote another best first Branch and bound using our priority queues which solves the TSP. It is clear that our TSP Branch and Bound is not the best one. But our main subject of interest is testing the concurrent priority queues and not solving the Travelling Salesman Problem.

Figure 6 shows the results obtained on a KSR1 machine. The programs solve the swiss42 of the TSPLibrary. This figure also presents the results with the Roa&kumar Concurent Heap. We can reiterate the same comments as those for the QAP results about the scalability of these priority queues. These tests are interesting because they show the importance of the DeleteGreater operation. It seems that the priority queues (Funnel and Splay) which have an efficient DeleteGreater operation give better results than the Heap results. We can also notice that the results are irregular. Anomalies are the cause of these irregular results. As the time to perform an operation on each of these priority queues is never the same, the order of the explorated nodes is not the same for two priority queues.

5 CONCURRENT DATA STRUCTURE FOR PARALLEL A*

In this section, we present an application of the partial locking protocol (cf. section3) on a *double criteria* data structure called *treap*. This data structure is particularly suitable for the implementation of parallel A* algorithm because

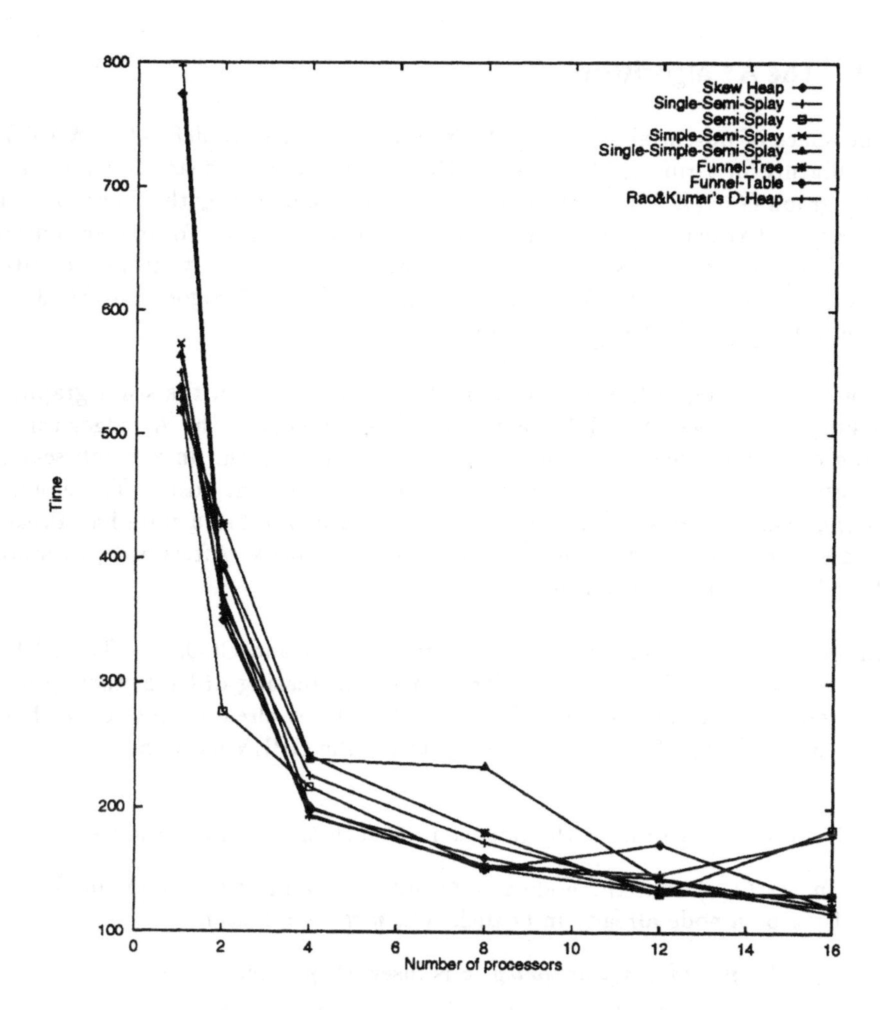

Fig. 6. Time results on KSR1 machine

there is no cross-referencing between two data structures as we will see later. Therefore, in the data structure management we save memory space and avoid deadlock problems. First, we briefly describe the principle of the A* algorithm from the data structure point of view.

5.1 The A* algorithm

The space of potential solutions of a problem is generally defined in Artificial Intelligence in terms of state space. Generally, the state space is represented by a graph $G = (E, V)$, where E is a set of edges labelled by the operators and the costs of transformation, and V a set of nodes/states. To find an *optimal* solution to a problem is equivalent to finding a least cost path from the initial state to a goal state in the associated graph. The A* algorithm [31, 33] is generally used to find such an optimal path.

When the A* algorithm is used to make a tree search on the state graph, it is similar to a best-first B&B algorithm [30]. However, the A* algorithm is more difficult to deal with when it makes a graph search. In a graph search, a state could be reached by several paths with different costs. Therefore, a comparison mechanism must be added to find out whether a state has already been explored or not and thus avoiding redundant work. Afterwards, we study the A* algorithm for graph search.

The A* algorithm manages two lists, named OPEN and CLOSED. The OPEN list contains the states to be explored in the increasing order of their priorities (f-values), while the CLOSED keeps the states already explored without priority order. The basic operations to access the OPEN list O are :

- $Search(O, state)$ finds the node x in O such that $x.state = state$;
- $Insert(O, x)$ adds the node x in O but $x.state$ must be unique in O; let y be a node already in O such that $y.state = x.state$,
  - if $y.priority < x.priority$, x is inserted, y is removed,
  - if $x.priority \leq y.priority$, x is not inserted;
- $DeleteMin(O)$ selects and removes the node x in O with the highest priority;
- $Delete(O, state)$ removes the node x from O such that $x.state = state$.

For the CLOSED list C, only three operations are necessary :

- $Search(C, state)$ finds the node x in C such that $x.state = state$;
- $Insert(C, x)$ adds the node x in C;

- *Delete*(C, *state*) removes the node x from C such that $x.state = state$.

Several parallelizations of the A* algorithm have been proposed for tree search, just as for the best-first B&B algorithm (see [16] for a large survey). But to the best of our knowledge, there are only a few specific parallel implementations of the A* algorithm [23, 8]. The reason is that the A* data structures seems to be difficult to parallelize [35]. The data structure used for OPEN is usually a combination of a priority queue (heap) and a hash table, and the one for CLOSED is a hash table. Furthermore, the parallelism increases some specific problems such as synchronization overheads, because operations have to be done in an exclusive manner. The combination of two data structures for the OPEN list also implies cross-references and thus deadlock problems and memory space overheads.

Thus, we propose a new data structure called *treap* which combines both criteria (the priority for *DeleteMin* and the state for *Search*) into one structure. This suppresses cross-references and limits synchronization overheads [7].

In the following sections we briefly discuss the Treap data structure (further details could be found in [7]).

5.2 Treap data struture

Treap was introduced by Aragon and Seidel [1]. The authors use it to implement a new form of binary search trees : Randomized Search Trees. McCreight [29] also uses it to implement a multi-dimensional searching. He uses the term *Priority Search Trees*[4].

Let O be a set of n nodes, a *key* and a *priority* are associated to each node. The keys are drawn from some totally ordered universe, and so are the priorities. The two ordered universes need not be the same.

[4] Vuillemin introduced the same data structure in 1980 and called it *Cartesian Tree*. The term *treap* was first used for a different data structure by McCreight, who later abandonned it in favor of the more commonly used *priority search tree* [29].

A *treap* for O is a rooted binary tree with a node set O that is arranged in In-order [5] with respect to the keys and in Heap-order [6] with respect to the priorities.

It is easy to see that for any set X such a treap exists. The node with the largest priority is in the root node.

The A* algorithm uses an OPEN list where a *key* (a state of the problem) and a priority (f-value of this state) are associated to each node. Thus, we can use a treap to implement the OPEN list of the A* algorithm.

Let T be the treap storing the node set O. The operations presented in the literature that could be applied on T are *Search*, *Insert* and *Delete*. We add one more operation *DeleteMin* and modify the *Insert* operation to conform to the basic operations of A* [7].

Each node contains a key and a priority. Thus, the set occupies $O(n)$ words of storage.

The time complexity of each operation is proportional to the depth of the treap T. If the key and the priority associated to a node are in the same order, the structure is a linear list. However, if the priorities are independent and identically distributed continuous random variables, the depth of the treap is $O(\log\ n)$ (the treap is a balanced binary tree). Thus, the expected time to perform one of these operations, is still $O(\log\ n)$ (n number of nodes in T) [29].

To get a balanced binary treap in an implementation for the A* algorithm, the problem is *reversed.* The priority order cannot be modified (cf. section 5.1). However, we can find an arbitrary bijective function to encode the ordered set of keys into a new set of randomized ordered keys. The priority order and the key order are then different.

Each operation goes through the Treap in the Top-Down Direction. Thus, our partial Locking boolean protocol can be applied to the Treap data structure.

[5] *In-order* means that for any node x in the tree $y.key \leq x.key$ for all y in the left subtree of x and $x.key \leq y.key$ for all y in the right subtree of x.

[6] *Heap-order* means that for any node x with parent z the relation $x.priority \leq z.priority$ holds.

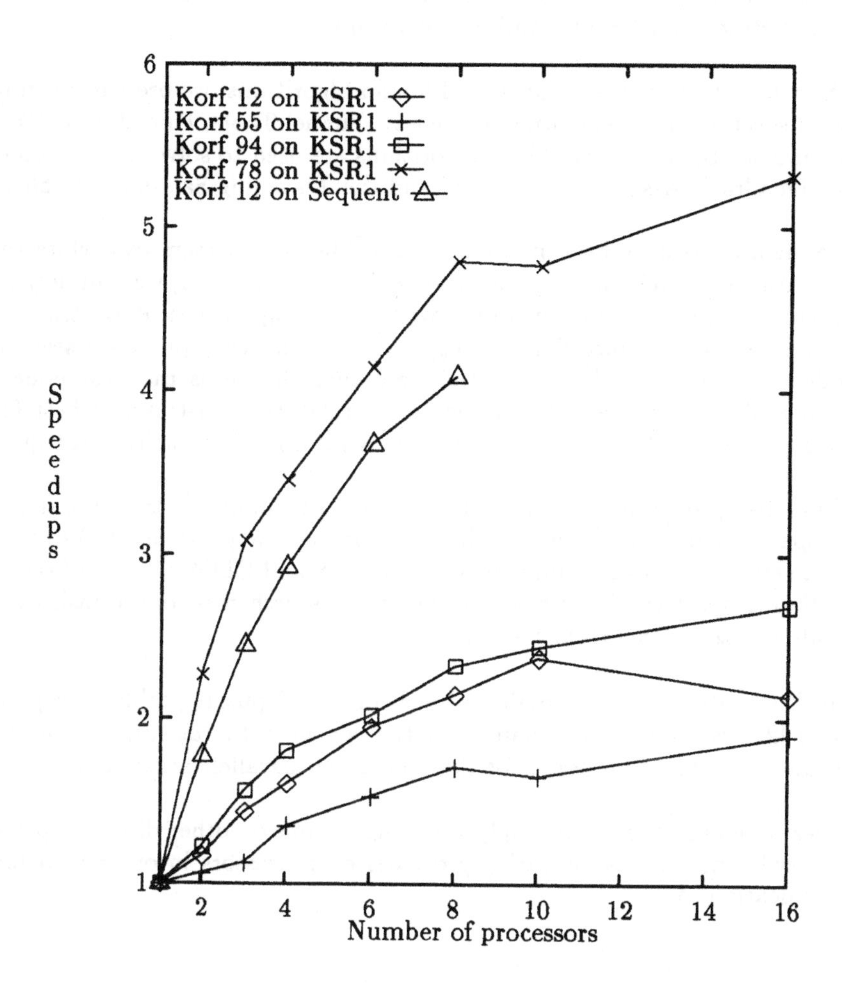

Fig. 7. Preliminary results on KSR1.

6 CONCLUDING REMARKS AND FUTURE WORKS

In this article, we have proposed an optimization of the partial locking method, which reduces the parallelism overhead in comparison with previous works. We also present the usefulness of our partial locking boolean protocol applying to data structures which are suitable for parallel Tree search algorithms such as best first Branch and Bound and A* algorithm

With regards to Best-first branch and bound algorithms, we present a comparison of several concurrent priority queues : Funnel-Table, Funnel-Tree, Skew-Heap and the Splay family. These algorithms are used to solve some instances of the Quadratic Assignment Problem and the Travelling Salesman Problem.

These algorithms have been implemented on the shared memory architecture KSR1. As the KSR1 machine has a virtual shared memory, we will decrease the memory contention and then improve the speed-up of these data structures relaxing the notion of best first strategy. At this time, each processor selecting a node with the DeleteMin operation, are sure that it is the best node in the global data structure. We can accept to relax the constraint of best first Strategy, as in a distributed memory model with a load balancing strategy.

We have also proposed a new asynchronous parallel implementation of the A* state space algorithm. It mainly differs from the few previous works [23] in the data structures used to implement the OPEN and CLOSED lists of the A* algorithm, respectively a priority queue with a search tree and a hash table. Instead, we use a *concurrent treap* and a hash table.

Applied on some relatively small instances of the 15 puzzle problem proposed by Korf [22], we can already claim that the *concurrent treap* surpasses all the other data structures proposed for the centralized parallelization of A*.

Another on-going study is to apply these algorithms on other discrete optimization problems such as allocating processes on processors in order to reduce communication cost.

REFERENCES

[1] Aragon (C.) et R (G. S.). – Randomized search trees. *FOCS 30*, 1989, pp. 540–545.

[2] Bitwas (J.) et Browne (J.). – Simultaneous update of priority structures. *IEEE international conference on parallel computing*, 1987, pp. 124–131.

[3] Brown (M.). – Implementation and analysis of binomial queue algorithmms. *SIAM Comput.*, vol. 7, 1978, pp. 298–319.

[4] Calhoun (J.) et Ford (R.). – Concurrency control mechanisms and the serializability of concurrent tree algorithms. *In : of the 3rd ACM SIGACT-SIGMOD Symposium on Principles of Database Systems*. – Waterloo Ontario, Avr. 1984. Debut de la theorie sur la serializability.

[5] Cun (B. L.), Mans (B.) et Roucairol (C.). – *Opérations concurrentes et files de priorité*. – RR n° 1548, INRIA-Rocquencourt, 1991.

[6] Cun (B. L.), Mans (B.) et Roucairol (C.). – *Comparison of concurrent priority queues for branch and bound Algorithms*. – RR n° 92-65, MASI Universite Pierre et Marie Curie, 1992.

[7] Cung (V.-D.) et Le_Cun (B.). – An efficient implementation of parallel a*. *In : Lecture Notes in Computer Science, Parallel and Distributed computing : Theory and Practice, Canada-France Conference on Parallel Computing*, pp. 153–168.

[8] Cung (V.-D.) et Roucairol (C.). – *Parcours parallèle de graphes d'états par des algorithmes de la famille A* en Intelligence Artificielle*. – RR n° 1900, INRIA, Avr. 1993. In French.

[9] Das (S.) et Horng (W.-B.). – Managing a parallel heap efficiently. *In : Proc. PARLE'91-Parallel Architectures and Languages Europe*, pp. 270–288.

[10] Deo (N.). – Data structures for parallel computation on shared-memory machine. *In : SuperComputing*, pp. 341–345.

[11] Deo (N.) et Prasad (S.). – Parallel heap: An optimal parallel priority queue. *Journal of Supercomputing*, vol. 6, n° 1, Mars 1992, pp. 87–98.

[12] Ellis (C.). – Concurrent search and insertion in 2-3 trees. *Acta Informatica*, vol. 14, 1980, pp. 63–86.

[13] Ellis (C.). – Concurrent search and insertion in avl trees. *IEEE Trans. on Cumputers*, vol. C-29, n° 9, Sept. 1981, pp. 811–817.

[14] Elshafei (A.). – Hospital layout as a quadratic assignment problem. *Operational Research Quarterly*, vol. 28, 1977, pp. 167–179.

[15] Fredman (M.), Sedgewick (R.), Sleator (D.) et Tarjan (R.). – The pairing heap: A new form of self-adjusting heap. *Algorithmica*, vol. 1, 1986, pp. 111–129.

[16] Grama (A. Y.) et Kumar (V.). – A survey of parallel search algorithms for discrete optimization problems. – Personnal communication, 1993.

[17] Huang (Q.). – An evaluation of concurrent priority queue algorithms. *In: Third IEEE Symposium on Parallel and Distributed Processing*, pp. 518–525.

[18] Jones (D.). – An empirical comparaison of priority queue and event set implementation. *Comm. ACM*, vol. 29, n° 320, Avr. 1986, pp. 191–194.

[19] Jones (D.). – Concurrent operations on priority queues. *ACM*, vol. 32, n° 1, Jan. 1989, pp. 132–137.

[20] Knuth (D.). – *The Art of Programming: Sorting and Searching*. – Addison-Wesley, 1973volume 3.

[21] Koopmans (T.) et Beckman (M.). – Assignment problems and the location of economic activities. *Econometrica*, vol. 25, 1957, pp. 53–76.

[22] Korf (R. E.). – Depth-first iterative-deepening : An optimal admissible tree search. *Artificial Intelligence*, no27, 1985, pp. 97–109.

[23] Kumar (V.), Ramesh (K.) et Rao (V. N.). – Parallel best-first search of state-space graphs : A summary of results. *The AAAI Conference*, 1987, pp. 122–127.

[24] Kung (H.) et Lehman (P.). – Concurrent manipulation of binary search trees. *ACM trans. on Database Systems*, vol. 5, n° 3, 1980, pp. 354–382.

[25] Lehman (P.) et Yao (S.). – Efficient locking for concurrent operation on b-tree. *ACM trans. on Database Systems*, vol. 6, n° 4, Déc. 1981, pp. 650–670.

[26] Mans (B.) et Roucairol (C.). – *Concurrency in Priority Queues for Branch and Bound algorithms*. – Rapport technique n° 1311, INRIA, 1990.

[27] Mans (B.) et Roucairol (C.). – *Performances des Algorithmes Branch-and-Bound Parallèles à Stratégie Meilleur d'Abord*. – RR n° 1716, Domaine de Voluceau, BP 105, 78153 Le Chesnay Cedex, INRIA-Rocquencourt, 1992.

[28] Mautor (T.) et Roucairol (C.). – A new exact algorithm for the solution of quadratic assignment problems. *Discrete Applied Mathematics*, vol. to appear, 1993. – MASI-RR-92-09 - Université de Paris 6, 4 place Jussieu, 75252 Paris Cédex 05.

[29] McCreight (E. M.). – Priority search trees. *SIAM J Computing*, vol. 14, n° 2, Mai 1985, pp. 257–276.

[30] Nau (D. S.), Kumar (V.) et Kanal (L.). – General branch and bound, and its relation to a* and ao*. *Artificial Intelligence*, vol. 23, 1984, pp. 29–58.

[31] Nilsson (N. J.). – *Principles of Artificial Intelligence.* – Tioga Publishing Co., 1980.

[32] Nugent (C.), Vollmann (T.) et Ruml (J.). – An experimental comparison of techniques for the assignment of facilities to locations. *Operations Research*, vol. 16, 1968, pp. 150–173.

[33] Pearl (J.). – *Heuristics.* – Addison-Wesley, 1984.

[34] Rao (V.) et Kumar (V.). – Concurrent insertions and deletions in a priority queue. *IEEE proceedings of International Conference on Parallel Processing*, 1988, pp. 207–211.

[35] Rao (V. N.), Kumar (V.) et Ramesh (K.). – *Parallel Heuristic Search on Shared Memory Multiprocessors : Preliminary Results.* – Rapport technique n° AI85-45, Artificial Intelligence Laboratory, The University of Texas at Austin, Juin 1987.

[36] Stasko (J.) et Vitter (J.). – *Pairing Heap: Experiments and Analysis.* – Rapport technique n° 600, I.N.R.I.A., Fév. 1987.

[37] Tarjan (R.) et Sleator (D.). – Self-adjusting binary search trees. *Journal of ACM*, vol. 32, n° 3, 1985, pp. 652–686.

[38] Williams (J.). – Algorithm 232: Heapsort. *CACM*, vol. 7, 1964, pp. 347–348.

8

A DISTRIBUTED IMPLEMENTATION OF ASYNCHRONOUS PARALLEL BRANCH AND BOUND

Ricardo Corrêa* and Afonso Ferreira**

** LMC – IMAG*
46, av. Félix Viallet, 38031
Grenoble Cedex, France
Ricardo.Correa@imag.fr

Partially supported by a brazilian CNPq fellowship, grant 201421/92-5 (BC)

*** CNRS – LIP – ENS-Lyon,*
46, allée d'Italie, 69364,
Lyon Cedex 07, France,
ferreira@lip.ens-lyon.fr

This work was partially supported by DRET and the project Stratagème of the French CNRS. It was done in relation to the ESPRIT III - HCM project SCOOP – Solving Combinatorial Optimization Problems in Parallel.

ABSTRACT

Branch-and-bound is a popular method for searching an optimal solution in the scope of discrete optimization. It consists of a heuristic iterative tree search and its principle lies in successive decompositions of the original problem in smaller disjoint subproblems until an optimal solution is found. In this paper, we propose new models to implement *branch-and-bound* in parallel, where several subproblems are concurrently decomposed at each iteration. In our ***shared data-object model***, processors communicate through a ***distributed shared memory***, that allows subproblems to be ***totally*** or ***partially*** ordered. We show that a variation from the ***synchronous*** to ***asynchronous*** parallel *branch-and-bound* allows optimizations of the operations over the data stru-

A. Ferreira and J. D. P. Rolim (eds.), Parallel Algorithms for Irregular Problems: State of the Art, 157–176.

cture that improve their performance and, possibly, the overall performance as well. Finally, some experiments are described that corroborate our theoretical previsions.

1 INTRODUCTION

The use of parallel algorithms for solving NP-hard problems becomes attractive as parallel systems consisting of a number of powerful processors offer large computation power and memory storage capacity. Even though the parallelism will not be able to overdue the assumed worst case exponential complexity of those problems (unless an exponential number of processors is used), the average time complexity of heuristic search algorithms for many NP-hard problems is polynomial. Consequently, those parallel systems, possibly with hundreds or thousands of processors, give us the perspective of solving relatively large instances of NP-hard problems. Hence, it is of major importance the study of ways to parallelize existing sequential methods used for solving NP-hard problems.

Among the universe of NP-hard problems, we consider in this report a *discrete optimization problem*, stated as follows. Let n be a positive integer, $f : S \rightarrow R$ be a function from a finite domain $S \subset \mathbb{N}^n$ onto a completely ordered set R, $x \in S$ be a *feasible solution* and $f(x)$ be the *value* of x. A feasible solution x is *better* than $x' \in S$ if $f(x) < f(x')$. We search an *optimal solution* x^* (we denote $f^* = f(x^*)$), where $f^* \leq f(y)$, for all $y \in S$. Maximization problems can be defined and dealt with similarly.

The search for optimal solutions constitutes the most important problem in the scope of discrete optimization. Improving the search efficiency is of considerable importance since exhaustive search is often impracticable, especially if the problem is NP-hard. The method called *branch-and-bound* (noted B&B) is often used as an intelligent search in this context, and parallelism has also been widely studied as an additional source of improvement in search efficiency. This paper focuses on *parallel best-first branch-and-bound algorithms* for *distributed memory* parallel machines. These machines are composed of a set of processors, each one with its local memory and with no physically shared memory. The communications among the different processors are implemented through the exchange of messages.

The principle of a B&B algorithm lies in successive decompositions of the original problem in smaller disjoint subproblems until an optimal solution is found.

A *state space tree* $T = (\mathcal{V}, \beta)$ describes these subproblems (the set of vertices $\mathcal{V}$, rooted at S) and the decomposition process (the set of edges β). It consists of a heuristic iterative search in T that avoids visiting some subproblems which are known not to contain an optimal solution. The principle of the high level parallelization considered in this paper consists of concurrently decomposing several subproblems at each iteration. This parallel B&B is traditionally considered as a non-regular parallel algorithm due to the fact that the structure of the state space tree is not known beforehand [1]. As a consequence, several special techniques have been developed to manage the problems related to the irregularity of the state space tree or of the parallel tree searching process, essentially related to dynamic workload sharing [2, 6, 7, 10, 14, 18].

In this work we study another approach, where the structure of the state space tree is not taken into account, avoiding the troubles inherent to irregularity. For this purpose, the two parallel versions we consider, namely *synchronous* and *asynchronous*, assume a model, called *shared data-object model (SDOM)*, that allows processors to communicate through a *global data structure* ($\mathcal{D}$) [4, 12]. The *global data structure* keeps subproblems generated but not decomposed of T. This model permits $\mathcal{D}$ to be consistently shared by all processors, i.e., whenever any processor invokes an operation on $\mathcal{D}$, any other processor that subsequently operates on $\mathcal{D}$ will retrieve the result of the operation just performed. The motivation for using the SDOM for implementing parallel B&B stems from the fact that it provides a *global data structure*, in contrast with implementations based on distributed lists.

An SDOM consists of an abstract data type in which mutual exclusive accesses are provided implicitly and the processors communicate by means of applying the operations over the *global data structure*, with an associated latency time. The operations are executed indivisibly, i.e., the model guarantees serializability of operation invocations; if two operations are applied simultaneously, then the result is as if one of them was executed before the other. The order of invocation, however, is non-deterministic.

A traditional question related to parallel iterative algorithms, parallelized by separating it into several local algorithms operating concurrently at different processors, is comparative performances of their synchronous and asynchronous approaches [3]. For instance, in a parallel B&B, the subproblems are allocated to different processors and the local algorithms accomplish the subproblems decompositions. In the synchronous case, once an operation request has been issued to $\mathcal{D}$, the local algorithms remain idle due to the latency time associated to $\mathcal{D}$. On the other hand, the main characteristic of an asynchronous algorithm is that the local algorithms do not have to wait at predetermined points of the execution for predetermined data to become available. We thus allow some processors to operate over $\mathcal{D}$ more frequently than others, and we allow the latency time to be substantial and unpredictable.

Asynchronism may yield the following main advantages: a reduction of the synchronization penalty, overlap between communication and computation, and a potential speedup over synchronous algorithms in some applications. Further, it may provide a smaller number of accesses to $\mathcal{D}$. On the other hand, we may accomplish more unnecessary subproblem decompositions, and the detection of termination tends to be somewhat more difficult to implement.

In [4], we introduced a new asynchronous model under the SDOM and gave some guidelines for implementations. The goal of this paper is to present some experimentations of asynchronous parallel B&B including a *global data structure*, with the purpose of analyzing the suitability of the SDOM in parallel B&B. Some of the concepts presented in [4] were revisited and yielded efficient implementations of the models studied. The remainder of the paper is organized as follows. In sections 2 and 3, the definitions of, respectively, sequential, and synchronous and asynchronous B&B are presented, including the restrictions of the models employed. In sections 4 and 5, we describe the implementations of each model. Section 6 consists of some experimental results on the *knapsack problem* and, finally, section 7 consists of conclusions and perspectives for further research.

2 PRELIMINARIES ON SEQUENTIAL B&B

A sequential B&B algorithm consists of a sequence of *iterations*, in which some operators are applied over a *data structure* $\mathcal{D}$. This *data structure* keeps a priority list of subproblems. The operators choose a subproblem from $\mathcal{D}$, decompose it and eventually insert new subproblems into the *data structure*, chosen among those generated by the decomposition. A decomposition may generate singletons, and the one with the best feasible solution so far is kept, as well as its value. At some points of the execution, some subproblems shown not to contain a feasible solution better than the current best known feasible solution are eliminated [5, 15].

We call *open subproblems* the subproblems that contain a feasible solution and that, at any point of the execution, were generated but neither decomposed nor eliminated. At each iteration, the *data structure* $\mathcal{D}$ is composed of a collection of open subproblems, of an ***upper bound*** U, which is the best value found so far, and of a feasible solution whose value is U. The execution of the algorithm starts in the initial state $\mathcal{D}_0 = (\{S\}, U_0)$, where $U_0 \geq f^*$ is a constant (possibly infinite) corresponding to the initial upper bound, until the final state $(\emptyset, f^*)$. The operators applied at each iteration are four, namely, ***selection***, ***evaluation***, ***insertion*** and ***decomposition***. We say that a processor ***selects***, ***decomposes*** and ***inserts*** when it performs, respectively, a ***selection***, a ***decomposition*** and an ***insertion*** on $\mathcal{D}$.

The *selection* function, applied to a collection of subproblems V, is any δ such that if $V \neq \emptyset$ then $\delta(V) \neq \emptyset$, indicating that if there is an open subproblem at the beginning of an iteration, then at least one such a subproblem will be chosen for decomposition, and $\delta(\emptyset) = \emptyset$. Let l be a ***lower bound*** function that associates to each open subproblem v in the state space tree a value smaller than or equal to the value of the best feasible solution in v, $l(v) \leq \min_{x \in v}\{f(x)\}$. Let h be a ***heuristic priority function*** that applied to an open subproblem determines its priority for decomposition. The subproblems in $\mathcal{D}$ are available in a nondecreasing order of h. We concentrate on *best-first* search, where $h(v) = l(v)$ and δ selects the subproblem with smallest priority.

After the selection of a subproblem v, a ***decomposition*** creates a set of new subproblems corresponding to a partition of v (ensuring that T is a tree). At an iteration, the decomposition operation is the only operation that may generate feasible solutions. Thus, the new upper bound is always generated just after a decomposition. Furthermore, the lower bound l of each subproblem is calculated when this subproblem is generated by a decomposition.

An *evaluation* eliminates all subproblems v such that $l(v) \geq U$, allowing an intelligent search in T that avoids considering subproblems that not lead to an optimal solution of the original problem. Under the SDOM, an evaluation is performed whenever the upper bound is updated.

Finally, by *insertion* we mean that an open subproblem is included in $\mathcal{D}$ in such a way that the set of open subproblems remains sorted. In particular, in the sequential version, all the open subproblems generated by a decomposition are inserted.

3 MODELING SYNCHRONOUS AND ASYNCHRONOUS PARALLEL B&B

Several papers concerning models for synchronous parallel B&B can be found in the literature [2, 5, 11, 13, 17, 18]. In this section, we first recall its definition, then we describe a new and general asynchronous model of parallel B&B, both models under the SDOM (figure 1). The synchronous model, which outlines the role of $\mathcal{D}$, is only useful for efficient implementations in special cases, but it is used as a reference of efficient *qualitative workload sharing*. The asynchronous approach employed keeps the modularity of the method and allows implementations that satisfy the following important features in parallel B&B (some of these features were mentioned in [19]):

F1. they should be modularly expandable to ensure scalability up to a very large number of processors;

F2. it should use distributed control so that a centralized controller would not become the bottleneck in future system expansions;

F3. they should present efficient *qualitative workload sharing*.

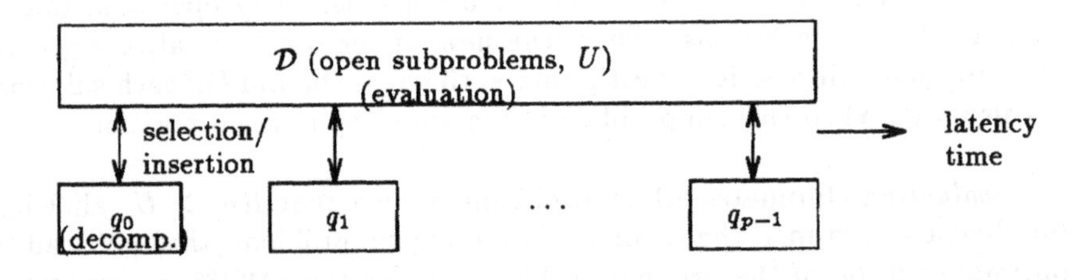

Fig. 1. Overall description of the SDOM with p processors.

The items F1 and F2 refer essentially to the implementation of $\mathcal{D}$. The item F3 applies to the selection operations. Let us consider as *useful work* the decomposition of the subproblems v such that $l(v) \leq f^*$. By *efficient qualitative workload sharing* we mean that each processor should accomplish approximately the same amount of useful work.

3.1 Synchronous parallel B&B model

The synchronous formulation adopted in this work consists of a set of p processors working iteratively and synchronously, where each iteration consists of five synchronous steps:

S1. each processor selects (non-empty selections, unless there are not enough open subproblems in $\mathcal{D}$);

S2. each processor decomposes its selected subproblem independently from the others (eventually generating feasible solutions);

S3. the processors that generated feasible solutions in step S2 update the new upper bound together (some subproblems may be eliminated);

S4. each processor inserts all subproblems generated in step S2 and not eliminated in step S3.

We note that this synchronous model determines that *a selection is not performed before all the insertions of the previous iteration have finished* (see figure 2). Such a behavior implies the wasted times, indicated in figure 2, determined by the waiting states for insertion and selection, and for synchronization (idle states). Therefore, synchronous B&B severely restricts the set of possible optimizations of operations over $\mathcal{D}$, yielding high latency times. On the other hand, the open subproblems are *totally ordered*, i.e., the several subproblems selected at each iteration are guaranteed to be the current open subproblems with smallest priority function values h. This fact ensures the best workload sharing, in terms of the heuristic function values, at each iteration.

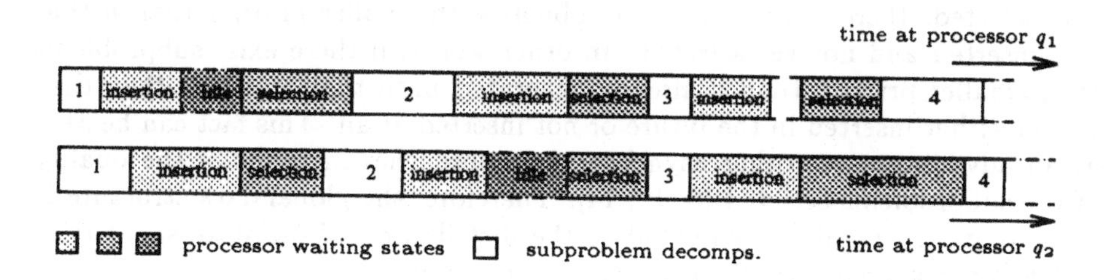

Fig. 2. An example of a synchronous execution with 2 processors.

3.2 Asynchronous parallel B&B model

In the asynchronous model, each of the p processors works in its own pace, only limited by the time needed to operate the data structure, carrying out their sequences of operations independently from the other processors. This model avoids synchronization restrictions and represents executions allowing one or more selections from a given processor q_i after another processor q_j has selected but before it has accessed $\mathcal{D}$ in order to perform the corresponding evaluation or a corresponding subproblem insertion. We reformulate the asynchronous model in [4] so that subproblems generated but not inserted in $\mathcal{D}$ can be decomposed (item A3 below). This modification in the model simplifies the implementations (see section 5), while maintaining the theoretical results presented in [4].

Although each processor works in a sequence of iterations, there is no synchronization among processors. In this case, an iteration is characterized by:

A1. each processor selects (at least one non-empty selection);

A2. each processor decomposes its selected subproblem independently from the other processors;

A3. sequential execution on non-inserted open subproblems until some condition holds (e.g., until there are no non-inserted open subproblems);

A4. each processor having found a feasible solution updates the upper bound;

A5. each processor inserts some of its best open subproblems (possibly none).

In spite of the asynchronism, the model guarantees that the open subproblems are *partially ordered*. By *partially ordered* we mean that whenever a subproblem v is selected, there is no other subproblem with smaller priority than v that was inserted and not yet selected. In other words, if there exist subproblems with smaller priority than v already generated until that moment, then they will be either inserted in the future or not inserted at all. This fact can be seen as an advantage over other asynchronous approaches in terms of the quality of the subproblems selected [8, 10, 14]. Therefore, the global data structure is used to avoid irregularities related to the distribution of the state space tree, which is not known beforehand, over the processors.

The example in figure 3 illustrates the capacity of the asynchronous model for avoiding waiting states. In this case, in contrast with the synchronous example in figure 2, all the processors remain working on subproblem decompositions

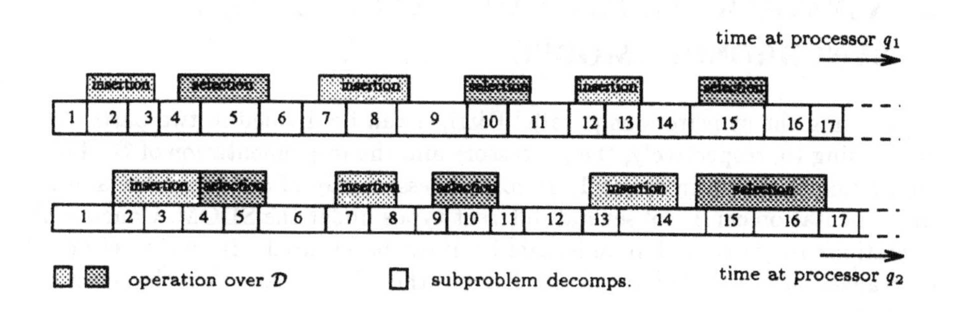

Fig. 3. An example of an asynchronous execution with 2 processors.

while operating on $\mathcal{D}$. This supposes that each processor keeps a number of local subproblems not inserted in $\mathcal{D}$. This asynchronous behavior stems from the inherent non-determinism of the operations of a B&B algorithm, i.e., different time durations for operations in $\mathcal{D}$ as well as different time durations for decomposition and computation of lower bounds.

Table 1 summarizes the positive features of each model, when compared to parallel B&B formulations based on distributed lists and no global data structure.

parallel B&B under SDOM	
SYNCHRONOUS	**ASYNCHRONOUS**
the open subproblems are totally ordered, inducing at each interaction the best workload sharing	open subproblems are partially ordered
	reduction of the synchronization penalty
	overlap of communication and computation
	hidden memory latency times
possibly less decompositions	possibly less operations on $\mathcal{D}$

Table 1. Synchronous and asynchronous positive features.

4 A DISTRIBUTED IMPLEMENTATION OF THE SYNCHRONOUS MODEL

Concerning the synchronous parallel B&B, it can be divided in two parts, corresponding to, respectively, the processors and the implementation of $\mathcal{D}$. Their connection is shown in figure 1. $\mathcal{D}$ consists essentially of a priority list and its implementation can be of several different types under the SDOM. Concurrent operations may occur, but indivisibility must be ensured. It can be either a centralized or a distributed priority list and the critical element is the selection operation.

The use of this model with a centralized strategy was investigated by several authors [2, 6, 10]. We adopt a distributed strategy where each processor q_i maintains a local data structure $\mathcal{D}_i$ containing open subproblems and a local upper bound U_i (see figure 4). In this strategy, an algorithm that provides a distribution among the p processors of the p open subproblems with smallest priority, taken over all local data structures, called ***best-nodes-exchange***, is used at each iteration. At each iteration of best-nodes-exchange, the global upper bound is updated and global termination is tested.

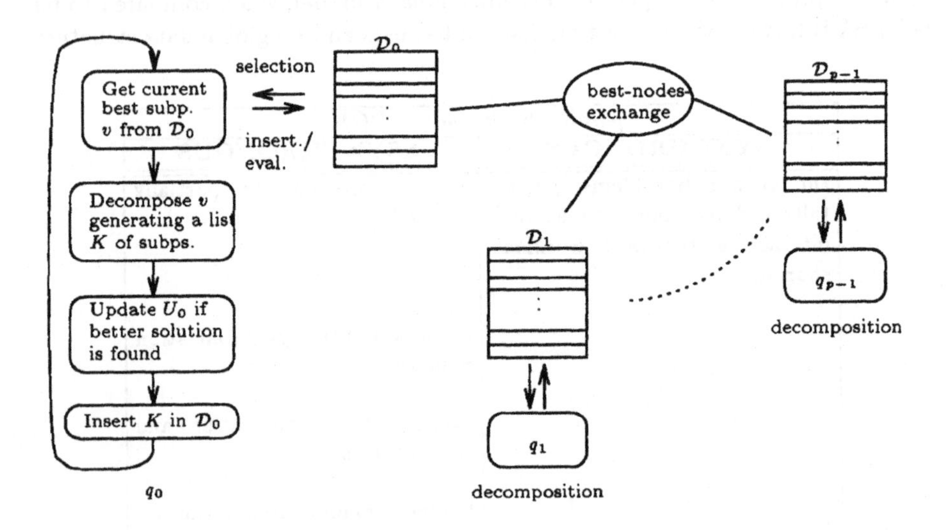

Fig. 4. A general scheme for the synchronous parallel B&B.

5 A DISTRIBUTED IMPLEMENTATION OF THE ASYNCHRONOUS MODEL

In this section, we describe a distributed implementation of an asynchronous parallel B&B. It is based on the synchronous implementation of the previous section, using the same distributed strategy adopted in that section. The asynchronous nature is implemented by a slight modification of the interaction between each processor q_i and its $\mathcal{D}_i$, which is shown in figure 5. The principles described in the sequel are also valid in the case of a centralized strategy, except that $\mathcal{D}$ is implemented differently. We keep in mind that the conditions A1-A5 from subsection 3.2 must be satisfied.

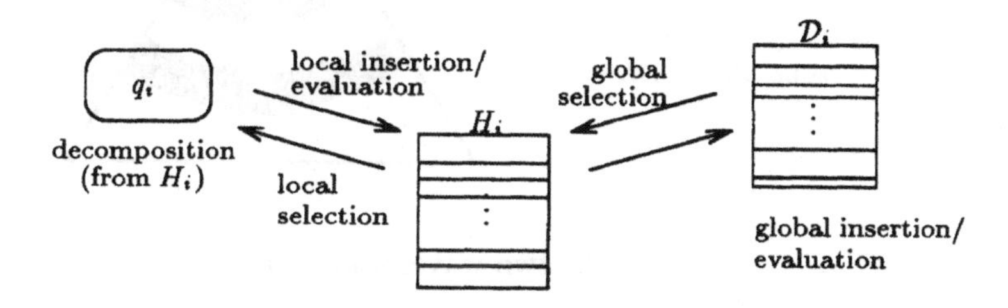

Fig. 5. Asynchronous B&B as a slight modification of synchronous B&B.

The operations from each processor q_i over $\mathcal{D}$ are addressed first to the corresponding *local data structure* H_i (also a priority list). Such operations are not really accomplished until they are performed over $\mathcal{D}$. Consequently, seen from the processors, the operations over the global data structure have the latency time determined by the accesses to the local data structure, hiding the latency time associated to $\mathcal{D}$. In algorithmic terms, two techniques are used:

- *delaying the effect of an insertion:* too many insertion operations over $\mathcal{D}$ can degenerate their individual and overall performance. Thus, some of these operations are delayed in order to keep a given performance;
- *pre-selections:* in order to be able to maintain low latency times and to keep each processor q_i busy, a subproblem v can be selected from H_i before it has been inserted in $\mathcal{D}$.

With these techniques, the search in T can be seen as two waves propagating from the root to the leaves (see figure 6) [4]. The deepest wave represents the local open subproblems. The pre-selections move the deepest wave away from the other one, that represents the inserted and not selected open subproblems (the execution over $\mathcal{D}$), while the operations performed in $\mathcal{D}$ bring the two waves closer to each other. The latency time of the operations over $\mathcal{D}$ is used to decompose pre-selected subproblems that, otherwise, could be decomposed in the future. These decompositions overlap the communications related to the execution of the best-nodes-exchange.

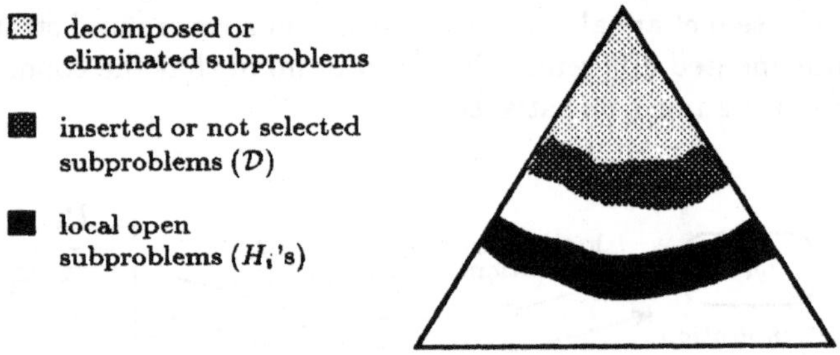

decomposed or eliminated subproblems

inserted or not selected subproblems ($\mathcal{D}$)

local open subproblems (H_i's)

Fig. 6. The two waves search in T.

Decompositions of pre-selected subproblems eventually find feasible solutions and, consequently, some unnecessary operations over $\mathcal{D}$ can be avoided. First, if all feasible solutions of a pre-selected subproblem v were found, then v and all its descendants can be eliminated before insertion in $\mathcal{D}$. Second, a feasible solution found in a decomposition of a pre-selected subproblem can be update the upper bound U and, then, eliminate either some subproblems in the local data structure, avoiding future insertions in $\mathcal{D}$, or some subproblems in $\mathcal{D}$, avoiding future selections.

The asynchronism can be controlled from the extreme case where there are no pre-selections to the other extreme, where no operations are performed over $\mathcal{D}$. This control balances synchronous order constraints, against hiding memory latency time, and depends both on the frequency of operations over $\mathcal{D}$ and on the quantity of subproblems reserved to insertion in $\mathcal{D}$ from each H_i. We adopt a strategy, based on the ***blackboard communication strategy*** [10], described as follows. For convenience, when H_i or $\mathcal{D}_i$ is empty, we consider that the corresponding smallest lower bound (and priority) is infinite.

Let us consider an execution of the algorithm in figure 7. Initially, we define a *threshold* for each processor q_i, which is used to control the distance between the two waves mentioned above. Let *difference* designate the difference between the lower bound of the smallest priority subproblems in H_i and in $\mathcal{D}_i$. Then, during the asynchronous execution, while the *difference* is smaller than the threshold, q_i proceeds its sequence of operations on H_i, decreasing the threshold after each selection from H_i. The threshold is decreased at a fixed rate α until it becomes smaller than or equal to the *difference*. Then, if $\mathcal{D}_i$ is not empty, the smallest priority subproblem in $\mathcal{D}_i$ is selected and decomposed, and the new generated subproblems are inserted in H_i.

1. **begin**
2.   **while** $\mathcal{D}$ is not empty **do**
3.   **while** the *difference* is smaller than threshold **do**
4.   operate from H_i;
5.   decrease threshold by α;
6.   **if** $\mathcal{D}_i$ is not empty **then**
7.   get the smallest priority subproblem from $\mathcal{D}_i$, decompose it and put the new generated subproblems into H_i;
8.   get the s' smallest priority subproblems from H_i and put them into $\mathcal{D}_i$;
9.   set threshold to its initial value;
10.   best-nodes-exchange in parallel with decompositions from H_i;
11. **end**

Fig. 7. Asynchrounous B&B.

Let now s be a constant parameter, $0 < s < p$, that represents the maximum number of subproblems transferred from a H_i to $\mathcal{D}_i$ at an iteration, and s'' be the number of subproblems of H_i satisfying the following two conditions:

a. the lower bound is smaller than the lower bound of the smallest priority subproblem in $\mathcal{D}_i$, and

b. the difference of its lower bound to that of the smallest priority subproblem in $\mathcal{D}_i$ is greater than or equal to the threshold (possibly none).

Then, the $s' = \min\{s, s''\}$ smallest priority subproblems of H_i are transferred to $\mathcal{D}_i$. After this transfer, the threshold is reassigned its initial value (see figure 7). Finally, the line 10 represent the simultaneous execution of the best-nodes-exchange and decompositions of pre-selected subproblems. All this procedure is repeated until $\mathcal{D}$ becomes empty.

The objective when tolerating some gap between lower bounds in $\mathcal{D}$ and in local data structures is to limit the number of operations on $\mathcal{D}$. As a consequence, we also limit the number of best-nodes-exchange executions since it is executed only if an operation on some $\mathcal{D}_i$ is performed. This tolerance assumes the simple and reasonable principle that, immediately after a best-nodes-exchange execution, the workload is evenly distributed over the processors. Thus, we set the threshold to its initial (maximum) value (figure 7, line 9). On the other hand, as each processor works between two best-nodes-exchange executions (increasing asynchronism), the probability of workload unbalance increases. Thus, we decrease the threshold by multiplying it by a constant rate $\alpha < 1$, increasing the probability of an operation on $\mathcal{D}$.

Two optimizations on the asynchronous algorithm are possible. First, it is not required that each $\mathcal{D}_i$ holds more than p subproblems at any point of the execution, since only p subproblems are involved in a best-nodes-exchange. Then, if more than p subproblems are present in a $\mathcal{D}_i$ after an insertion, some of the greatest priority subproblems can be transferred from $\mathcal{D}_i$ to H_i, deleting their previous insertions. Second, let v be the subproblem with the smallest priority among the s' subproblems in line 8. Its priority can be compared to the smallest priority in $\mathcal{D}_i$. If the former is smaller, v is not transferred to $\mathcal{D}_i$. Hence, v is decomposed before being inserted in $\mathcal{D}$. Let us analyze the advantage of this mechanism. If v was placed among the p open subproblems with smallest priority over all the processors, than it would be selected, by some processor, in the next iteration. Thus, the latency time associated to $\mathcal{D}$ is prevented when selecting it directly. Otherwise, it would be inserted, but not selected. Then, there is no loss of useful information in $\mathcal{D}$ in the next iteration.

6 EXPERIMENTAL RESULTS AND COMPARISONS

The theoretical models and previsions provided in previous sections are corroborated by the experimental results presented in this section. We adopt, besides the setting of the distributed implementation of $\mathcal{D}$ in sections 4 and 5, the ***knapsack problem*** defined below. The major objective of these implementations is not to provide a good algorithm for the ***knapsack problem***, but rather to validate the previous models and features in table 1, through extensive testing.

Given a set of n objects, to which are associated ***weights*** w_i and ***profits*** c_i, $i = 1, 2, \cdots, n$, the ***0-1 knapsack problem*** consists of finding a subset M of objects that maximizes[1] $\sum_{i \in M} c_i$, subject to $\sum_{i \in M} w_i \leq C$, where C is the ***knapsack capacity***. All the weights, profits and knapsack capacity are positive integers. In our experiments, the values of w_i and c_i are randomly generated, considering a correlation factor $r = 0.0001$, as $w_i = rand(10^6)$ and $c_i = w_i - 10^6 r + rand(2 \cdot 10^6 r)$, such that $0 \leq c_i < 10^6$. The function $rand(a)$ generates randomly a real number between 0 and $a - 1$. The knapsack capacity is set to $C = 10^6 n/4$ and the sequential B&B algorithm used is the one described in [16].

The results shown in tables 2, 3 and 4 correspond to averages over 20 randomly generated instances, with $n = 180$, on an Intel Paragon parallel computer with 32 processors. We set $threshold = 0.5$, $\alpha = 0.95$ and $s = 2$. The average ***sequential*** time is 3.5 min. In table 2, ***#it. ratio*** is the ratio between the number of iterations in sequential and the corresponding synchronous implementation. In table 3, ***#subp. ratio*** is the ratio between the number of subproblems decomposed in the asynchronous case and in the sequential implementation.

An important point to be noted is that the B&B algorithm for the knapsack problem is fine grained, in the sense that a subproblem decomposition and its lower bound computation are simple. This is the major reason for disappointing time speedups obtained with the synchronous implementation. However, we note that the number of iterations decreases linearly with p, due to the global data structure approach (see table 2). Thus, for this synchronous algorithm applied to the knapsack problem, the best qualitative workload sharing obtained at each iteration implies a very good workload sharing as a whole. It implies that the execution of the synchronous algorithm, in terms of subproblems decomposed and order of decompositions, can be used as a reference to measure the ***quality*** of an asynchronous algorithm; if an algorithm asynchronously decomposes the same subproblems, in the same order, then this algorithm explores the advantages of both synchronous and asynchronous models.

In spite of the fact that the application (the knapsack problem) is fine grained, the asynchronous implementation attains reasonable time speedups. In table 3, the ***speedup*** is calculated as the ratio between the time in parallel and the time in sequential. Along with the time speedup values, it is shown the number of best-nodes-exchange invocations, which are much more frequent in the synchronous implementations (one invocation per iteration). Comparing to the synchronous B&B, this implies less communications in the asynchronous implementation, although a loss in the qualitative workload sharing and an increasing in the

---

[1] This maximization problem can be easily transformed in a minimization one by setting $c'_i = -c_i$.

# proc.	# it.	# it. ratio
1	1694466	1.00
2	847235	2.00
4	423621	4.00
8	211822	8.00
16	106037	15.98
32	53725	31.54

Table 2. Comparison between synchronous and sequential, $n = 180$ and $r = 0.0001$.

number of subproblems decomposed are verified. Clearly, there is a trade-off to be found between best-nodes-exchange invocations and the number of subproblems decomposed.

# proc.	# subp.	# subp. ratio	speedup	# global op.
1	1694466	1.00	1.00	-
2	1694466	1.00	1.95	25634
4	1711411	1.01	3.76	15979
8	1723079	1.02	7.36	8934
16	1816415	1.07	14.24	4659
32	1849750	1.09	27.2	2529

Table 3. Comparison between asynchronous and sequential, $n = 180$ and $r = 0.0001$.

We recall that the number of best-nodes-exchange invocations compared to the number of subproblems decomposed gives an idea of the latency time associated to $\mathcal{D}$, which is very large when compared to the time for decomposition and lower bound computation. One of the most important features of our proposed and implemented asynchronous model (see table 1) is the ability to hide this latency time by efficiently using the local data structures (see tables 3 and 4). In table 4, *#subp. (max)* is the maximum, among all the processors, of the number of decomposed subproblems in the asynchronous case. The column *#subp. ratio* gives the ratio between the maximum number and the total number of subproblems decomposed (table 3). The column *%from sync.* shows the increase on the maximum number of subproblems decomposed by a processor in an asynchronous execution compared to the corresponding synchronous execution.

# proc.	# subp. (max)	# subp. ratio	% from sync.
1	1694466	1.00	-
2	849145	0.501	0.2
4	428895	0.2506	1.24
8	220633	0.1217	4.16
16	117306	0.058	10.6
32	59866	0.027	11.4

Table 4. Comparison between synchronous and asynchronous, $n = 180$ and $r = 0.0001$.

We finally remark that the asynchronism does not substantially increase the number of subproblem decompositions when the number of processors employed is relatively small (32). Nonetheless, in all asynchronous cases, some useless work is accomplished. There are two main complementary reasons for this, both related to the application instances: their fine grained and hardness natures. First, the memory latency time is large, yielding too many subproblem decompositions during this time. Second, we considered hard instances of the knapsack problem, where the difference of lower bound among the subproblems is very small. Then, the occurrence of subproblems with nearly the same lower bound is frequent, yielding the possibility of "bad" selections [5, 11]. Larger instances should minimize this effect.

7 CONCLUSIONS AND FURTHER RESEARCH

In this paper, we studied an approach for parallel best-first B&B as a generalization of a known sequential model, based on a shared data-object model (SDOM). It allows the formulation of the two models of parallel B&B, synchronous and asynchronous. These two models were implemented, using the knapsack problem as an application. The key point in these implementations is the shared data-object model (and its distributed global data structure). The analyses and comparisons of the results obtained with the implementations corroborated the positive features of the models (table 1, page 165), with respect to the important features in parallel B&B described in section 3 (F1-F3). Generally speaking, the SDOM is a structured, easily expandable approach, which makes it suitable for parallel B&B tools development [9]. The only tasks left to the user would be the implementation of the decomposition and the lower bound computation routines, as well as the setup of some parameters (*threshold*, α and s, section 6).

The synchronous model is important because it attains a very good quality workload sharing for best-first B&B. However, its time speedup is poor because of synchronization constraints and because of the fine grained application considered. On the other hand, the asynchronous model is flexible and allowed reasonable time speedup values in our implementations, mainly due to two factors, when compared to the synchronous implementation. First, synchronization penalties were eliminated. Second, less global data structure operations were performed. These facts did not increase the number of decomposed subproblems substantially. Thus, the major feature of the synchronous model, namely the good quality workload sharing, was partially maintained.

The flexibility of the asynchronous model was not completely explored in our implementations (the results presented in section 6 correspond to an ongoing work), and we are improving the implementation so that better speedups and scalability can be achieved. These improvements include variations in the degree of asynchronism according to the problem to be solved, in order to balance the advantages of each model (synchronous and asynchronous), keeping in mind F1-F3. Also, a coarse grained application should be considered, as well as other algorithms for best-nodes-exchange (sections 4 and 5). Based on the experimental results presented in this paper, we claim that the models described are suitable for good implementations of parallel B&B.

REFERENCES

[1] *IRREGULAR '94 - International Workshop on Parallel Algorithms for Irregularly Structured Problems*, Geneva, August 1994.

[2] G. Ananth, V. Kumar, and P. Pardalos. Parallel processing of discrete optimization problems. *Encyclopedia of Microcomputers*, 13:129–147, 1993.

[3] D. Bertsekas and J. Tsitsiklis. *Parallel and Distributed Computation: Numerical Methods.* Prentice-Hall International Editions, 1989.

[4] R. Corrêa and A. Ferreira. Modeling parallel branch-and-bound for asynchronous implementations. In *'94 DIMACS Workshop on Parallel Processing.* DIMACS Center, Rutgers University. To appear.

[5] R. Corrêa and A. Ferreira. On the effectiveness of synchronous branch-and-bound algorithms. *Submitted for publication*, 1994.

[6] J. Eckstein. Control strategies for parallel mixed integer branch and bound. In IEEE Computer Society Press, editor, *Supercomputing '94*, 1994. To appear.

[7] J. Eckstein. Parallel branch-and-bound algorithms for general mixed integer programming on the CM-5. Technical Report TMC-257, Thinking Machines Corp., 1994. To appear in *SIAM Journal on Optimization*, 1994.

[8] R. Karp and Y. Zhang. Randomized parallel algorithms for backtrack search and branch-and-bound computations. *Journal of the ACM*, 40(3):765–789, July 1993.

[9] G. Kindervater. Towards a parallel branch and bound machine. Presentation at IRREGULAR '94, Geneva, August 1994.

[10] V. Kumar, A. Grama, A. Gupta, and G. Karypis. *Introduction to Parallel Computing: Design and Analysis of Algorithms.* The Benjamin/Cummings Publishing Company, Inc., 1994.

[11] T. Lai and S. Sahni. Anomalies in parallel branch-and-bound algortithms. *Communications of the ACM*, 27:594–602, 1984.

[12] W. Levelt, M. Kaashoek, H. Bal, and A. Tanenbaum. A comparision of two paradigms for distributed shared memory. *Software – Practice and Experience*, 22(11):985–1010, November 1992.

[13] G. Li and B. Wah. Coping with anomalies in parallel branch-and-bound algorithms. *IEEE Transactions on Computers*, C-35(6):568–573, June 1986.

[14] R. Lüling and B. Monien. Load balancing for distributed branch & bound algorithms. In *IPPS*, pages 543–549, 1992.

[15] L. Mitten. Branch-and-bound methods: General formulation and properties. *Operations Research*, 18:24–34, 1970.

[16] G. Nemhauser and L. Wolsey. *Integer and Combinatorial Optimization.* John Wiley and Sons Interscience, 1988.

[17] P. Pardalos and X. Li. Parallel branch-and-bound algorithms for combinatorial optimization. *Supercomputer*, pages 23–30, September 1990.

[18] H. Trienekens. *Parallel Branch and Bound Algorithms.* PhD thesis, Erasmus University, Rotterdam, 1990.

[19] B. Wah and Y. Eva. MANIP - A multicomputer architecture for solving combinatorial extremum-search problems. *IEEE Transactions on Computers*, C-33(5):377–390, May 1984.

9

EXPERIMENTS WITH A PARALLEL SYNCHRONIZED BRANCH AND BOUND ALGORITHM

Claude G. Diderich and Marc Gengler

Swiss Federal Institute of Technology - Lausanne
Computer Science Department
CH-1015 Lausanne, Switzerland

ABSTRACT

In this paper we present an efficient parallel synchronized branch and bound (PSBB) algorithm. This parallelization of a sequential branch and bound best-first algorithm is based on the concept of alternating computation and synchronization or macro-communication steps. The computational steps simplify the problem to solve whereas the synchronization phases solve the problem of load balancing and data distribution. We will describe the implementation of heuristics for solving mixed integer linear programs. Experimental results will show the efficiency of the proposed PSBB algorithm when executed on a massively parallel Cray T3D machine.

1 INTRODUCTION

Combinatorial optimization problems form a common class of problems. Examples can be found while producing time tables, when searching for the shortest round-trip through a given set of cities or when solving various scheduling problems. All these problems have been shown to belong to the class of **NP**-hard problems. There exist three basic approaches to tackle such difficult problems. First, one may be interested in the exact solution of the problem, admitting that finding it may take exponential time in the size of the problem. Second, one may use polynomial time approximation algorithms. And third, one may use a heuristic approach to find a good, but not necessarily optimal solution using methods like local search, taboo or simulated annealing.

A. Ferreira and J. D. P. Rolim (eds.), Parallel Algorithms for Irregular Problems: State of the Art, 177–193.

We have chosen the approach of solving the problems exactly which belong to the class **NP**-hard by developing a parallel algorithm based on the sequential branch and bound method [6, 7]. The key idea of our parallel synchronized branch and bound algorithm, also called PSBB algorithm, is to alternate computation and communication phases such as to minimize useless work and to maximize computation versus communication time.

In order to study the efficiency of the PSBB algorithm, we implemented a general combinatorial optimization problem, the mixed integer linear programming problem (MIP) [1] on a Cray T3D machine.

The basic idea of any parallel branch and bound algorithm is to expand multiple problems in parallel. One can distinguish between two types of approaches. The set of unexpanded problems, also called the actif pool of problems, may be managed either in a centralized or in a distributed way. Parallel branch and bound algorithms dealing with a centralized active pool usually follow the farmer/worker approach. They are well suited for shared memory machines. The other approach consists in having each processor maintain its own list of unexpanded problems. Various techniques have been invented in order to avoid idle time of some of the processors and to minimize useless work. One possible scheme is to use an implicit problem distribution algorithm which guarantees that the work load is always evenly balanced. There also exist algorithms in which idle processors ask their neighbors for work. Another approach is to execute load balancing algorithms at a given time or after certain conditions are verified. Such a load balancing scheme may concern all the processors (global load balancing) or only a subset of them (local load balancing). An overview of these techniques can be found in [8].

This paper is subdivided into five sections. In Sec. 2, we will describe the parallel synchronized branch and bound algorithm, its computation and macro-communication phases as well as its actual implementation. Section 3 describes the heuristics used for solving the MIP problem. In Sec. 4 we describe various experimental results on a Cray T3D massively parallel system [2]. Finally, in Sec. 5 we will summarize the advantages of our approach and give some ideas for further work.

2 THE SYNCHRONIZED PARALLEL BRANCH AND BOUND ALGORITHM

In order to simplify the description of our parallel synchronized branch and bound algorithm we will start by briefly introducing the sequential best-first branch and bound algorithm. By viewing the branch and bound algorithm as a fringe enumeration method, we will then derive the PSBB algorithm. For the sake of simplicity, we will only consider minimization problems.

2.1 The sequential branch and bound algorithm

The branch and bound algorithm traverses a tree where each node is a subproblem of the initial problem in order to find a feasible leaf node with minimal value. For each non leaf node, the solution of at least one of its sons corresponds to the solution of that node. To each node P in the tree, we associate a lower bound $l(P)$ and an upper bound $u(P)$. The nodes in the search tree are explored in increasing order according to their lower bound $l(P)$. This has as a consequence that the first feasible solution that is considered for branching is the optimal solution to the initial problem. A more in depth description of the sequential branch and bound algorithm and its axiomatization may be found in [4, 6, 7]. It may be described by the following pseudo code, in which OPEN is a list of unexpanded problems sorted in increasing order by their lower bound $l(P)$ or optimistic evaluation.

1. **begin** initialize all data structures $insert(\text{root}, l(\text{root}), \text{OPEN})$, $u \leftarrow +\infty$ **loop** $P \leftarrow get\_first(\text{OPEN})$ **if** P is a feasible solution **then return** P is the optimal solution subdivide P into b subproblems $P_1, \ldots, P_b$ compute $l(P_i)$ and $u(P_i)$ for each P_i, set $u \leftarrow \min\{u, u(P_1), \ldots, u(P_b)\}$ **if** $l(P_i) \leq u$ **then** $insert(P_i, l(P_i), \text{OPEN})$ **end loop**
2. **end**

2.2 The concept of fringe enumeration

The branch and bound algorithm described in Sec. 2.1 may bee seen as a *fringe enumeration algorithm*. A *fringe* F of problems in the branch and bound algorithm is a set of subproblems P_i in the OPEN list having all the same lower bound or optimistic evaluation, and therefore being equivalent for the branch and bound algorithm. The notion of a fringe is illustrated in Fig. 1. It is easy to see that all the problems P_i belonging to a single fringe F_j have to be explored or expanded, except for the case when the fringe F_j contains the solution problem. We extend the notion of fringes by defining a *macro fringe* F as the union of all the problems in a set of fringes.

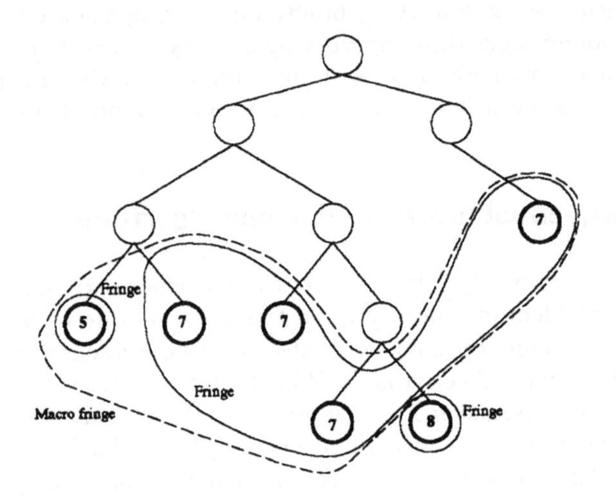

Fig. 1. The concepts of fringes and macro fringes.

The concept of macro-fringe is also denoted by the term of *fringe fusion* or *fringe merging* and is illustrated on Fig. 1. The sequential best-first branch and bound algorithm explores the fringes of the search tree in increasing order, according to their lower bounds. An in depth description of the fringe concept and its relation to the branch and bound algorithm through its axiomatization may be found in [4].

2.3 The parallel synchronized best-first branch and bound algorithm

The idea behind our parallel synchronized best-first branch and bound algorithm (PSBB) is to explore the nodes belonging to a certain macro-fringe F in parallel. In fact, if all the subproblems of F have a lower bound smaller than the optimal value, then they must be explored. The major problem is to construct a macro-fringe F such that it is as large as possible but does not contain nodes having a lower bound larger than the solution value. In a global communication operation, we determine the fringes to merge such that the number of subproblems on all processors together belonging to the macro-fringe is about $c \cdot p$, where p is the number of processors and c a constant. Then the $c \cdot p$ problems are evenly redistributed among all processors. The constant c is chosen in such a way that the computation time always dominates the synchronization and data redistribution time.

We assume a MIMD-DM (Multiple Instructions Multiple Data – Distributed Memory) massively parallel machine and use the SPMD (Single Program Multiple Data) programming model. Each processor maintains its own data structures. Our PSBB algorithm is described by the following pseudo-code.

1. **begin**
   initialize all data structures each processor explores a portion of the search tree until it owns a given number of subproblems, set $u \leftarrow +\infty$ **loop** ⟨Computation phase⟩ ⟨Synchronization phase⟩ **if** ⟨a solution was found⟩ **and** ⟨no better solution can be found⟩ **then** **stop** all the processors **and return** the found solution **end if end loop**
2. **end**

The computation phase is described by the following pseudo code.

1. **begin**
   ⟨Computation phase⟩ **loop** **if** ⟨I have no more work⟩ **or** ⟨I have done enough work⟩ **then** **broadcast** please synchronize **exit loop** **end if** **exit loop when** someone asked *please synchronize* **or** ⟨I have already done some work⟩ $P \leftarrow get\_first(\text{OPEN})$ **if** P is an feasible solution **then** **broadcast** please synchronize **else** subdivide P into b subproblems $P_1, \ldots, P_b$ compute $l(P_i)$ and $u(P_i)$ for each P_i $u \leftarrow \min\{u, u(P_1), \ldots, u(P_b))\}$ **if** $l(P_i) \leq u$ **then** $insert(P_i, l(P_i), \text{OPEN})$ **end if end loop**
2. **end**

In the computation phase of the PSBB algorithm, subproblems are expanded until one of the following conditions holds.

- ⟨I have no more work⟩: This condition is required to avoid starvation.
- ⟨I have done enough work⟩: Enough work has been done when all the problems in the current macro-fringe, as determined by the previous macro-communication or synchronization phase, have been branched.

- If someone has found a feasible solution, a synchronization phase is initiated in order to find out if the found solution is the optimal solution to the problem, in which case the algorithm may be terminated.

A synchronization request is only accepted if some work has already been done. This is necessary in order to avoid empty computation phases, which might occur when some processors do not have any subproblems to expand (because there are not enough subproblems for everyone).

The pseudo code of the macro-communication or synchronization phase, corresponding to the dimension-ordered token distribution algorithm, as described in Sec. 2.4, is given below.

1. **begin**
   ⟨Synchronization phase⟩ Exchange the currently best known upper bound u and update the local OPEN list accordingly (OPEN $\leftarrow \{P: P \in \text{OPEN} \wedge l(P) \leq u\}$
   Determine the set of fringes to explore (value and size) and evenly redistribute all the subproblems belonging to the selected fringes
2. **end**

In order to determine the set of fringes to explore all processors exchange information about how may problems they own in each of their fringes (this is done is step 1, described in Sec. 2.4). As the maximal size of a macro-fringe is give as a parameter to the algorithm[1], it is easy to compute the set of fringes to merge.

2.4 The macro-communication phase

We will assume that the network topology of the considered parallel machine can be described as a $(k_1, \ldots, k_d)$-ary d-cube network or a d-dimensional torus. An in depth description of the macro-communication or synchronization algorithm, also called *Dimension Ordered Token Distribution* algorithm may be found in [3]. The load balancing part of the synchronization phase of our parallel branch and bound algorithm proceeds in three steps: 1) All processors exchange information about the number of subproblems they own with their immediate neighbors, 2) each processor computes the final distribution l' of subproblems and the message passing needed in order to lead from the

[1] In a further version of our implementation of the PSBB algorithm, we plan to dynamically adjust the maximal size of a macro-fringe.

initial distribution l to the distribution l', and 3) all the processors realize the redistribution by exchanging subproblems according to the scheme determined by step 2.

Processors exchange information during step 1, while they transfer subproblems during step 3. In order to avoid costly routing algorithms, communications are between physical neighbors only. The computation of step 2 is entirely local and does not need any communication between processors. Thus, the black processor communicates only with processors along the different rings it belongs to (Fig. 2.b).

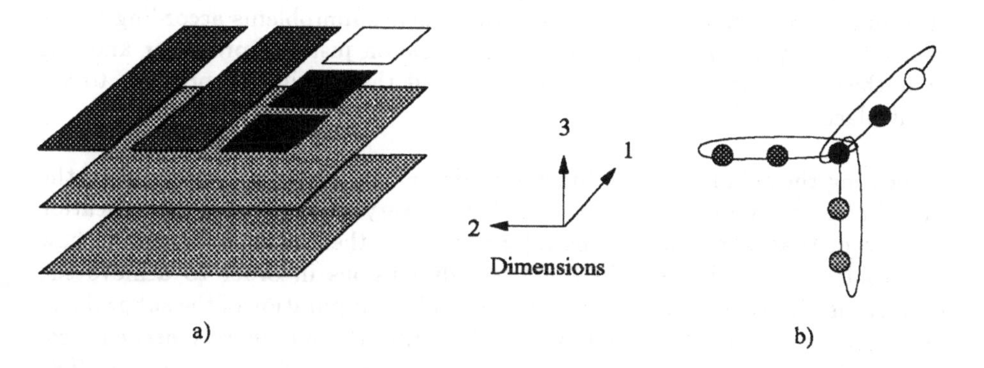

Fig. 2. Representation of the subproblem distribution known to one processor.

The information exchange we use can most easily be understood from figure 2.a). The black processor (which could be any other processor since a **k**-ary d-cube is a regular graph) knows the number of subproblems owned by its neighbor processors along dimension one, shown in dark grey and white on Fig. 2.a). Along dimension two, the black processor does not know any detailed information about the number of subproblems owned by an individual processor. It only knows the sum of subproblems owned collectively by all the processors that belong to a same ring of dimension one. The processors involved in such a sum are shown as mid grey blocks. This same information is of course also available to the dark-grey and white processors. The decomposition continues this way and all individual processors of the upper plane see the sums of subproblems owned collectively by the processors of the lower planes (light grey).

Let us now look at the way subproblems are exchanged so as to determine the routing information computed in step 2. Subproblems are successively routed through given dimensions, the dimensions being visited in some fixed order.

Suppose that the upper plane has one extra subproblem while the plane in the middle lacks one subproblem and the lower plane is balanced in the number of subproblems. Knowing both the sum of subproblems available on the lower plane and the final distribution wished by these processors, the middle and above plane conclude that they will not exchange any subproblems with the lower plane. With the same informations, they also know that they must themselves exchange exactly one subproblem. The difficulty comes from the fact that the processors of the middle plane distribute their local subproblems such that there will be one processor that gets the *hole*, i.e. that lacks one subproblem. The location of this hole must be known to the above processors so that they can put their extra subproblem on their corresponding processor. The middle processors just distribute their local subproblems according to the distribution strategy and thereby put the hole on just any processor knowing that the exceeding plane will manage to send the missing subproblem to the right place.

Inspecting the subproblem exchanges needed along a given dimension gives the distribution of subproblems that is needed on every plane before communication is done in that dimension. This informations is then used to determine how subproblems must be sent in the lower dimensions in order to achieve the correct distribution on the different planes. The computation of the subproblem exchanges is a recursive procedure that computes the successively needed local distributions and communications starting with the highest dimension. The effective exchange of subproblems is done during step 3 of the algorithm.

2.5 Some implementation details

In this section we will briefly describe the basic structure of our implementation of the algorithm described in Sec. 2.3. One of our main goals was to be as modular as possible.

Our implementation is subdivided into two major modules, the branch and bound interpreter implementing the computation phase and the synchronization and subproblem distribution module realizing the synchronization phase.

Except for the fact that the algorithm currently only solves minimization problems, the implementation is completely independent of the problem to be considered. The algorithm can be applied to any **NP**-hard minimization problem for which the functions described in table 1 can be implemented.

Function	**Description**
subdivide_problem	Subdivide a problem into zero or more mutually exclusive independent subproblems
is_admissible_sol	Test if a given subproblem is a, not necessarily optimal, admissible solution
lower_bound	Compute a lower bound of a given subproblem
upper_bound	Compute an upper bound of a given subproblem

Table 1. Functions required for any problem to be solved by the parallel synchronized best-first branch and bound algorithm.

In order to communicate between processors, we defined a machine independent interface containing the basic communication primitives we need. This was necessary to be independent of any particular communication library[2]. The functions required by our application are sending and receiving messages as well as a mechanism for sending signals to all processors to indicate that a given event occurred. The synchronization and token distribution algorithm assumes that the underlying hardware topology is a **k**-ary d-cube network.

3 THE MIXED INTEGER PROGRAMMING PROBLEM

A mixed integer programming (MIP) problem, in its most general form, is described by the following system.

$$\begin{aligned} \text{minimize}_{\mathbf{x}\in\mathbf{R}^n} \quad & \mathbf{c}^\mathsf{T}\mathbf{x} \\ \text{subject to} \quad & A\mathbf{x} \leq \mathbf{b} \\ & x_j \in \mathbf{Z} \text{ for } j \in \mathcal{B} \end{aligned} \tag{9.1}$$

where $\mathbf{c}$ is a cost vector of size n, A a matrix of size $m \times n$, $\mathbf{b}$ a vector of size m and $\mathcal{B} \subseteq \{1..n\}$, a set of distinguished indices which identify integer variables (see also [9]).

[2]We did not consider using a standard message passing interface like PVM or MPI as there do not (yet) exist efficient implementation of these interfaces for all parallel machines.

3.1 Computation of lower bounds

To compute a lower bound of a MIP subproblem, we solve the relaxed linear programming problem associated, i.e. the MIP subproblem without the integrity constraints. Depending on the MIP problem structure, the gap between the lower bound and the optimal solution may be large.

3.2 Computation of upper bounds

Although knowing good upper bounds does not diminish the overall execution time[3], they are important in order to reduce memory consumption, especially when using a best-first branch and bound strategy. We implemented a simplified strategy which is based on rounding the value of integer variables having non integer values in the solution of the relaxed problem and iteratively removing inadmissibilities occurring due to these roundings.

3.3 Branching strategy

Typically a branching strategy subdivides a given feasible MIP problem into one or more feasible and disjoint subproblems. The branching strategy hopefully increases the lower bound for each of the newly generated subproblems, but an increase cannot be guaranteed.

In the case of a MIP subproblem P, we chose a variable x_j which has a non integer value $\hat{x}_j$ in the solution of the relaxed problem of P and which must be integer, i.e. $j \in \mathcal{B}$. Now the branching strategy subdivides the problem P into two disjoint subproblems P_1 and P_2 such that $P_1 = P \cup \{x_j \leq \lfloor \hat{x}_j \rfloor\}$ and $P_2 = P \cup \{x_j \geq \lceil \hat{x}_j \rceil\}$

The selection of the variable used in the branching of a subproblem is based on the estimated degradation of the objective function and the notion of pseudo-costs. Let $H(t)$ denote the set of all subproblems P_i for which a lower bound has already been computed before or at time t. Let $l(P)$ denote the lower bound corresponding to problem P, $\hat{\mathbf{x}}(P)$ the corresponding solution of the relaxed problem, P_1 and P_2 its associated subproblems. Then the up pseudo-costs is

[3] This is not entirely true as smaller internal data structures which have a non constant access time certainly reduce the overall execution time.

defined by eqn. (9.2).

$$C_j^{\mathsf{U}}(t) = \text{avg}\left\{ \frac{l(P_2) - l(P)}{\lceil \hat{x}_j(P) \rceil - \hat{x}_j(P)} : P, P_2 \in H(t) \right\} \tag{9.2}$$

The down pseudo-cost $C_j^{\mathsf{D}}(t)$ is defined similarly. These quantities attempt to measure the average rate at which the objective value increases as one forces the variable x_j either up or down. The quantities $C_j^{\mathsf{U}}(t)$ and $C_j^{\mathsf{D}}(t)$ are computed iteratively and therefore there is no need to store $H(t)$ in memory.

In our implementation, we select the variable x_j as a branching variable such that the quantity S_j, defined by eqn. (9.3), is minimized.

$$S_j = \alpha_1 \cdot \min\{d_j, u_j\} + \alpha_2 \cdot \min\{C_j^{\mathsf{U}}(t) \cdot u_j, C_j^{\mathsf{D}}(t) \cdot d_j\} \tag{9.3}$$

where $u_j = \lceil \hat{x}_j \rceil - \hat{x}_j$, $d_j = \hat{x}_j - \lfloor \hat{x}_j \rfloor$ and α_1 and α_2 two constants.

Furthermore, as the quantities C^{U} and C^{D} depend on previously subdivided problems, they represent a local information and are different on each processor in the PSBB algorithm. This has as a consequence that for different numbers of processors, the branching variable chosen for a given subproblem P may change. This phenomenon leads to different search trees of different sizes being explored when using different numbers of processors.

4 EXPERIMENTAL RESULTS

In this section we will describe the experimental results obtained from executing our PSBB algorithm on a Cray T3D massively parallel computer [2]. The machine which we used for our tests was composed of 128 DEC Alpha processors running at 150Mhz. Each processor had 16Mb of local memory. The 128 processors were interconnected as a three dimensional torus. Although our algorithm takes advantage of a **k**-ary d-cube network topology, the current implementation of the system software does not allow the mapping of a logical **k**-ary d-cube onto a physical one.

In order to test the efficiency of our parallel algorithm, we solved various small to medium sized mixed integer program problems from the MIPLIB [1]. In table 2, we give a list of the considered problems as well as their characteristics and their application domain, when known. The test problems were selected in order to have on the order of one thousand to one hundred thousand search nodes, although even larger trees are probably tractable. For all our tests we used $\alpha_1 = 1$ and $\alpha_2 = 1000$ in the branch-scoring function (9.3). The tree sizes reported in table 2 give the number of nodes branched when running the PSBB algorithm on a single processor.

Problem	Cnstr.	Var./I.V.	Tree size	Application/Characterization
bell3a	123	133/71	27044	Fiber-optic network design
egout	98	141/55	75579	Drainage Network design
flugpl	18	18/11	5443	Airline model
pipex	25	48/48	1619	pure 0/1 IP
p0033	16	33/33	10293	pure 0/1 IP [5]
p0201	133	201/201	347	pure 0/1 IP [5]

Table 2. Summary of test problems from the MIPLIB.

4.1 Determination of an efficient macro-fringe size

It is very important to determine an efficient macro-fringe size as indicated by the following considerations:

- If the macro-fringe size is small, then the number of synchronization phases required is large and the overhead due to the synchronization phases becomes important.
- If the macro-fringe size is large, then, at the end of the algorithm, an increasing number of problems may be subdivided which need not be explored to find the optimal solution.
- The choice of how to subdivide a problem into subproblems depends on problems previously subdivided.

In order to chose an efficient macro-fringe size, we have solved two medium sized problems p0033 and flugpl, for various number of problems per macro-fringe. All the tests were run on 64 processors logically interconnected in a (8, 4, 2)-ary 3-cube. The results are shown in Fig. 3.

As we can see from Fig. 3, the number of synchronizations required sharply decreases with the increase of the macro-fringe sizes in order to stabilize itself around 20 problems per macro-fringe. At that point, the synchronization phases

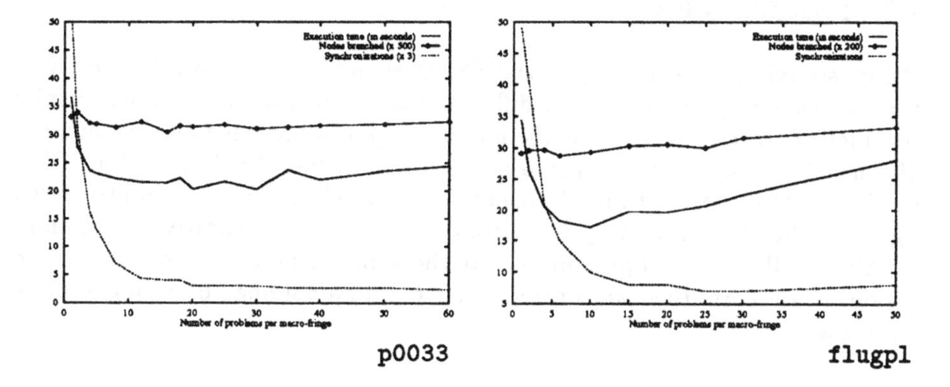

Fig. 3. Representation of the execution time, the number of subproblems generated and the number of synchronization phases for various sizes of the macro-fringes solving the problems **p0033** and **flugpl**.

take on the average over all processors about 10 to 20 percent of the total execution time[4].

The number of problems expanded increases with the number of problems per macro-fringe, although a fluctuation can be observed for small macro-fringe sizes. This fluctuation is due to the implementation of the branching heuristic. In fact, the more ancestors of a given node have been expanded on the same processor, the better the branching heuristic seems to work.

Finally, we see that the execution time rapidly decreases for small increasing number of problems per macro-fringe. Then a local minimum can be seen. The macro-fringe size at which this minimum occurs is problem dependent but lies around 10 to 20 problems per macro-fringe. For larger numbers, the execution time gradually increases again.

From the considerations above, we decided to use a macro-fringe size of 20 for all further experiments, except when stated otherwise.

[4] This value dramatically increases as soon as the problem size becomes too small in comparison with the number of processors available. In such a situation, also called starvation, there are not enough problems to form a macro-fringe of a given size.

4.2 Speedup study

Before studying the efficiency of the PSBB algorithm, we would like to stress the fact that on the Cray T3D, different executions give very similar results considering the execution time, the number of nodes branched and the number of synchronizations. The only non determinism in the PSBB algorithm is due to the variable arrival time of signals. On the Cray T3D, we implemented signal handling using the very fast virtual shared memory primitives. Therefore signals usually arrive approximately at the same instant of execution during various runs. Form this, we conclude that the data obtained by a single run is representative.

Some hard to solve problems

In Fig. 4 we show the speedup obtained when solving three very difficult problems, namely `bell3a`, `p0033` and **`egout`**[5]. The sequential execution times were obtained on a SPARCserver 600 system and were multiplied by 1.1 in order to obtain timings for one processor on the Cray T3D.

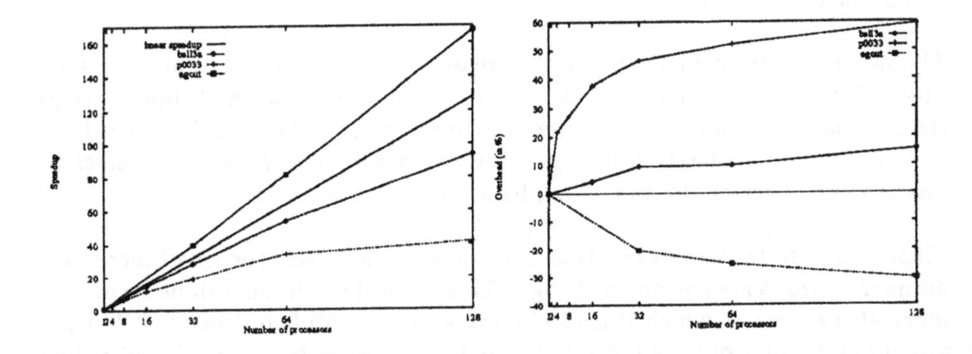

Fig. 4. Overhead in the number of problems branched for various problems.

As we can see from Fig.4, we obtain an efficiency of 73% on 128 processors when solving the `bell3a` problem. About 15% more nodes were explored by the parallel algorithm than by the sequential one. From this we can deduce that about 23.3% of the overall execution time is lost in synchronization and waiting. There are two major reasons for this behavior.

[5]The notion of difficult problem is very subjective. It highly depends on the heuristics used during the search. For us a problem is difficult if the number of nodes branched by our implementation is very hard.

First, at the beginning and at the end of the execution, there do not exist enough subproblems to form large enough macro fringes. This problem is namely due to the structure of the problem solved and is independent of any particular parallel branch and bound algorithm.

Second, if one processor has no more work, it has to wait until the last processor has received its notification of starvation and has finished branching its current subproblem. This problem may be efficiently solved by using an interrupt based message passing system. Such a mechanism is not available on the Cray T3D machine.

15% of the total execution time is spent in synchronization and waiting. We estimate that about 30% of this time is effective synchronization time.

Interestingly, for the problem **egout**, we measure a superlinear speedup. This is due to the fact that the size of the tree explored decreases with the number of processors used to solve it. This consists a rather special case as the tree size explored tends to increase with the number of processors used (see Fig. 4).

The **p0033** problem is completely different from **egout**. The number of subproblems branched considerably increases with the number of processors. On a 128 processor execution, 1.6 times the number of nodes branched by the sequential algorithm are expanded. This has a direct impact on the efficiency, which is in this case about 30%, as can be seen on Fig. 4.

Solving easy problems

In Fig. 5, we have represented the results obtained from solving the tree easy MIP problems **fluglp**, **pipex** and **p0201**. As can be seen, the speedup obtained is nearly linear up to a given number of processors (32, 16 and 8). At that point, the problem size becomes too small and there is not enough work to occupy all the processors.

5 OPEN PROBLEMS AND CONCLUSION

In this paper we have described a parallel branch and bound algorithm based on the strategy of using alternations between synchronization and computation steps. We have shown the simplicity of the resulting algorithm as well as its

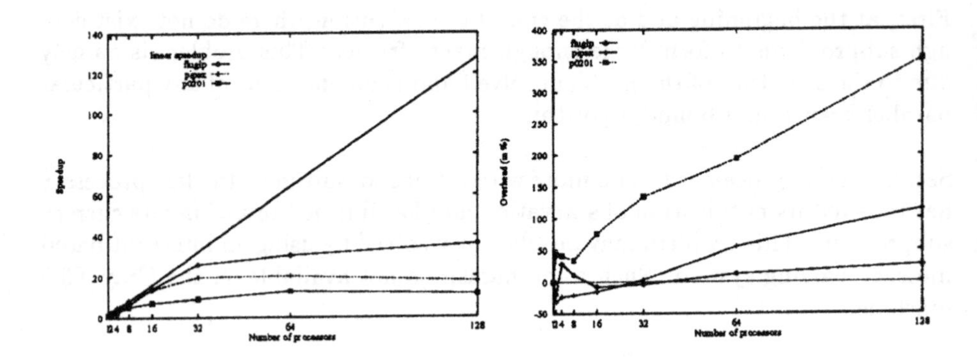

Fig. 5. Overhead in the number of problems branched for various problems.

independence of any specific problem. Finally we have shown that the obtained results are very promising.

In further work, we will apply the PSBB algorithm to other problems, like the traveling salesman problem, and use other parallel machines, like the Intel Paragon. We will also try to apply these principles to more structured trees, like game trees.

REFERENCES

[1] R. E. Bixbiy, E. A. Boyd, and R. R. Indovina, *MIPLIB: A test of real-world mixed-integer programming problems*, SIAM News **25** (1992), 16.

[2] Cray Research, *Cray T3D system architecture overview*, Cray Research, Chippewa Falls, WI, Sep. 1993.

[3] C. G. Diderich, M. Gengler, and S. Ubéda, *An efficient algorithm for solving the token distribution problem on k-ary d-cube networks*, Proc. ISPAN'94 (S. Horiguchi and D. F. Hsu, eds.), Dec. 1994.

[4] M. Gengler and G. Coray, *A parallel best-first branch and bound algorithm and its aximatization*, Parallel Algo. and Appl. **2** (1994), 61–80.

[5] M. Padberg H. Crowder, E. L. Johnson, *Solving large-scale zero-one linear programming problems*, Operat. Res. **31** (1983), no. 5.

[6] E. Lawler and D. Wood, *Branch-and-bound methods: A survey*, Operations Research **14** (1966), 699–719.

[7] L. Mitten, *Branch and bound methods: General formulation and properties*, Operations Research **18** (1970), 24–34.

[8] C. Roucairol, *Parallel branch and bound algorithms — An overview*, Parall. and. Dist. Algo. (1989), 153–163.

[9] S. Schrijver, *Theory of linear and integer programming*, Wiley and Sons, Chichester, UK, 1984.

10

PARALLEL LOCAL SEARCH AND JOB SHOP SCHEDULING

M.G.A. Verhoeven, E.H.L. Aarts*

Eindhoven University of Technology,
Department of Mathematics and Computing Science,
P.O. Box 513, 5600 MB Eindhoven, The Netherlands

* *Philips Research Laboratories, P.O. Box 80000, 5600 JA Eindhoven*

ABSTRACT

We discuss parallel local search approaches to the job shop scheduling problem, based on edge reversal neighborhoods. Speed-up is achieved by parallel exploration of neighborhoods and parallel computation of the P longest paths in a disjunctive graph. A complexity analysis shows that the resulting parallel local search algorithm has a speed-up of $O(P/\log\log P)$ on a PRAM machine with P processors, and a speed-up of $O(P/\log P)$ on a distributed-memory MIMD machine. Furthermore, we show that the problem to verify local optimality with respect to the 1-opt neighborhood for the job shop scheduling problem is in $\mathcal{NC}$.

1 INTRODUCTION

In a combinatorial optimization problem one must find an optimal solution among a finite, possibly large, number of alternatives [Papadimitriou & Steiglitz, 1982]. Many combinatorial optimization problems are NP-hard [Garey & Johnson, 1979], with the direct consequence that optimal solutions cannot be obtained in reasonable amounts of computation time for the larger instances. Local search algorithms can be used to handle hard combinatorial optimization problems [Reeves, 1993]. Extensive empirical performance analyzes have revealed that local search algorithms can find good quality solutions within acceptable running times for a wide variety of problems. However, for large instances of NP-hard problems, the running time can still be considerable. One way to reduce running times is by using parallelism. This paper is one in a series of papers in which we investigate the use of parallelism in local search for various combinatorial optimization problems. In [Verhoeven, Aarts, Van

A. Ferreira and J. D. P. Rolim (eds.), Parallel Algorithms for Irregular Problems: State of the Art, 195–212.

de Sluis & Vaessens, 1992] we studied the traveling salesman problem. In this paper we study the job shop scheduling problem.

Job shop scheduling is an important model in the scheduling theory [Lawler, Lenstra, Rinnooy Kan & Shmoys, 1993], which serves as a proving ground for new algorithmic ideas and provides a starting point for the more complicated practically relevant scheduling models. Furthermore, the problem is NP-hard and local search has proved to belong to the best approximation algorithms for it [Vaessens, Aarts & Lenstra, 1994].

The organization of the paper is as follows. In Section 2, we briefly summarize the most important aspects of local search. Section 3 presents the job shop scheduling problem, and Section 4 discusses the application of local search to this problem. In Section 5 we discuss some general issues of parallel local search and in Section 6, we discuss parallel local search for the job shop scheduling problem on a PRAM machine.

2 LOCAL SEARCH

The use of local search presupposes the specification of an instance of a combinatorial optimization problem and a neighborhood structure.

Definition 2.1 *An instance of a combinatorial minimization problem is denoted as a pair $(\mathcal{S}, f)$, where the* solution space *$\mathcal{S}$ is a set of feasible solutions and $f : \mathcal{S} \rightarrow \mathbb{R}$ is a function that gives the cost of a feasible solution. The problem is to find a solution with minimal cost. A* neighborhood structure *$\mathcal{N} : \mathcal{S} \rightarrow \mathcal{P}(\mathcal{S})$ assigns to each feasible solution a set of solutions that can be directly obtained from it.*

Neighbors are obtained by applying some transition function to a solution. Most transition functions are based on replacements or exchanges of a small number of elements in a solution.

Definition 2.2 *A solution $s \in \mathcal{S}$ is a* local minimum *of $\mathcal{N}$, if $f(s') \geq f(s)$ for all $s' \in \mathcal{N}(s)$.*

Local search algorithms constitute a class of approximation algorithms that are based on exploring neighborhoods of solutions. The simplest form of local

```
proc Iterative_Improvement(s ∈ S)
W := ∅ ;
while N(s) \ W ≠ ∅ do
    s' :∈ N(s)\ W ;
    if f(s') ≥ f(s) then W := W ∪ {s'}
    else s, W := s', ∅
od /* s is a local minimum of N */
```

Fig. 1. An iterative improvement algorithm.

search is iterative improvement in which a solution is continually replaced by a solution in its neighborhood with lower cost, until no such solution can be found anymore. In that case a local minimum has been found. Figure 1 presents an iterative improvement algorithm. There are many variations on this basic local search algorithm known in the literature, such as simulated annealing, tabu search, genetic local search, and some types of recurrent neural networks. For an overview we refer to [Aarts, Korst & Zwietering, 1993]. Here, we restrict our discussion to the basic iterative improvement algorithm presented in Figure 1.

An unresolved issue in local search is the complexity to find a local optimum. To this end, Johnson, Papadimitriou & Yannakakis [1988] introduced a new complexity class $\mathcal{PLS}$, which contains the problems for which local optimality can be verified in polynomial time.

Next, a PLS-completeness notion has been introduced, and an important unresolved issue is whether PLS-complete problems are also NP-hard. Consequently, it is not possible to give a non-trivial upper bound on the number of iterations needed to find a local optimum, and instances of PLS-complete problems may exist for which no local optimum can be found within a running time that can be bounded by a polynomial in the size of the instance.

3 JOB SHOP SCHEDULING

An instance of the job shop scheduling problem consists of a set of jobs and a set of machines. Each machine can handle at most one job at a time. Each job consists of a chain of operations, which need to be processed in that order during an uninterrupted time period of a given length on a given machine. The problem is to find a schedule, which is defined as an assignment of the operations to time intervals, such that the total length of the schedule, is minimal [French, 1982]. More formally, the problem can be defined as follows.

Definition 3.1 *Given are finite sets J of jobs, M of machines, and O of N operations. For each operation $a \in O$, there is a job $j(a) \in J$ to which it belongs, a machine $m(a) \in M$ on which it must be processed, and a processing time $d(a) \in \mathbb{N}$. There is a binary relation $\prec$ on O that decomposes O into chains corresponding to the jobs; more specifically, if $a \prec b$, then $j(a) = j(b)$ and there is no $c \in O \setminus \{a, b\}$ with $a \prec c$ and $c \prec b$. The problem is to find a non-negative start time $s(a)$ for each operation $a \in O$ such that the cost function*

$$\max_{a \in O}(s(a) + d(a))$$

is minimal, subject to

(1) $s(b) - s(a) \geq d(a)$ *for all* $a, b \in O$ *with* $a \prec b$, *and*
(2) $s(b) - s(a) \geq d(a) \vee s(a) - s(b) \geq d(b)$ *for all* $a, b \in O$ *with* $m(a) = m(b)$.

The cost function gives the schedule length. The constraints corresponding to (1) are job precedence constraints, and those corresponding to (2) are machine capacity constraints.

4 LOCAL SEARCH AND JOB SHOP SCHEDULING

To apply local search to the job shop scheduling problem, it is most appropriate to use the disjunctive graph representation of Roy & Sussmann [1964].

Definition 4.1 *An instance of the job shop scheduling problem of Definition 3.1 is represented by a vertex weighted disjunctive graph $G = (V, A, E)$ with node set $V = O$, arc set $A = \{(a, b) | a \prec b\}$, and edge set $E = \{\{a, b\} | m(a) = m(b)\}$. A weight $d(a)$ is associated with each vertex a in V. A feasible solution is represented as a minimal subset E' of orientations of the edges in E, such that E' gives for each machine a linear ordering of the operations that have to be processed on it, and such that the resulting digraph $D = (V, A \cup E')$ is acyclic. The start time of an operation $v \in V$ is the length of the longest path up to and without v, and the cost $f(D)$ of a feasible solution D is the longest path in D.*

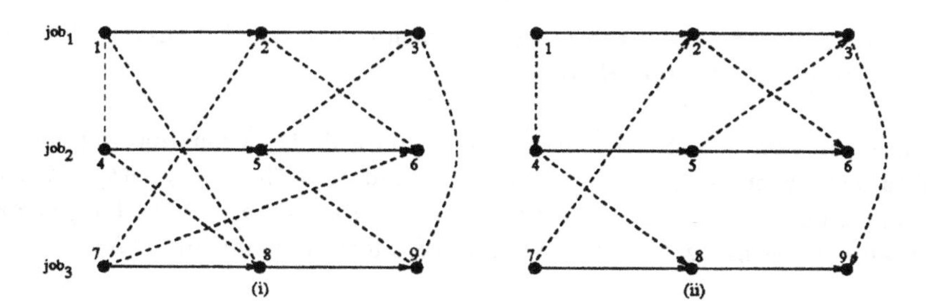

Fig. 2. Example of an instance (i) and a digraph of a solution to this instance (ii). Solid arcs represent job precedence constraints, dashed edges and arcs machine capacity constraints.

The (directed) arcs represent the job precedence constraints; the (undirected) edges represent the machine capacity constraints. Figure 2 gives an example of this representation for an instance with three machines and three jobs each consisting of three operations, viz. $\text{job}_1 = \{1,2,3\}, \text{job}_2 = \{4,5,6\}$, and $\text{job}_3 = \{7,8,9\}$. A digraph D corresponding with a feasible solution, contains only oriented edges between direct successors and direct predecessor on a machine, since the orientation of the remaining edges in E are uniquely determined by E'. So, each vertex has at most two incoming edges and at most two outgoing edges, viz. at most one successor and at most one predecessor in its job and on its machine. Given any such orientation E', we can determine feasible start times by setting each start time $s(a)$ equal to the length of a longest path in D up to a. The cost f of a solution with N operations can be computed in $O(N)$ time with Bellman's labeling algorithm [Lawler, 1976], since the degree of each vertex in D is at most four. The problem is now to find an orientation of the edges in E that minimizes the longest path length in D.

Definition 4.2 *Given are a digraph $D = (V, E)$ and a vertex $a \in V$. Then, the successor and predecessor on the machine and in the job of operation a are given by sm_a, pm_a, sj_a, and pj_a, respectively. We use $sj_a = \diamond$ or $sm_a = \diamond$ to denote that a vertex a has no successor in its job or on its machine, respectively. $st(a, D)$ is the length of the longest path up to and without a. $rt(a, D)$ is the length of the longest path starting from and including a. $l(a, D)$ is the length of the longest path through a. The start time $s(a)$ is equal to $st(a, D)$.*

Most neighborhoods for the job shop scheduling problem are based on reversing machine capacity arcs in the digraph, i.e., reversing the order in which operations are processed on a machine. Van Laarhoven, Aarts & Lenstra [1992] give the following 1-opt neighborhood $\mathcal{N}_1$.

Given a solution $s \in \mathcal{S}$ represented by a digraph D, a neighboring solution is obtained by choosing two operations a and b, with $j(a) \neq j(b)$, that are adjacent on some machine m and for which the arc (a, b) is on a longest path in D, and reversing (a, b). Next, we give of a formal definition of $\mathcal{N}_1$.

Definition 4.3 *Let $s \in \mathcal{S}$ with digraph $D = (V, E)$, and let $a, b \in V$ with $b = sm_a$, then the transition function rev is defined as*

$$rev(D, a, b) = (V, (E \setminus \{(a, b), (pm_a, a), (b, sm_b)\}) \cup \{(pm_a, b), (b, a), (a, sm_b)\}).$$

The 1-opt neighborhood $\mathcal{N}_1$ is then defined as

$$\mathcal{N}_1(D) = \{rev(D, a, sm_a) \mid a, sm_a \in V \wedge (a, sm_a) \text{ on a longest path in } D\}.$$

Figure 3 gives an example of an arc reversal. The size of a neighborhood is $O(N)$. For this neighborhood the following properties hold [Van Laarhoven, Aarts & Lenstra, 1992].

a. The reversal of (a, sm_a) on a longest path results in an acyclic digraph D', corresponding again to a feasible solution s'.

b. Reversals of arcs on a longest path are the only arc reversals that can —but need not— result in a digraph with a shorter longest path. Furthermore, a reversal of an arc not on a longest path can lead to an infeasible solution.

c. For any digraph D, corresponding to an arbitrary solution s, it is possible to construct a sequence of arc reversals leading from D to a digraph corresponding to a globally minimal solution. This is a necessary and sufficient condition for asymptotic convergence of simulated annealing.

Further, we have the following result which generalizes an observation made by Taillard [1994].

Theorem 4.1 *Given are a digraph D and functions st and rt. Let $D' = rev(D, a, b)$ with $b = sm_a$ and (a, b) on a longest path in D, and let $mx(D', a, b)$ denote the length of the longest path through a or b in D'. Furthermore, let l_2 be the length of the second longest path in D —which can be equal to $f(D)$— and k the length of the longest path in D not through a or b. Then,*

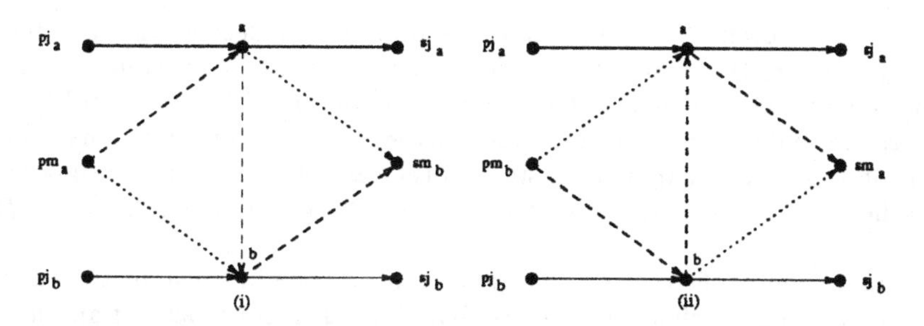

Fig. 3. Part of a digraph before (i) and after (ii) reversal of an arc (a, b). Dotted arcs are redundant machine capacity arcs.

(1) $mx(D', a, b) \geq f(D) \Rightarrow f(D') = mx(D', a, b)$,
(2) $l_2 < mx(D', a, b) < f(D) \Rightarrow f(D') = mx(D', a, b)$, *and*
(3) $l_2 \geq mx(D', a, b) \Rightarrow f(D') = \max\{k, mx(D', a, b)\}$.

Moreover, $mx(D', a, b)$ is computable in constant time.

Proof. From Figure 3 we infer that

$$\begin{aligned} st(b, D') &= \max\{st(pm_a, D) + d(pm_a), st(pj_b, D) + d(pj_b)\}, \\ st(a, D') &= \max\{st(b, D') + d(b), st(pj_a, D) + d(pj_a)\}, \\ rt(a, D') &= \max\{rt(sj_a, D), rt(sm_b, D)) + d(a)\}, \text{ and} \\ rt(b, D') &= \max\{rt(a, D'), rt(sj_b, D)) + d(b)\}. \end{aligned}$$

This gives $l(a, D') = st(a, D') + rt(a, D')$ and $l(b, D') = st(b, D') + rt(b, D')$. Thus, $mx(D', a, b) = \max\{l(a, D'), l(b, D')\}$ and $f(D') \geq mx(D', a, b)$. Paths through a or b in D are the only paths in D which possibly do not exist in D' as a consequence of the reversal of (a, b). So, in case (1) when $mx(D', a, b) \geq f(D)$ then $f(D') = mx(D', a, b)$, because all paths in D' not through a or b have lengths of at most $f(D)$. In case (2) when $l_2 < mx(D', a, b) < f(D)$ then $f(D') = mx(D', a, b)$, because all paths in D' not through a or b have lengths of at most l_2 with $l_2 < mx(D', a, b)$, and paths in D' through a or b have lengths of at most $mx(D', a, b)$ and at least one path has length $mx(D', a, b)$. In case (3) when $mx(D', a, b) \leq l_2$ then the length of the longest path in D' is given by $\max\{k, mx(D', a, b)\}$. Furthermore, $mx(D', a, b)$ is given by expressions that are all computable in constant time. □

The reversal of (a, b) in D leads to a neighbor D' with $f(D') \geq f(D)$ in case (1) of Theorem 4.1, $f(D') < f(D)$ in case (2), and $f(D') \leq f(D)$ in case

(3). So, Theorem 4.1 shows that the problem to decide whether $f(D) > f(rev(D, a, sm_a))$ for any a on a longest path in D, is solvable in constant time. Furthermore, it shows that the lower bound $mx(D', a, b)$ for $f(D')$ makes it possible to decide in constant time whether to accept or reject a neighboring solution in simulated annealing, since in the case of a cost deteriorating transition the exact cost difference between D' and D is given by $mx(D', a, b) - f(D)$.

If computation of the cost of a neighbor can be done in constant time and the size K of a neighborhood is polynomially bounded in the size of an instance, the verification of local optimality is in $\mathcal{NC}$, because then a parallel algorithm with K processors that uses general function parallelism can decide in $\log K$ time whether a solution is a local minimum [Verhoeven & Aarts, 1994]. This result together with Theorem 4.1 leads to the result that verification of local optimality is in $\mathcal{NC}$. — $\mathcal{NC}$ is the class of problems solvable in polylogarithmic time using a polynomial number of processors [Cook, 1981].—

Corollary 4.1 *Given are a digraph D, mappings st, rt, $nv : V \to \mathbb{N}$, and $na : E \to \mathbb{N}$ with $nv(a)$ the number of paths with length $f(D)$ through vertex a and $na(e)$ the number of paths with length $f(D)$ through arc e. Then, the problem to verify whether D is a local minimum with respect to $\mathcal{N}_1$ is in $\mathcal{NC}$.*

Proof. According to Theorem 4.1 $f(D') \geq f(D)$ if $mx(D', a, b) \geq f(D)$ with $D' = rev(D, a, b)$. If $mx(D', a, b) < f(D)$ then $f(D') = f(D)$ when there are paths in D with length $f(D)$ not through a or b. The number of paths in D with length $f(D)$ through a or b is equal to $nv(a) + nv(b) - na(a, b)$. Let nl be the number of paths with length $f(D)$ in D, then $nl - (nv(a) + nv(b) - na(a, b)) > 0$ implies $f(D') = f(D)$. The mappings nv, na and nl can be computed in constant time during the computation of st. □

The condition that the mappings st, rt are known is not a restriction, since these mappings are already computed to determine the cost of a digraph D. The following theorem gives an upper bound on the cost reduction that can be obtained by a certain class of transitions.

Theorem 4.2 *Given are a digraph D and a vertex a with a weight $d(a)$. If the longest path in D passes through (pm_a, a) and (a, sm_a), then*

$$f(D) - f(rev(D, a, sm_a)) \leq d(a).$$

Proof. Let $D' = rev(D, a, sm_a)$. If the longest path also passes through (b, sm_b) with $b = sm_a$, then the $f(D') = f(D)$. If the longest path passes through (b, sj_b), then the longest path in D' is at most $d(a)$ shorter than the longest path in D. □

5 PARALLEL LOCAL SEARCH

In the study of parallel local search algorithms we distinguish between algorithms with function and data parallelism [Verhoeven & Aarts, 1994]. In an algorithm with data parallelism, data is distributed over a number of processes, and each process executes the same local search algorithm. In an algorithm with function parallelism one step of a sequential local search algorithm is executed in parallel. Furthermore, we distinguish between general and tailored parallelism. A general approach can be applied to a wide variety of problems, whereas a tailored approach strongly depends on the problem at hand.

In a local search algorithm with tailored function parallelism, a problem dependent step is executed in parallel. In an algorithm with general function parallelism the step in which the neighborhood of a solution is explored to find a solution with lower cost is parallelized. However, speed-up obtained with general function parallelism is often small, since in a concrete implementation of local search a neighbor with lower cost is often found in constant time, although worst case it takes $O(K)$ time with K the size of the neighborhood. So, although the speed-up of an algorithm with general function parallelism is $O(P)$ for P processors, it is often small in practice.

Local search with general data parallelism requires a population of solutions that is distributed among a number of processors. Tailored data parallelism can be incorporated into local search by partitioning a solution into a number of disjoint partial solutions. Next, the neighborhoods of all partial solutions are explored simultaneously by a local search algorithm. In this paper, we concentrate on tailored data and function parallelism.

6 PARALLEL LOCAL SEARCH FOR JOB SHOP SCHEDULING

Taillard [1994] presents a local search algorithm with tailored function parallelism for the job shop scheduling problem. In this algorithm, the longest path in a digraph is computed using function parallelism. Taillard states that his algorithm has a limited scalability. Our goal is to investigate the applicability of tailored data parallelism to introduce scalable parallelism in local search for the job shop scheduling problem.

6.1 Towards a parallel algorithm

In Section 4 we have argued that only reversals of arcs on a longest path can lead to digraphs with shorter longest paths. Furthermore, infeasible digraphs can be constructed when arcs not on a longest path are reversed. So, a local search algorithm with data parallelism must be based on reversals of arcs on longest paths. However, a reversal of an arc on a longest path leads to a graph that has a different longest path, and consequently only a single arc can be reversed on a given longest path.

Our approach to introduce data parallelism in local search for the job shop scheduling problem is based on Theorem 4.1. This theorem shows that the longest path in a neighbor D' of a digraph D with $D' = rev(D, a, b)$, can be determined in constant time, provided that the second longest path in D is known; more specifically, the longest path through a or b in D', which can be determined in constant time, is the longest path in D' if it is longer than the second longest path in D, otherwise the second longest path in D is the longest path in D', provided that this path does not pass through a or b. Hence, the second longest path in D is the longest path in many neighbors D' of D. A similar argument shows that the third longest path in D is the longest path in many neighbors D'' of D'. A repetition of this argument shows that the p-th longest path in D is often the longest path in a digraph E which can be obtained from D by p subsequent cost decreasing arc reversals.

Our parallel local search algorithm is based on computing the P longest paths in a digraph D and proposing arc reversals on each of these paths. To formalize our approach we give the following definition.

Definition 6.1 *Let p be an integer between 1 and P, and $D = (V, E)$ a digraph. Then, for a vertex $c \in V$, $st_p(c, D)$ equals the length of the p-th longest path up to and without c, and $rt_p(c, D)$ equals the length of the p-th longest path from and including c. Furthermore, $\lambda_p(D)$ denotes the p-th longest path in D. Let $a_p, b_p \in V$ with (a_p, b_p) on $\lambda_p(D)$. Then, we recursively define D_p as follows: $D_1 = D$, and $D_{p+1} = rev(D_p, a_p, b_p)$ for $1 \leq p < P$.*

Notice that $st_p(c, D)$ and $rt_p(c, D)$ are descending in p. D_p is the graph obtained by effectuating the reversals on the paths $\lambda_q(D)$ with $1 \leq q < p$. We have argued that $\lambda_p(D)$ is often the longest path in D_p. This gives rise to the following approach to introduce data parallelism in local search for the job shop scheduling problem. Given are a digraph D and P processors.

a. Compute the P longest paths in D and assign $\lambda_p(D)$ to a process p with $1 \leq p \leq P$.

b. Process p proposes an arc reversal $rev(D, a_p, b_p)$ with (a_p, b_p) on $\lambda_p(D)$. A proposal is accepted if $mx(D', a_p, b_p) <$ length$(\lambda_p(D))$.

c. Effectuate the proposed arc reversals by successively constructing each digraph D_{p+1} obtained by reversal of (a_p, b_p) in D_p, provided that $f(D_{p+1}) \leq f(D_p)$.

d. Replace D by D_ρ with ρ the maximum p for which D_p is feasible. Repeat the above steps until $f(D_\rho) > f(D)$.

Notice that $f(D_{p+1}) \leq f(D_p)$ for $1 \leq p < \rho$. Figure 4 presents a formal description of the above approach for a local search algorithm with tailored data parallelism. In the next sections we discuss the steps of this algorithm in more detail.

---

proc PLS(D:digraph; P : $\mathbb{N}$)
z_1 :=true;
while z_1 **do** /* $D = d$ */
    Init_st(D, P);**Init_rt**(D, P);
    /* $\forall_{a \in V}\ s(a,p) = st_p(a, D) \ \wedge\ r(a,p) = rt_p(a, D)$ */
    Propose(D, z);
    /* $z_p \Rightarrow$ process p proposes $rev(D_p, a_p, b_p)$ */
    Effectuate(D)
    /* $D = d_\rho \ \wedge\ \rho \geq 1$ */
od /* D is a local minimum of $\mathcal{N}_1$ */

Fig. 4. A parallel local search algorithm for the job shop scheduling problem.

```
proc Init_rt(D:digraph; P : ℕ)
V₀, V₁ := ∅, ∅ ;
for a ∈ V do
    if sm_a = ◇ ∧ sj_a = ◇ then V₁ := V₁ ∪ {a}
rof;
while V₁ ≠ ∅ do
    a :∈ V₁;
    if sm_a = ◇ ∧ sj_a = ◇
    then r(a,1) := d(a); r(a,2), ···, r(a,P) := −∞, ..., −∞
    else if sm_a = ◇
    then r(a,1), ···, r(a,P) := r(sj_a,1) + d(a), ···, r(sj_a,P) + d(a)
    else if sj_a = ◇
    then r(a,1), ···, r(a,P) := r(sm_a,1) + d(a), ···, r(sm_a,P) + d(a)
    else r(a,1), ···, r(a,P) := Merge ([r(sm_a,p) + d(a) | 1 ≤ p ≤ P],
                                        [r(sj_a,p) + d(a) | 1 ≤ p ≤ P])[1, ···, P];
    if sm(pj_a) ∈ V₀ ∪ {◇} then V₁ := V₁ ∪ {pj_a};
    if sj(pm_a) ∈ V₀ ∪ {◇} then V₁ := V₁ ∪ {pm_a};
    V₀, V₁ := V₀ ∪ {a}, V₁ \ {a}
od /* ∀_{a∈V} r(a,p) = rt_p(a,D) */
```

Fig. 5. An algorithm that computes the P longest paths in a digraph D.

6.2 A parallel algorithm for the P longest paths in a graph

In this section, we present the algorithm **Init_rt** which computes the P longest paths from all vertices in a digraph. The algorithm **Init_st** which computes the P longest paths up to all vertices is similar. To compute rt_p with $1 \leq p \leq P$, we use the following invariants

$$\forall_{a \in V_0}\ r(a,p) = rt_p(a,D)\ \wedge\ \forall_{a \in V_1}\ \{sm_a, sj_a\} \subseteq V_0 \cup \{\diamond\},$$

where V_0 contains vertices for which rt_p is computed, and V_1 the vertices for which rt_p can be computed. Using these invariants, we obtain the algorithm given in Figure 5. In this algorithm a row a of k elements is denoted as $[a(i) \mid 1 \leq i \leq k]$ where $a[i,\ldots,j]$ with $i \leq j$, gives the elements $a(i),\ldots,a(j)$. The function **Merge** outputs, on input of two descending rows r_0 and r_1 of length P, one descending row which contains the elements of r_0 and r_1. The time complexity of **Merge** is $\Theta(P)$, and consequently the complexity of of **Init_rt** to compute the P longest paths from vertices in a digraph with N vertices is $\Theta(N \cdot P)$.

The time complexity of a single iteration of **PLS** is at least the time complexity to compute the P longest paths in the digraph D. So, the time complexity of this parallel algorithm is $\Omega(N \cdot P)$. In a single iteration of **PLS** at most P transitions are accepted. However, the time complexity of a sequential algorithm that finds P successively improving transitions in D is $\Theta(N \cdot P)$, because the time complexity to find a neighbor with lower cost, i.e., a single improving transition, is $\Theta(N) + O(N) \cdot O(1) = \Theta(N)$; more specifically, computation of the longest path requires $\Theta(N)$ time, enumeration of the neighborhood of D requires $O(N)$ time, decision to accept or reject a neighbor D' requires $O(1)$ time. Hence, **PLS** with tailored data parallelism only, has the same worst-case time complexity as a sequential algorithm.

In order to design an efficient parallel algorithm for the job shop scheduling problem, we need to design a parallel algorithm to compute the P longest paths in a digraph D with a time complexity smaller than $\Theta(N \cdot P)$, i.e., we have to apply tailored function parallelism in **Init_rt**. Borodin & Hopcroft [1985] give a parallel algorithm to merge two rows of length P. Their algorithm has a time complexity of $O(\log\log P)$ on a PRAM machine with P processors. Using this algorithm to parallelize the function **Merge** in the algorithm **Init_rt**, we obtain a parallel algorithm to compute the P longest paths in a digraph with a time complexity of $\Theta(N \cdot \log\log P)$ on a PRAM machine with P processors.

---

proc Propose(D:digraph; $z : 1, \ldots, P \rightarrow$ boolean)
par $1 \leq p \leq P$ **do**
  $z_p, W :=$ false, $\emptyset$;
  while $\lambda_p(D) \setminus A \setminus W \neq \emptyset$ **do**
    $(a_p, b_p) :\in \lambda_p(D) \setminus A \setminus W$; /* $D' = rev(D, a_p, b_p)$ */
    if $mx(D', a_p, b_p) \geq \text{length}(\lambda_p)$ **then** $W := W \cup \{(a_p, b_p)\}$
    else $z_p :=$ true
  od
rap /* $z_p \Rightarrow$ process p proposes $rev(D_p, a_p, b_p)$ */

Fig. 6. An algorithm that proposes arc reversals.

6.3 Parallel algorithms to propose and effectuate transitions

In this section we discuss in more detail the functions `Propose` and `Effectuate` from the algorithm `PLS` given in Figure 4.

Figure 6 presents an algorithm to implement function `Propose`. In this function the P longest paths of a digraph are distributed over P processors and each process proposes a transition which is expected to lead to a digraph with lower cost. The time complexity of this algorithm is $O(N)$, since the path $\lambda_p(D)$ contains at most N arcs.

In `Propose` we assume that edge (a_p, b_p) is on a longest path in the graph D_p. However, this need not be the case, if the longest path through (a_p, b_p) in D also passes through (a_q, b_q) in D whose reversal is proposed by process q with $1 \leq q < p$. This means that effectuating $rev(D_p, a_p, b_p)$ might lead to an infeasible solution. Furthermore, consolidating a proposed transition $rev(D_p, a_p, b_p)$ may result in a digraph with larger cost, when st or rt needed to compute $mx(D_{p+1}, a_p, b_p)$ are not equal to the values of st or rt used to compute $mx(D, a_p, b_p)$, i.e., reversals of (a_q, b_q) have resulted in a longer path to or from a vertex that precedes or succeeds a_p or b_p in D_p.

In order to deal with such situations, it has to be checked whether a proposed transition leads to a feasible digraph with lower cost, before it is effectuated. This is done in the function `Effectuate` of `PLS`. A transition $rev(D_p, a_p, b_p)$ can be effectuated, if (a_p, b_p) is on a longest path in D_p and $f(D_{p+1}) < f(D_p)$, i.e., the transition leads to a feasible digraph with lower cost. In to order to determine whether $rev(D_p, a_p, b_p)$ leads to a feasible digraph D_{p+1} with lower cost, the graph D_p is constructed and the longest path in D_p is computed

---

proc `Effectuate`(D:digraph)
par $1 \leq p \leq P$ **do** /* $D = d$ */
  construct D_p;
  `Init_st`$(D_p, 1)$;`Init_rt`$(D_p, 1)$;
  $ok_p := st_1(a, D_p) + d_a + rt_1(b, D_p) = f(D_p) \wedge mx(D_{p+1}, a_p, b_p) < f(D_p) \wedge z_p$
rap;
$\rho := \max\{p \mid 1 \leq p \leq P \wedge \forall_{1 \leq q \leq p}\, ok_q\}$; $D := D_\rho$ /* $D = d_\rho \wedge \rho \geq 1$ */

Fig. 7. An algorithm that effectuates the proposed transitions.

by `Init_rt`$(D_p$`,1)`. Figure 7 presents a parallel algorithm to effectuate the proposed transitions. The time complexity of this algorithm is $\Theta(P+N)$, since each digraph D_p is constructed in $O(P)$ time (cf. Definition 6.1) and the time complexity to compute st_1 and rt_1 is $\Theta(N)$. Furthermore, we observe that ρ transitions are effectuated with $1 \leq \rho \leq P$.

Combining the complexity results for the functions used in algorithm PLS in Figure 4 results in the following theorem.

Theorem 6.1 *Algorithm* PLS *has a speed-up of* $O(\frac{P}{\log\log P})$ *on a PRAM machine with* P *processors and* $P = O(N)$.

Proof. The time complexity of a single iteration of PLS is $\Theta(N \log\log P)$ $+O(N) + \Theta(N+P) = \Theta(N \log\log P)$ for $P = O(N)$. In a single iteration of PLS at most ρ transitions are accepted with $1 \leq \rho \leq P$. A sequential algorithm which accepts ρ transitions has time complexity $\Theta(\rho \cdot N)$. So, the speed-up is $\Theta(\frac{\rho}{\log\log P}) = O(\frac{P}{\log\log P})$. □

The actual speed-up obtained by PLS for a concrete instance strongly depends on ρ, which is the number of consecutive arc reversals that can be effectuated in a single iteration of PLS. A proposed arc reversal cannot be effectuated, when this arc is not on a longest path. Such situations can occur when, as a consequence of an arc reversal on path λ_p, a new path arises in D_{p+1} which is longer then λ_ρ with $p < \rho \leq P$. In this case, reversals proposed by processes q with $\rho \leq q \leq P$ cannot be effectuated. If for a proposed transition $rev(D_p, a_p, b_p)$ holds that $mx(D_{p+1}, a_p, b_p) <$ length(λ_P), then the above situation cannot occur, because then all paths through a_p or b_p are shorter than λ_q with $p < q \leq P$. So, to improve the speed-up obtained by PLS, $mx(D', a, b)$, the length of the longest path through a or b after reversal of (a, b), should be as small as possible. This can be achieved by preferring transitions for which (pm_a, a) is not on a longest path in D (cf. Theorem 4.2).

Furthermore, reversal of an edge (a_p, b_p) on the p-th longest path that is also on a q-th longest path with $p < q \leq P$, prohibits effectuation of a proposed reversal of (a_q, b_q) on this q-th longest path. Reversal of an edge (a_p, b_p) on a p-th longest path that is not on a q-th longest path, is therefore to be preferred in `Propose` in order to maximize ρ.

7 DISCUSSION

We have designed a parallel local search algorithm for the job shop scheduling problem with a speed-up of $O(P/\log\log P)$ on a PRAM machine with P processors. In this section we discuss the time complexity of our algorithm on a distributed-memory MIMD machine.

The worst-case complexity of the algorithm depends on the parallel merging algorithm from Borodin & Hopcroft [1985], which is used in the computation of the P longest paths. This parallel merging algorithm, however, does not show a good speed-up on a distributed-memory MIMD machine due to large communication overhead. Fortunately, it is not necessary to implement this parallel merging algorithm to obtain a good speed-up, as is shown in the remainder of this section.

In order to determine the speed-up on a distributed-memory MIMD machine, we discuss the complexity of a single iteration of PLS in more detail. The worst-case complexity of a single iteration of PLS is dominated by the complexity to compute the P longest paths in `Init_rt` and `Init_st`, but the overall time complexity also consists of the complexity of `Propose` and `Effectuate`. The complexity of `Init_rt` is given by $(c_1 + c_2 P)N$, where c_1 denotes the complexity of a single iteration of the repetition in `Init_rt` and c_2 the complexity to merge two elements of a row. For the constants c_1 and c_2 holds $c_1 \gg c_2$. The complexity of `Propose` is given by $c_3 N$, where c_3 denotes the complexity of a single iteration of the repetition in `Propose`, with $c_3 \approx c_1$. The complexity of `Effectuate` is $c_1 N + c_4 P$, where c_4 is the complexity to effectuate an arc reversal and $c_1 \gg c_4$. Furthermore, global communication has to take place between the consecutive stages of the algorithm. The complexity of global communication depends on the interconnection network of the parallel machine at hand, but can be done in $c_5 \log P$ time on most machines. Hence, the time complexity of a single iteration of **PLS** —in which ρ, $1 \le \rho \le P$, arc reversals are effectuated— is $2c_1N + c_2PN + c_3N + c_4P + c_5 \log P \approx 3c_1N + c_5 \log P$.

A sequential algorithm to find ρ consecutive cost decreasing edge reversals requires $\rho(c_1 + c_2 + c_3)N \approx 2\rho c_1 N$ time, because no call to **Effectuate** is needed in **PLS**, only **Init_rt**, **Init_st**, and **Propose** have to be invoked. So, implemented on a distributed-memory MIMD machine with P processors, the parallel algorithm has a speed-up equal to

$$\frac{2\rho c_1 N}{3c_1N + c_5 \log P} = O(\frac{\rho}{\log P}) = O(\frac{P}{\log P}).$$

We observe that the speed-up of the algorithm on a distributed-memory MIMD machine, without implementing the parallel merging algorithm, is not substantially lower than the speed-up on a PRAM machine. So, our parallel local search algorithm for the job shop scheduling problem can obtain an acceptable speed-up on a distributed-memory MIMD machine.

REFERENCES

AARTS, E., J. KORST, AND P. ZWIETERING [1993], Deterministic and randomized local search, in: P. Smolensky, M. Mozer, and D. Rumelhart (eds.), *Mathematical Perspectives of Neural Networks*, Erlbaum Associates.

item[AARTS, E. H. L. , J. H. M. KORST, AND P. J. ZWIETERING] [1993], Deterministic and randomized local search, in: P. Smolensky, M. Mozer, and D. Rumelhart (eds.), *Mathematical Perspectives of Neural Networks*, Erlbaum Associates.

BORODIN, A., AND J.E. HOPCROFT [1985], Routing, merging, and sorting on parallel models of computation, *Journal of Computer and Systems Sciences* **30**, 130–145.

COOK, S.A. [1981], Towards a complexity theory of synchronous parallel computation, *Enseignement Mathématique.* **27**, 99–124.

FRENCH, S. [1982], *Sequencing and Scheduling: An Introduction to the Mathematics of the Job-Shop*, Wiley & Sons, New York.

GAREY, M.R., AND D.S. JOHNSON [1979], *Computers and Intractability: A Guide to the Theory of NP-completeness*, W.H. Freeman and Company, New York.

JOHNSON, D.S., C.H. PAPADIMITRIOU, AND M. YANNAKAKIS [1988], How easy is local search?, *Journal of Computer and System Sciences* **37**, 79–100.

LAARHOVEN, P.J.M. VAN, E.H.L. AARTS, AND J.K. LENSTRA [1992], Job shop scheduling by simulated annealing, *ORSA Operations Research* **40**, 113–125.

LAWLER, E.L. [1976], *Combinatorial Optimization: Networks and Matroids*, Holt, Rinehart & Wilson, New York.

LAWLER, E.L., J.K. LENSTRA, A.H.G. RINNOOY KAN, AND D.B. SHMOYS [1993], Sequencing and scheduling: algorithms and complexity, in: S.C. Graves, A.H.G. Rinnooy Kan, and P. Zipkin (eds.), *Handbooks in Operations Research and Management Science 4*, North-Holland.

PAPADIMITRIOU, C.H., AND K. STEIGLITZ [1982], *Combinatorial Optimization: algorithms and complexity*, Prentice Hall.

REEVES, C.R. (ed.) [1993], *Modern Heuristic Techniques for Combinatorial Problems*, Blackwell.

ROY, B., AND B. SUSSMANN [1964], *Les problèmes d'ordonnancement constraints disjonctives*, Note DS 9 bis, SEMA, Paris, France.

TAILLARD, E. [1994], Parallel taboo search techniques for the job shop scheduling problem, *to appear in ORSA Journal on Computing.*

VAESSENS, R.J.M., E.H.L. AARTS, AND J.K. LENSTRA [1994], *Job shop scheduling by local search*, Memorandum COSOR 94-5, Eindhoven University of Technology, Netherlands.

VERHOEVEN, M.G.A., E.H.L. AARTS, E. VAN DE SLUIS, AND R.J.M. VAESSENS [1992], Parallel local search and the travelling salesman problem (extended abstract), *Parallel Problem Solving from Nature 2*, 543–552.

VERHOEVEN, M.G.A., AND E.H.L. AARTS [1994], A parallel Lin-Kernighan algorithm for the traveling salesman problem, *Parallel Computing: Trends and Applications*, 559–563.

11

A RANDOMIZED PARALLEL SEARCH STRATEGY

A. Clementi, L. Kučera* and J. D. P. Rolim

Centre Universitaire d'Informatique
University of Geneva, Switzerland

** Computer Science Department*
Charles University, Prague, Tchequie

ABSTRACT

The problem of computing shortest paths in graphs is of particular interest in combinatorial optimization. Unfortunately, known parallel shortest path algorithms perform significantly more work than the best known sequential algorithms. This gap holds for both *Single-Source* and *All-Pairs* versions of this problem. We present here two parallel algorithms for the *Single-Source Shortest Path* problem having an efficient *expected* performance for two important cases of *Random Graphs* $\mathcal{G}(n, P(edge) = p)$:
- a) p is a fixed constant. We give a parallel-randomized algorithm using n processors and time with "high probability" bounded by $\log n$.
- b) $p = \frac{c \log n}{n}$ (for any constant $c > 2$). We give a parallel algorithm using n processors and time with "high probability" bounded by $\log^2 n$. This algorithm remains efficient even for $p \in \Theta(\frac{\log^k}{n})$, for any constant $k > 1$.
These algorithms can be easily adapted in order to solve the *All-Pairs* version by using n^2 processors and the same expected time.

1 INTRODUCTION

The problem of computing shortest paths in graphs has numerous applications in many areas of combinatorial optimization like networks theory or, in general, when the possible solutions of a given problem can be represented by using a graph. Thus the algorithms, for computing shortest paths, are frequently used as a subroutine in the solution of other important problems.

A. Ferreira and J. D. P. Rolim (eds.), Parallel Algorithms for Irregular Problems: State of the Art, 213–227.

One version of the shortest paths problem consists in, given an undirected graph $G(V, E)$ and a vertex $r \in V$, find the shortest path (and thus the distance) between r and every vertex of G (we denote $n = |V|$). This variant is commonly called *Single-Source Shortest Path* problem (in short $SSSP$). The *All-Pairs* version (denoted as $APSP$), instead, requires to compute the shortest paths between all node pairs of G.

The inability to perform efficiently shortest paths computations in parallel represents a long standing open problem that is commonly dubbed in the literature "the transitive closure bottleneck" [12, 4]. In particular, satisfactory sequential algorithms exist for $SSSP$: the Dijkstra's algorithm (or its improvements) is almost optimal also for nonnegative weighted graphs, but, on the other hand, the parallel algorithms, introduced in literature, are not efficient: the best known algorithm has a work complexity (i.e. the product of number of processors and running time) $w(n) = O(n^3 \text{polylog}(n))$ which is significantly greater than that of the best sequential algorithm.

The above facts led us to consider less ambitious goals than to solve $SSSP$ *exactly* and within an efficient *worst-case* work complexity (clearly, this holds also for the $APSP$ version). An interesting approach to cope efficiently with this problem in parallel is to devise algorithms, having an efficient "expected" work complexity, which can, eventually, make random choices and thus allowing a "small" probability of error (for other interesting approaches see [4, 10, 11]).

The expected (average-case) complexity analysis of any computational problem requires the definition of a probabilistic distribution on the input space. Under this point of view, as far as graph problems are considered, one of the most frequently encountered probability spaces for graphs is the *random graph* model $\mathcal{G}(n, P(edge) = p(n))$ (see [2]), where edges are chosen independently and with probability $p(n)$.

In this paper, we present $SSSP$ parallel algorithms, for two different definitions of $p(n)$, both frequently considered in literature [2, 5, 6]: $p(n) = p$, where p is an absolute constant in the range $(0, 1)$ and for $p = \frac{c \log n}{n}$ (for any constant $c > 2$). Similar algorithms have been first introduced in [3]. Both algorithms use n processors and their simplicity makes easy the relative implementations on the CM-2 machine. As usual, the difficulties lie in the the probabilistic arguments determining the bounds on the errors and on the expected running-time. For p constant, we give a randomized CRCW-PRAM algorithm working in time with high probability (see definition 2.2) bounded by $\log n$ and with an error-probability bounded by the inverse of a suitable polynomial in n (i.e. $O(1/n^d)$, $d > 0$). To do this, we will adopt a technique for efficiently detecting the existence of an element in a set S of size $O(n)$, by performing only a $O(\log n)$ number of random accesses in S. Similar methods have been recently applied for obtaining important results in the theory of probabilistic checkable proofs and in the theory of approximation of optimization problems (see [1]).

For the second case (i.e. $p = \frac{c \log n}{n}$), we construct a CREW-PRAM-algorithm working in time with high probability bounded by $\log^2 n$, using no random choices and therefore without any errors.

Moreover, we will observe that the algorithm, given for the second case, works efficiently also when the probability function can be expressed as a function $p(n) = \frac{a \log^k n}{n} + b$ for any nonnegative constants a, b and c (in this case, the work-complexity is with high probability bounded by $n \log^{k+1} n$).

Finally, concerning with the *APSP* problem, we will easly show that one can run in parallel the proposed algorithms for every possible *source*-node: we thus are able to solve this problem by using n^2 processors and the same expected time.

2 PRELIMINARIES

In this paper we adopt the convention that, given an undirected graph $G(V, E)$ (with $V = \{1, \ldots, n\}$ and $|E| = m$) and a vertex $r \in V$ (i.e. the root), a randomized algorithm A for the *SSSP* problem is a procedure which assigns to each vertex $v \in V$ a value $d^A(r, v)$ which is with high probability equal to the real distance from node r to node v (i.e. $d(r, v)$). However, our algorithms, can be easly modified, in order to compute, for any vertex $v \in V$, a path from r to v of length $d^A(r, v)$ as well. The set of neighborhoods of a vertex $v \in V$ will be denoted as $N(v)$.

Let us now adapt the concepts of error probability and expected-time complexity to the *SSSP* problem according with the model $\mathcal{G}(n, P(edge) = p(n))$.

Definition 2.1 *We say that A is a* randomized *algorithm for SSSP if, given a graph $G(V, E)$ according to the model $\mathcal{G}(n, P(edge) = p(n))$ and a vertex $r \in V$, the error probability:*

$$Pr(d(r, v) \neq d^A(r, v))$$

tends to 0 as $n \to \infty$ for any $v \in V$, where $d^A(r,v)$ is the distance between r and v computed by A.

Definition 2.2 *Let A be an algorithm for SSSP with input probability distribution $\mathcal{G}(n, P(edge) = p(n))$, we say that the time-complexity of A (in short T^A) is* with high probability bounded *(in short w.h.p. – bounded) by the function f if three positive constant a, b and k exist such that:*

$$\lim_{n\to\infty}[Prob(T^A(n) > af(n) + b)]n^k < +\infty.$$

Thus, the probability that the asymptotical time-complexity is not bounded by the function f must tend to 0 at least as fast as the inverse of a polynomial.

Throughout the paper, for brevity's sake, claiming the following facts:

a. $\exists k > 0\ :\ \lim_{n\to\infty} \frac{f(n)}{g(n)} = k$:

b. $\exists\ \overline{n}\ : \forall n > \overline{n} : \Pi(\mathcal{G}(n, P(edge) = p))$ *is true*;

c. a constant $k > 0$ exists such that:

$$\lim_{n\to\infty}[Prob(\Pi(\mathcal{G}(n, P(edge) = p))\ is\ false)]n^k < +\infty$$

(where $f(n)$, $g(n)$ are two arbitrary functions and $\Pi(\mathcal{G}(n, P(edge) = p)$ is an arbitrary boolean predicate), will be sometimes replaced, respectively, with the notations:

a. $f(n) \sim g(n)$;

b. *almost always* $\Pi(\mathcal{G}(n, P(edge) = p))$ is true;

c. *with high probability* $\Pi(\mathcal{G}(n, P(edge) = p))$ is true.

Since all proofs in the paper are based on the well known Chernoff's bound to the tail of the binomial distribution, we present it in the following form:

Theorem 2.1 *Let $f_1, \ldots, f_r$ be independent $0,1$-random variables and $\epsilon \in (0,1]$ be a constant. If $p_i = Prob(f_i = 1)$, $i = 1, \ldots, r$, then:*

$$Prob(\sum_1^r f_i \ - \ E(\sum_1^r f_i) \geq \epsilon E(\sum_1^r f_i)) \leq \exp(-\frac{\epsilon^2}{2} E(\sum_1^r f_i)),$$

$$Prob(E(\sum_1^r f_i) \ - \ \sum_1^r f_i \geq \epsilon E(\sum_1^r f_i)) \leq \exp(-\frac{\epsilon^2}{2} E(\sum_1^r f_i)).$$

3 RANDOM GRAPHS WITH CONSTANT EDGE PROBABILITY

In this section we shall define a CRCW-PRAM randomized algorithm for random graphs according to the model $\mathcal{G}(n, P(edge) = p)$ where $p \in (0,1)$ is an absolute constant. The idea of the algorithm is the following one. Given a graph $G = (V, E)$ and a vertex (the root) $r \in V$, the algorithm constructs a partition of V according to the distance between the root and each vertex. The h-th set (i.e. the h-th level) of the partition (denoted as $L(h)$) includes all nodes having exactly distance h from r. The procedure generates the first level $L(1)$ in a deterministic way, within constant time by using n processors. Then, the algorithm defines a random subset S of $L(1)$: each element of $L(1)$ belongs to S with uniform and mutually-independent probability equal to $q = \frac{c \log n}{|L(1)|}$ (for a suitable constant $c > 0$). Then, every non-visited node i checks whether there is at least one of its neighbors belonging to S; if this test has a positive answer, then we include i in level $L(2)$. After this stage, the algorithm detects whether all nodes are included in the first three levels. If this is not the case, the algorithm performs a procedure (totally deterministic) which constructs the other levels in an exhaustive way. It is clear that a running of this last procedure has a linear time for each level $L(h)$ to be constructed, however we shall prove that every pair of nodes has, with high probability, a path of length not greater than two.

Algorithm $A1$

INPUT:

The adjacency matrix $A \in \{0,1\}^{n \times n}$ of a graph $G(V, E)$ according to the distribution $\mathcal{G}(n, P(edge) = p)$ (the source-node corresponds to the first row of A and its label is 1);

OUTPUT:
A function $d : V \to \{0, \ldots, n-1\}$ such that $d(i) = d(1, i)$ for $i = 1, \ldots, n$ (i.e. the distance of every node from node 1);

```
Begin
1) d(1) := 0; L(0) := {1}; L(h) := ∅, h = 1...,n-1; d(i) = +∞, i = 2,...,n;
   S := ∅; Flag := true;
2) for i := 2 to n do in parallel              (computing the first level)
   if A(1,i) = 1 then
   Begin
   d(i) := 1; L(1) := L(1) ∪ {i}; |L(1)| := |L(1)| + 1;
   End;
3) for i := 2 to n do in parallel              (computing a random subset of
   L(1))
   if d(i) = 1 then S := S ∪ {i} with probability q = c log n / |L(1)|;   (for a
   suitable c)
4) for i := 2 to n do in parallel              (computing the second level)
   if d(i) > 1 then
   Begin
   for any s ∈ S do
   if A(i,s) = 1 then
   Begin
   d(i) := 2; L(2) := L(2) ∪ {i};
   End;
   if d(i) > 2 then Flag := false;
   End;
5) if Flag then stop;
   else
   while ( not(Flag) and h < n ) do            (Computing levels L(h),
   h > 2)
   Begin
   Flag := true;
   for i := 2 to n-1 do in parallel
   Begin
   if d(i) > h then
   for any j ∈ L(h) do
   if A(i,j) = 1 then
   Begin
   d(i) := h+1; L(h+1) := L(h+1) ∪ {i};
   End;
```

```
    if d(i) > h + 1 then Flag := false;
    End;
    h := h + 1;
    End;
End.
```

Note 1. The algorithm $A1$ can be easly modified in order to compute also the shortest paths; indeed, for any $h > 0$ and for each node $i \in L(h)$, a path of length h, from 1 (i.e. the source) to i, can be recover by simply implementing the node $j_i \in L(h-1)$ (computed in steps 4 and 5) such that $A(i, j_i) = 1$.

Theorem 3.1 *The algorithm A1 is a CRCW-PRAM randomized algorithm for SSSP using n processors. Morover, the time complexity T^{A1} is w.h.p.-bounded by* $\log n$.

Proof.

The proof is based on the following facts: **a)** the size of L(1) is with high probability greater than βpn for some constant $\beta > 0$; **b)** the set S has with high probability a size $|S|$ such that $\gamma_1 \log n \leq |S| \leq \gamma_2 \log n$ where $0 < \gamma_1 < \gamma_2$; **c)** from the first two points, it will be proved that the probability that a vertex $i \in L(2)$ exists such that $(N(i) \cap S) = \emptyset$ tends to 0 as $n \to \infty$; **d)** the probability that G has diameter greater than 2 tends to 0 as $n \to \infty$, according to definition 2.2. Let us prove the above facts.

a) According to the model $\mathcal{G}(n, P(edge) = p)$, the expected size of $L(1) = N(1)$ is

$E(|L(1)|) = p(n-1)$. Thus, by Chernoff's inequality, we have:

$$[Prob(|L(1)| - p(n-1) > \frac{1}{2}p(n-1))]n^k \leq [\exp(-\frac{1}{8}p(n-1))]n^k$$

and

$$[Prob(p(n-1) - |L(1)| > \frac{1}{2}p(n-1))]n^k \leq [\exp(-\frac{1}{8}p(n-1))]n^k,$$

for any constant $k > 0$. Hence, for $n \to \infty$, both the above expressions tends to 0 and so, with high probability, the size of the first level lies between the following bounds:

$$\frac{1}{2}p(n-1) < |L(1)| < \frac{3}{2}p(n-1).$$

b) Since the random sampling in $L(1)$ is made independently and with probability $q = \frac{c \log n}{|L(1)|}$, the expected size of set S is

$$E(|S|) = q|L(1)| = c \log n.$$

Similarly to step a), by applying Chernoff's inequality, we have

$$[Prob(|S| - c \log n > \frac{1}{2} c \log n)]n^k \leq [\exp(-\frac{1}{8} c \log n)]n^k$$

and

$$[Prob(c \log n - |S| > \frac{1}{2} c \log n)]n^k \leq [\exp(-\frac{1}{8} c \log n)]n^k.$$

By choosing any k in the range $(0, (1/8)c]$, we have, with high probability:

$$\frac{1}{2} c \log n \leq |S| \leq \frac{3}{2} c \log n.$$

c) Let us consider a vertex $i \in L(2)$ and a node j different from both i and 1 (i.e. the root) then we have: $Prob(j \in L(1)) = Prob(j \in N(i)) = p$ and, by step 3 of the algorithm (i.e. the random sampling in $|L(1)|$), $Prob(j \in S \mid j \in L(1)) = q$.
By the independence of these three events, the probability that there are no elements in the set $N(i) \cap S$ is bounded by:

$$Prob((N(i) \cap S) = \emptyset) \leq (1 - p^2 q)^{n-2} \leq (\exp(-p^2 q))^{n-2};$$

by introducing the values of q, we obtain the following bound:

$$Prob((N(i) \cap S) = \emptyset) \leq n^{-cp^2}.$$

Hence, for any fixed $i \in L(2)$, the probability that $(N(i) \cap S) \neq \emptyset$ is not smaller than $(1 - n^{-cp^2})$; then, the probability $Prob(\epsilon)$ such that there is *at least one* node $i \in L(2)$ for which $(N(i) \cap S) = \emptyset$ satisfies the following inequality (here we use the obviuos inequality $|L(2)| < n$):

$$Prob(\epsilon) \leq (1 - (1 - n^{-cp^2})^n)$$

which tends to 0 as $n \to \infty$, when $c > (1/p)^2$ (note that this bound also determines the choice of c in step 3 of $A1$).

d) Our goal is to prove that the diamater of G is with high probability not greater than two[1]. Let us consider a fixed pair of vertices (i, j), since in the model $\mathcal{G}(n, P(edge) = p)$ the existence of an edge is independent from the existence of any other, then the probability that $d(i, j)$ is greater than two satisfies the following inequalities:

$$Prob(d(i,j) > 2) \leq (1 - p^2)^{n-2} \leq exp(-p^2(n-2));$$

hence, the probability that *at least one* vertex pair (i, j) exists such that $d(i, j) > 2$ satisfies the following inequality:

$$[Prob(\exists(i,j) \ : \ d(i,j) > 2)]n^k \leq [1 - (1 - exp(-p^2(n-2)))^{n^2}]n^k.$$

The last expression tends to 0 as $n \to \infty$, for any $k > 0$, hence, with high probability, every pair of vertices has a path of length not greater than two.

Informally, level one (i.e. $L(1) = N(1)$) is deterministically determined in constant time by using n processors. Moreover, since the size of subset $S \subseteq L(1)$ is bounded by $3/2 \log n$ (with "high" probability - point b), each processor $i = 2, \ldots, n$ can detect whether $(N(i) \cap S) \neq \emptyset$ in time $w.h.p.$-bounded by $O(\log n)$. Hence, the time for constructing the set $L(2)$ is $w.h.p.$-bounded by $\log n$ using n processors. From point c) of the proof, the error probability, in computing $L(2)$, tends to 0 as $n \to \infty$. Finally from point d), the probability that, in step 5 of $A1$, the boolean variable $Flag$ is false tends to 0 as $n \to \infty$, thus the choice of the exhaustive procedure in step 5 does not affect the bound of the expected running time of $A1$ according to definition 2.2.

□

4 OTHER EDGE PROBABILITY FUNCTIONS

In this section, we investigate the cases in which the probability functions are, respectively, $p = \frac{c \log n}{n}$ for an arbitrary constant $c > 2$ and $p \sim \frac{\log^k n}{n}$, $k > 1$. We show that the exhaustive algorithm, for these cases, has an efficient expected work-complexity.

[1] a different proof of this fact can be also found in [2]; here we give a simplified version

Algorithm $A2$

INPUT:
The adjacency matrix $A \in \{0,1\}^{n \times n}$ of a graph $G(V, E)$ according to the distribution $\mathcal{G}(n, P(edge) = q(n))$; (the label of the source-node is 1 and corresponds to the first row of A)

OUTPUT:
A function $d : V \to \{0, \ldots, n-1\}$ such that $d(i) = D(1, i)$ for $i = 1, \ldots, n$; (i.e. the distance of every node from the node 1)

Begin

1) $d(1) := 0$; $d(i) = \infty$, $i = 2, \ldots, n$; $Flag := false$; $h := 1$;

2) **while** (**not**($Flag$) **and** $h < n$) **do**
  Begin
  $Flag := true$;

3) **for** $i := 2$ **to** n **do in parallel** (computing level $h+1$)
  Begin
  if $d(i) > h$ **then**
  for any $j \in N(i)$
  if $d(j) = h$ **then** $d(i) := h + 1$;
  if $d(i) > h + 1$ **then** $Flag = false$;
  End;

4) h:=h+1;
  End;

End.

This algorithm is totally deterministic and thus the error probability is always equal to zero. On the other hand, it remains to prove that the time-complexity is $w.h.p.$-bounded by a polylogarithmic function.

Theorem 4.1 *The algorithm A2 is a PRAM algorithm for SSSP using n processors. Morover, the time complexity* T^{A2} *is w.h.p.-bounded, respectively, by* $\log^2 n$ *for* $p = \frac{c \log n}{n}$ *and by* $\log^{k+1} n$ *for any function p such that* $p \sim \frac{\log^k n}{n}$, $k > 1$.

Proof. The iterating process, determined by step 2 of $A2$, ends when every processor i has computed the correct distance between node i and the root. Moreover, since in the h-th iteration all nodes at distance $h+1$ are determined, the total number of iterations is equal to the maximum distance from the source and the other nodes. Finally, the parallel-time of step 3 is bounded by $max\{|N(i)| : i = 1, \ldots, n\}$. From these facts, it is easy to verify that the time of the algorithm $A2$ is completely determined by the following two parameters of the input graph: **i)** the maximum number of neighborhoods of a vertex; **ii)** its diameter. Hence, the proof will consist of determining the expected values of them. Moreover, we will study only the case in which $p = \frac{c \log n}{n}$ since the other case (i.e. when $p \sim \frac{\log^k n}{n}$, $k > 1$) can be analysed in the same way.

i) Let us consider a vertex $i \in V$, according with our probability model, we have the following expected number of neighbors: $E(|N(i)|) = pn = \frac{c \log n}{n} n = c \log n$.

Then, by using the Chernoff's bound, it follows that:

$$p(i) \stackrel{def}{=} Prob((|N(i)| - c \log n) > c \log n) \leq exp(-\frac{1}{2} c \log n) \leq n^{-\frac{1}{2}c}$$

Since the edges are chosen independently, the probability $\pi(n)$ that a vertex $i \in V$ exists such that $|N(i)| > 2c \log n$ satisfies the following inequality $\pi(n) \stackrel{def}{=}$

$Prob(\exists\, i : |N(i)| > 2c \log n) = (1 - \prod_{i=1}^{n}(1 - p(i))) \leq (1 - (1 - n^{-\frac{1}{2}c})^n)$.

Finally, for $c > 2$, we have that

$$\lim_{n \to \infty} [\pi(n)] n^k = 0$$

for any constant $k \in [0, 1)$ (note that for $p \sim \frac{\log^k n}{n}$, $k > 1$, we can easy verify that $E(|N(i)|) \sim \log^k n$: this is the only difference, between the two considered cases, which has a significant consequence in the expected running-time of $A2$).

ii) Let us consider a vertex $i \in V$; for any $h = 1, \ldots n - 1$, let $L_i(h)$ be the set of vertices in V having distance h from node i, (i.e. the h-th level of the BFS-tree T_i rooted on node i) and consider only that first levels whose size is not greater than $(2n)^{1/2}$. Then, we now prove that for any constant $\alpha > 0$ and for any $h \geq 0$, satisfying the condition stated above, we have:

$$|L(h + 1)| \geq \alpha |L(h)|$$

with high probability.

The proof will be performed by induction on h. From part i) of the proof, the above fact is clearly true for $h = 0$. Thus, let $h > 0$ such that, for any $k = 0, \ldots h : |L(k)| \le (2n)^{1/2}$; then we have that:

$$s \stackrel{def}{=} |V \setminus \bigcup_{k=0}^{h} L(k)| \sim n. \tag{11.1}$$

Moreover, given a vertex $i \in (V \setminus \bigcup_{k=0}^{h} L(k))$ the probability that i is in level $L(h+1)$ is:

$$Prob(i \in L(h+1)) = Prob(\exists\, j \in L(h)\ s.t.\ j \in N(i)) = 1-(1-p)^{|L(h)|} =$$

$$= 1 - (1 - |L(h)|p + \binom{|L(h)|}{2} p^2 - \ldots + p^{|L(h)|}) \sim p|L(h)|.$$

Hence, the following inequality holds almost always:

$$Prob(i \in L(h+1)) \ge \frac{1}{2} p|L(h)| \tag{11.2}$$

By expressions 11.1 and 11.2, the expected size of $|L(h+1)|$ satisfies

$$E(|L(h+1)|) \ge \frac{1}{2} sp|L(h)|)$$

and, therefore, by using again Chernoff's bound, we have:

$$Prob((\frac{1}{2} sp|L(h)| - |L(h+1)|) > \frac{1}{4} sp|L(h)|) \le \exp(-\frac{1}{8} sp|L(h)|)$$

By rewriting the above inequalities in a slightly different way, we obtain:

$$\pi_1(n) \stackrel{def}{=} Prob(\ |L(h+1)| < \frac{1}{2} sp|L(h)|\ ) \le \exp(-\frac{1}{8} sp|L(h)|).$$

Since $sp \sim c \log n$ $(c > 2)$, then the expression

$$[\pi_1(n)](n^k)$$

tends to 0 as $n \to \infty$, for any $k \in (0, 1/4)$.

Let us now consider a vertex pair i and j. From the above results, with high probability, the size of the first levels $L^i(h)$ and $L^j(h)$ of both BFS-trees T^i and T^j, having no more than $(2n)^{1/2}$ elements, grows within exponential rate with respect to h (i.e. $|L_i(h+1)| \geq \alpha |L_i(h)|$ with $\alpha > 1$, the same holds for j). This implies that, with high probability, an integer $h^m > 0$, with $h^m = O(\log n)$, exists such that:

$$|L^i(h^m)| \geq (2n)^{\frac{1}{2}} \ and \ |L^j(h^m)| \geq (2n)^{\frac{1}{2}}.$$

Let us now consider the two subset of vertices $L^i(h^m)$ and $L^j(h^m)$, two cases may arise depending on whether $L^i(h^m) \bigcap L^j(h^m) = \emptyset$ or not. Let us consider the first case, that is when such two subsets are disjoint; then, the probability $Prob(\overline{edge})$ that there are no crossing edges between these two sets is:

$$Prob(\overline{edge}) = (1-p)^{2n} \leq \exp(-2np) \leq n^{-2c}$$

where in the second equivalence we have used the relation $np = n\frac{c \log n}{n} = c \log n$. Thus, the probability that there is at least one crossing edge between these two sets for any pair of vertices is:

$$(1 - Prob(\overline{edge}))^{\binom{n}{2}}.$$

Finally, the probability $\Pi(n)$ that a pair of vertices i and j exists such that their corresponding BFS-trees T^i and T^j have no common vertices after level $h^m + 1$ is:

$$\Pi(n) = (1 - (1 - Prob(\overline{edge}))^{\binom{n}{2}}).$$

Since $c > 2$, we have that $\Pi(n)(n^k)$ tends to 0 as $n \to \infty$, for any positive $k < c - 2$. Hence, given any pair of vertices of G, a path between them exists whose length is $w.h.p$-bounded by $\log n$ (note that for $p \sim \frac{\log^k n}{n}$, $k > 1$, the convergence to 0 of the "error" $\Pi(n)$ is faster than that for the case $p = \frac{c \log n}{n}$).

□

Note 2. For both edge-probability functions studied in this paper, we can adapt their relative algorithms ($A1$ and $A2$) in order to solve the $APSP$

problem. Without lost of generality, let us consider the first case (i.e. p constant): the algorithm for the $APSP$ problem will consist of n independent-parallel runs of $A1$, each of them starting with a different source-node. Since the probabilistic analysis, made in the proof of theorem 3.1, is based uniquely on the general combinatorial properties (i.e. diameter and expected size of node-neighborhoods) of the model $\mathcal{G}(n, P(edge) = p)$, it can be easily extended to the new algorithm for $APSP$. In particular, the choice of the parameter c (see point (c) of the proof of theorem 3.1) must now satisfy $c > (1/p)^2 + 1$. The only relevant difference is in the number of processor required by the new algorithms which is now augmented by a factor of n.

5 CONCLUSIONS

The simple structure of the algorithms and the use of two probability distributions which can be efficiently simulated on the computer have permitted us to make several experimental tests on a CM-2 machine by using random graphs with up to a few thousand nodes. The obtained results totally support our theoretical analysis.

Among interesting future works, we note especially the following; shortest path computations are often used as subroutine in the solution of other important combinatorial optimization problems like optimal scheduling in networks. Thus it could be interesting to apply our new algorithms for such problems.

REFERENCES

[1] L. Babai, Fortnow L. Levin L and M. Szegedy, "Checking computation in polylogarithmic time", Proc. of 23th Annual ACM STOC 21-, 1991.

[2] B. Bollobas, *Random Graphs*, Academic Press, 1985.

[3] A. Clementi, L. Kučera and J. Rolim, "A Note on Parallel Randomized Algorithms for Searching Problems", DIMACS - Workshop on Parallel Processing of Discrete Optimization Problems, DIMACS Center, New Jersey - USA, 1994.

[4] E. Cohen, "Polylog-time and near-linear work approximation scheme for undirected shortest paths", Proc. of the 26th Annual ACM STOC, 16-26, 1994.

[5] D. Coppersmith, P. Raghavan and M. Tompa, "Parallel Graph Algorithms that are Efficient on Average", Proc. of the 28th Annual IEEE FOCS, 260-269, 1987.

[6] M.E. Dyer and A. Frieze, "The Solution of Some Random NP-hard Problems in Polynomial Expected Time", Journal of Algorithms, 10, 451-489, 1989.

[7] R. Greenlaw, "Polynomial Completeness and Parallel Computation", in Synthesis of Parallel Algorithms, Ed. J. Reif, Morgan-Kaufmann, Chapter 21, 1993.

[8] R. Karp and V. Ramachandran, "Parallel Algorithms for Shared-Memory Machines", in Handbook of T.C.S., Ed. J. van Leeuwen, Elsevier Science, Vol. A, Chapter 17, 1990.

[9] R. Karp, "An introduction to randomized algorithms", Discr Appl Math, 34, 165-201, 1991.

[10] P.N. Klein and S. Sairam, "A Parallel Randomized Approximation Scheme for Shortest Paths", Proc. of the 24th Annual ACM STOC, 750-758, 1992.

[11] P.N. Klein and S. Sairam, "A linear-processor polylog-time algorithm for shortest paths in planar graphs", Proc. of the 34th Annual IEEE FOCS, 259-270, 1994.

[12] J. Ullmann and M. Yannakakis, "High Probability Parallel Transitive Closure algorithms", SIAM Journal of Computing, 20, 100-125, 1991.

12

A PARALLEL APPROXIMATION SCHEMA FOR SYMMETRIC LINEAR ASSIGNMENT PROBLEM

Erik Urland

Centre Universitaire d' Informatique
Université de Genève, 24 rue Général Dufour
CH-1211 Genève 4, Switzerland

ABSTRACT

In this paper we discuss the Jacobi version of the Auction algorithm and its implementation on a Connection Machine CM-2. We show the advantages as well as some shortcomings of such a massively parallel approach to the dense symmetric linear assignment problem. An approximation schema is proposed and empirically evaluated on some modified size random instances.

1 INTRODUCTION AND DEFINITIONS

A general *assignment* problem can be stated as follows. Let M be a set of objects x, N a set of objects y (where $|M|$ and $|N|$ are not necessary equal) and a cost $c_{u,v}$ associated with the assignment of object $u \in M$ to object $v \in N$. We can imagine that objects x represent persons and objects y are jobs. Let us assume now that $|M| \leq |N|$, then the classical *linear* assignment problem, shortly LAP, is to find such an assignment of each person to exactly one job that maximizes (or minimizes) the total cost.
From graph theoretical point of view, the LAP is to find a maximum matching with optimum weight in a bipartite graph. If the bipartite graph is complete then the LAP is called *dense*. Moreover, if $|M| = |N|$ then we have *symmetric* LAP and in this case one would like to find a perfect matching of optimum weight in the given bipartite graph.
Different type of problems arise from network flow theory which can be interpreted as LAP. Consider the following network.

A. Ferreira and J. D. P. Rolim (eds.), Parallel Algorithms for Irregular Problems: State of the Art, 229–242.

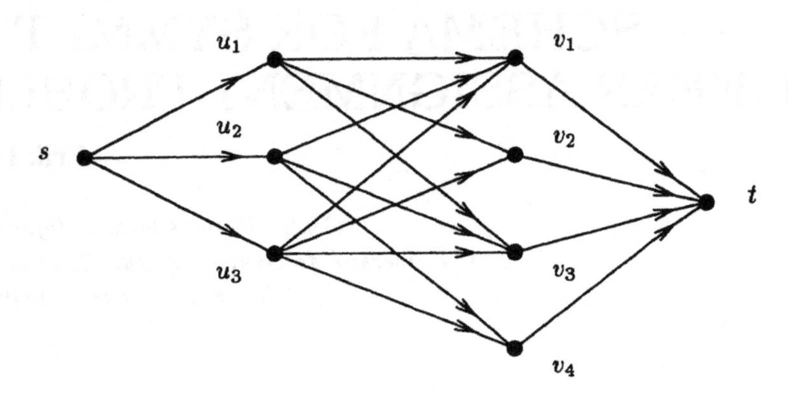

Figure 1.

We set the cost of the arcs connecting the *source* node s with nodes u_i, $i = 1, ..., 3$ and the arcs connecting the *sink* node t with nodes v_j, $j = 1, ..., 4$ to be zero. Further we assume that the capacity of each arc in the network is one. Now it is not difficult to see that the LAP is equivalent to the problem of finding an optimum (usually minimum) cost flow from node s to node t in such a network.
Other applications of the LAP can be found in such areas like project planning, capital budgeting, routing and scheduling, network design and in other areas which involve optimum assignment.
From the above mentioned examples follows that the LAP can be considered as one of the most fundamental and basic problem in combinatorial optimization. In addition, we can often find the LAP as an important part in the solution of various complex optimization problems.
As the symmetric LAP is a special case of the optimum cost max-flow problem it can be stated as the following linear program.

$$\text{Optimize} \quad \sum_{(i,j)\in E} c_{i,j} x_{i,j} \tag{1}$$

$$\text{subject to} \quad \sum_{(i,j)\in E} x_{i,j} = 1 \quad \text{for all } i = 1, ..., n,$$

$$\sum_{(i,j)\in E} x_{i,j} = 1 \quad \text{for all } j = 1, ..., n,$$

$$\text{where} \quad x_{i,j} \geq 0 \quad \text{for all } i, j = 1, ..., n,$$

where $c_{i,j}$, $1 \leq i, j \leq n$, are the entries of the assignment cost matrix. If the optimization problem is to maximize the sum (1) then the corresponding dual problem in the vectors $\pi = (\pi_1, ..., \pi_n)$ and $p = (p_1, ..., p_n)$ is

$$\text{Minimize} \quad \sum_{i=1}^{n} \pi_i + \sum_{j=1}^{n} p_j, \tag{2}$$

$$\text{subject to} \quad \pi_i + p_j \geq c_{i,j} \quad \text{for all } (i,j) \in E.$$

Since we can formulate the LAP using language of linear programming and network flow theory it is not too surprising that most algorithms for the LAP can be viewed as modifications of simplex method or methods for minimum cost flow problem. More precisely, we have the following classification [13].

- Maximum flow
- Primal simplex
- Dual simplex
- Shortest augmenting path
- Auction algorithm

Most of the maximum flow algorithms are based on *Hungarian* method which was introduced by Kuhn [14]. Some modifications and improvements of this method are presented in [7] and [11]. A specialization of primal simplex method, called *alternating basis* algorithm, was developed by Barr, Glover and Klingman [2]. Hung in [10] used similar approach to design a polynomial algorithm for LAP. A dual polynomial simplex method, known as *signature* method, is due to Balinsky [1]. Further extension and analysis of this method can be found in [9]. The shortest augmenting path technique is adapted by algorithms presented in [7] and [8]. Jonker and Volgenant [12] did an extensive study of algorithms based on shortest augmenting path. Moreover, they developed an algorithm [12] which seems to be very efficient in this category. Some computational results for large scale LAP instances can be found in [13]. The last of the above mentioned algorithms for LAP is the Auction algorithm which was introduced by Bertsekas [3] and [4]. The original version of the algorithm was partially based on Hungarian method and augmenting path technique and later improved by using the idea of ϵ-relaxation and adaptive scaling [6]. We give a more detailed description of this algorithm in the following section. The parallel implementation of the Auction algorithm is described in section 3. Section 4 deals with computational results obtained on a Connection Machine CM-2 while in the last section we present an approximation schema which is a variation of the original Auction algorithm.

2 AUCTION ALGORITHM

Suppose that there are n persons and n objects. We would like to find such an assignment A of each person to exactly one object that $\sum_{A(i)=j} c_{i,j}$ is maximized, where $A(i) = j$ means that person i is assigned to object j. Before introducing the algorithm itself we mention some facts from linear programming theory which help us to understand the principle of the algorithm. We have introduced in (2) two vectors π and p. We refer to dual variables p_j, $j = 1, ..., n$, as the *price* of object j, to scalar $\pi_i = \max_{(j=1,...,n)}\{b_{i,j}\}$, where $b_{i,j} = c_{i,j} - p_j$, as the *profit margin* of person i corresponding to price vector p and to $b_{i,j}$ as the *benefit* of assigning person i to object j. Now one can see that for a given vector p, the dual problem (2) is minimized when all $\pi_i = \max\{b_{i,j}\}$, for $j = 1, ..., n$. In this case if each person i is assigned to different object $A(i) = j$ such that $b_{i,j} = \pi_i$ (person i is assigned to his most benefitable object j) then the assignment A is globally optimal for (1). We know that an assignment $A = \{c_{i,A(i)}; i = 1, ..., n\}$ and a price vector p are simultaneously primal and dual optimal respectively iff

$$\pi_i = \max_{(j=1,...,n)}\{b_{i,j}\} = b_{i,A(i)} \text{ for } i = 1, ..., n. \tag{3}$$

This condition is known as *complementary slackness* condition and such an assignment A which satisfies (3) is called *optimal*. A weaker *ϵ-complementary slackness* (ϵ-CS) condition for an assignment A and a price vector p is

$$\pi_i - \epsilon = \max_{(j=1,...,n)}\{b_{i,j}\} - \epsilon \leq b_{i,A(i)} \text{ for } i = 1, ..., n, \tag{4}$$

and an assignment satisfying (4) is called *ϵ-optimal*.
Suppose that we fix $\epsilon > 0$. Let us choose an assignment A and a price vector p satisfying (ϵ-CS) condition, then the Auction algorithm can be described as follows.

STEP 1. Auction Phase

- For each person i that is unassigned under A compute $b_{i,j}$ for all $j = 1, ..., n$;
- Find the *best* object j such that $w_{i,best} = \max_{(j=1,...,n)}\{b_{i,j}\}$;
- Find the *next* object j such that $w_{i,next} = \max_{(j=1,...,n),j\neq best}\{b_{i,j}\}$;
- Compute the bid of person i to its favorite object *best* as $bid_{i,best} = p_{best} + w_{i,best} - w_{i,next} + \epsilon$.

STEP 2. Assignment Phase

- Let $P(j)$ be the set of persons bidding for object j. Each bid-upon object j determines its highest bidder $h = x$, where $bid_{x,j} = \max\{bid_{i,j}, i \in P(j)\}$ and increase its price to $p_j = bid_{h,j}$;
- Each bid-upon object j assigns itself to person h and deassigns the person i to whom it was previously assigned (if any).

STEP 3. Assignment Verification

- If any person is unassigned under A goto STEP 1;
- We have obtained an ϵ-optimal assignment A. If ϵ is sufficiently small then stop. Assignment A is globally optimal.
- Otherwise decrease ϵ. Deassign persons who no longer satisfy (4) and goto STEP 1.

The algorithm proceeds iteratively (STEP 1 and 2) and terminates at STEP 3 when an ϵ-optimal assignment is obtained for sufficiently small ϵ. At the beginning of a generic iteration we have an assignment A and a price vector p satisfying ϵ-CS. At the end of the iteration, A and p are updated under the ϵ-CS condition [5]. This is important because of the fact that if an optimal assignment exists then the algorithm will find it in a finite number of iterations. Another important question is how to choose the initial value ϵ in order to obtain a good performance of the algorithm. In this case a scaling approach can be used which is called $\epsilon-$*scaling* method [5] or *adaptive* scaling [13], and it can be shown (e.g. [3], [6]) that if $C = \max\{|c_{i,j}|; i,j = 1,...,n\}$ then an optimal assignment (more precisely, an assignment satisfying ϵ-CS condition for $\epsilon < \frac{1}{n}$) can be obtained after $O(log(nC))$ scaling phases. The worst case analysis for the above described algorithm is following. Assume that the LAP is symmetric with cardinality n and is fully dense which corresponds to finding a maximum weight perfect matching in bipartite graph. To find $w_{i,best}$ and $w_{i,next}$ for a person i who is unassigned in STEP 1 takes $O(n)$ time. There can be at most n persons unassigned, thus we have $O(n^2)$ time complexity for this step. In the sequential fashion of the Auction algorithm one unassigned person bids at a time. The cardinality of $P(j) = 1$ and there is exactly one object j which receives a bid, so the STEP 2 can be performed in $O(1)$ time, what does not have any influence on $O(n^2)$. As the prices cannot rise more

than $O(n)$ times per object (see [5], [6]), one can obtain $O(n^3)$ computational complexity for a scaling phase. From this follows that the total cost of the Auction algorithm is $O(n^3 log(nC))$. If C is a constant independent from n, then from the asymptotical point of view we have $O(n^3 logn)$ worst case sequential complexity, which is reasonable for this class of algorithms.
According to STEP 1 we distinguish two versions of the pure Auction algorithm. In the *Jacobi* version all unassigned persons bid simultaneously on their favorite objects before the prices are adjusted. In the second, $Gauss - Seidel$ version, a single unassigned person bids for an object and the updated price of the object is taken into the account when the next bid by an unassigned person is performed. Next we will concentrate on the Jacobi version of the Auction algorithm which has a greater potential for exploring parallelism.

3 PARALLEL APPROACH

The design of any parallel algorithm is done according to the architecture of target machine. From this point of view one can intuitively see that the Jacobi version of the Auction algorithm is well suited for $SIMD - MC^2$ parallel machine. Let us assume that the target parallel machine has the following mesh architecture.

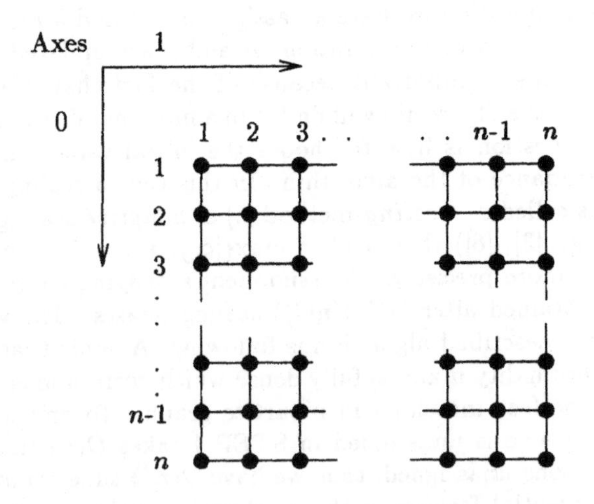

Figure 2.

An $SIMD-MC^2$ parallel machine performs two different kind of interprocessor communications. One is *general* communication, when a particular processor interacts with an arbitrary processor in the network, and the second is *grid* communication, when the processors communicate along any axis in the grid. It is clear that the operations which involve general communication are computationally much more expensive and we try to avoid them. The following algorithm uses only operations based on grid communication. Moreover, it does not have any part which involves a sequential approach.

STEP 1. Initialization

- Each processor in the network contains an element of the following parallel variables of shape $n \times n$: **eps**, it corresponds to ϵ; **price**, the price of objects j (for all processors along axis 0 it will be the same value); **post_ass**, where the element (x, y) is 1, if person x is assigned to object y or 0 in opposite case; **pers_ass**, another boolean variable indicating whether a particular person is assigned or not (similar to variable **price** but here all the processors along axis 1 will contain the same value 1 or 0); and finally variable **cost**, where each element (i, j) corresponds to $c_{i,j}$ in the cost assignment matrix.
- By a global-max operation set **eps**:=$(n + 1) * C$;
- Let **cost**:=**cost***$(n + 1)$;
- Set **post_ass**, **pers_ass** and **price** to 0;

STEP 2. Assignment Verification

- Use a global-and operation to check if all persons are assigned. If no, then proceed STEP 3;
- If each person i is assigned to some object j, $1 \leq i, j \leq n$, then check the value **eps**;
- Take an element of **eps** and if its value is < 1 then the algorithm stops.
- Otherwise **eps**:=**eps**/2;
- Take two parallel variables **benefit** and **help** of shape $n \times n$. Let us compute **benefit**:=**cost**−**price**. Next we perform a grid-spread-max operation along axis 1 on **benefit** to obtain in all elements of variable **help**

along axis 1 the maximum value of corresponding elements of the shape of **benefit**. Let **help**:=**help**−**eps**. Set the context to **post_ass**=1 and under this context compare variables **benefit** and **help**. Those elements where **benefit** < **help** set to 0 in **post_ass**. Finally, perform a grid-spread-max operation along axis 1 on **post_ass** and the result put to **pers_ass**. One can see that a processor containing value 1 in the element of **pers_ass** belongs to such a row i in the grid, that person i satisfies (4).

STEP 3. Auction Phase

- Set **benefit**:=**cost**−**price**;
- Perform a grid-spread-max along axis 1 on **benefit** and put the result to **help**;
- Take a parallel variable **test**. Set the context to **benefit**=**help** and assign 1 to those elements of **test** which satisfy the context. Other elements set to 0. Verify by a grid-spread-add operation along axis 1 on **test** if there is more than one best object for each person i, $i = 1, ..., n$. If yes, then determine one of them leaving the corresponding element in **test** without change while setting the other elements to 0 along axis 1;
- Set the context to **test**≠0 and perform a grid-spread-max along axis 1 on **benefit** and put the result to parallel variable **next**. Now, without any context, perform once more a grid-spread-max along axis 1 on **next**, and leave the result in **next**;
- Set the context to (**test**=1) *and* (**pers_ass**=0) and then perform the assignment **bid**:=**cost**−**next**+**eps**;
- Under the same context do a grid-spread-max along axis 0 on **bid** and the result put to parallel variable **max_bid**. One can see that **max_bid** contains the values of maximal bids from unassigned persons to all bid-upon objects (along axis 0). Now, similarly as in the case of finding the best object for a person, if there are more persons offering the highest bid to one particular object, we have to determine only one of them.

STEP 4. Assignment Phase

- Perform a grid-spread-max along axis 0 on **max_bid** without any context and the result leave in **max_bid**;

- Set the context to **max_bid** > 0 and assign **max_bid** to **price**. Note that only the prices of bid-upon objects are updated (because of the context);
- Let parallel variable **one_bid** contains value 1 in those elements which indicate the highest bidder for bid-upon object. The other elements are set to 0. Let **post_ass:=post_ass+one_bid**. We perform a grid-spread-add operation along axis 0 on **post_ass** and the result we assign to **help**. The context **help** > 1 guarantees that **post_ass:=one_bid** indicate the new person-object assignment after a parallel generic iteration.
- Perform a grid-spread-max along axis 1 on **post_ass** and the result put to **pers_ass**. Set **max_bid:=0** and goto STEP 2.

The worst case analysis for parallel version of the Auction algorithm is very similar to the sequential one. Consider the fully dense symmetric LAP with assignment cost matrix $c_{i,j}$, $1 \leq i,j \leq n$. Initialization phase takes $O(logn^2)$ time which we need for computing the initial value of **eps**. Other operations in STEP 1 can be performed in $O(1)$ time. All operations in STEP 2 take constant time except the computing a new ϵ-optimal assignment which needs $O(logn)$ time. It takes $O(logn)$ time to determine the best and the next object in STEP 3 for all unassigned persons. In the worst case, when all n persons bid to one particular object, we need the time $O(logn)$ to determine the maximal bid. As we mentioned before, the price of each object can be raised at most $O(n)$ times, thus the case when all persons bid to one object in each iteration leads to a complexity $O(n^2logn)$ for a scaling phase. There are $O(log(nC))$ scaling phases. Assume that C is a constant not depending on n, then the overall asymptotical complexity of the algorithm is $O(n^2log^2n)$ using n^2 processors. In the following section we deal with some modifications of the pure Auction algorithm which reduce the probability of the worst case and improve the performance of the algorithm.

4 COMPUTATIONAL RESULTS

We have implemented the Jacobi version of the Auction algorithm on a Connection Machine CM-2 which is a massively parallel *SIMD* machine. Although the physical processors of CM-2 are connected in hypercube interconnection network they can be arranged as an MC^2 architecture. Moreover, a number of *virtual* processors can be simulated by a single physical processor. The mapping of virtual processors to the physical processors of the machine is called

geometry. In our computations we use such a geometry that the virtual processors along two axes in the grid are uniformly mapped to the CM-2 physical processors.
As the CM-2 is designed for massively parallel computation we can perform all the bids at once in an auction phase. Some problems appear in STEP 3 when we have to choose one object from a set of best objects. This happens very often, especially when the cost range of the LAP is very small (0-100). Similar problem is to determine the highest bidder. To avoid the worst case of the algorithm some deterministic choices are not appropriate. In this case we can use efficiently randomness and thus obtain a *sherwood* fashion of the original algorithm. The following table shows the computational results for some modified size instances. The results are averaged over five random examples.

n	Cost	Bids	Iterations	Parallelism	Time(s)
128	0-100	4292	814	5.27	14.79/14.38
	0-1000	4794	859	5.58	15.39/15.18
	0-10000	5276	866	6.09	15.49/15.32
	0-100000	5882	1103	5.33	20.14/19.52
256	0-100	10820	2043	5.29	78.67/76.04
	0-1000	11041	1701	6.49	66.53/64.01
	0-10000	12335	1747	7.06	69.16/66.16
	0-100000	12944	1742	7.43	69.71/65.61
512	0-100	29606	4751	6.23	553.09/522.48
	0-1000	25609	3605	7.01	415.97/395.66
	0-10000	27231	3085	8.82	348.87/338.44
	0-100000	28487	3040	9.30	336.73/328.39

Table 1.

Note that the total and CM time in the last column of Table 1 are nearly the same. This corresponds to the fact that the algorithm does not contain sequential operations. During the experiments we have observed that most persons are assigned to objects very quickly and the rest of the computation is spent for finding the assignment of the last few persons. This *tail* effect reduces the possibility of parallelism, where by parallelism we mean the average number of persons bidding in one auction iteration (see Table 1).

A modification to avoid this overhead is proposed in [15]. The idea is to truncate the tail in such a way that instead of running each scaling phase until a complete ϵ-optimal assignment is achieved, the scaling phase is terminated when k% of the persons are assigned. As ϵ is decreasing, k is increasing until $k = 100$ in the last scaling phase. We have tested this modification with an initial assignment bound 70%, 80% and 90%. The best results have been achieved when $k = 80$ at the beginning of the first scaling phase. Table 2 shows the obtained results.

n	Cost	Bids	Iterations	Parallelism	Time(s)
128	0-100	3480	191	18.21	4.00/3.77
	0-1000	3726	159	23.43	3.32/3.12
	0-10000	4184	120	34.86	2.63/2.42
	0-100000	4459	153	29.14	3.29/3.01
256	0-100	10060	1460	6.89	62.40/59.01
	0-1000	9120	737	12.37	31.54/29.91
	0-10000	9445	269	35.11	11.91/11.43
	0-100000	9834	121	81.27	5.65/5.05
512	0-100	33479	4204	7.96	513.84/497.08
	0-1000	21352	1124	18.99	138.29/133.32
	0-10000	22226	1043	21.30	126.54/122.43
	0-100000	23153	934	24.78	114.93/111.02

Table 2.

One can see that in general the truncation of the tail leads really to an improved performance of the pure Auction algorithm code. The speedup is quite significant, especially for large cost ranges. Moreover, the parallelism is better explored, even if it is essentially less than the size of the instances. On the other hand, this modification is not very useful when n is increasing and the cost range is small (0-100). The results are more less the same as for the original algorithm. In this case we propose an approximation schema to improve the running time of the Auction algorithm.

5 APPROXIMATION SCHEMA

Suppose that during the auction phase more persons bid to the same object. As we mentioned before, this situation occurs very often for small cost ranges. The idea is to give somehow a preference to that person, to whom the particular object is much more important than for other bidders. This can be recognized by the fact that such a person can offer significantly lower bids for all other objects. Thus, before computing the bid in auction phase for person i, we find the *second* next object j such that $w_{i,second} = \max_{(j=1,\ldots,n),j\neq best,next}\{b_{i,j}\}$. Then $bid_{i,best} = (c_{i,best} - w_{i,next}) + (c_{i,best} - w_{i,second}) + \epsilon$. However, usage of the approach to compute the bid, has a shortcoming. After the last scaling phase one can obtain a price vector p and an assignment A which are not necessary dual and primal optimal, respectively. Thus we have proposed an approximation to the optimal solution. The following table shows the efficiency of this approach.

n	Cost	Bids	Iterations	Parallelism	Time(s)	Difference(%)
128	0-100	4323	197	21.94	5.08/4.79	0.24
	0-1000	5112	251	20.36	6.48/6.05	0.22
	0-10000	6207	301	20.62	7.72/7.33	0.23
	0-100000	7004	311	22.52	8.12/7.58	0.25
256	0-100	9906	247	40.10	13.50/12.68	0.10
	0-1000	11495	261	44.04	14.19/13.46	0.11
	0-10000	13720	294	46.66	15.78/15.19	0.11
	0-100000	15389	363	42.39	19.68/18.73	0.13
512	0-100	25932	932	27.82	144.32/140.35	0.01
	0-1000	25798	494	52.22	77.55/74.92	0.06
	0-10000	30458	490	62.15	75.02/72.85	0.05
	0-100000	33481	556	60.21	88.94/84.82	0.05

Table 3.

The difference in the table means the percentage of difference from an optimum assignment. As one can see, the speedup achieved by using this approximation schema is about a factor from 3 to 5 with comparison to the pure Auction algorithm code. Moreover, the difference from an optimum assignment is very small.

Acknowledgements

The author would like to thank Prof. Jose Rolim for fruitful discussions concerning combinatorial optimization and Prof. Peter Horak for several useful comments on a draft of this paper.

REFERENCES

[1] Balinsky, M.L.: Signature Methods for the Assignment Problem, Oper. Research 33, pp. 527-536 (1985).

[2] Barr, R.-Glover, F.-Klingman, D.: The Alternating Basis Algorithm for Assignment Problem, Math. Programming 13, pp. 1-13 (1977).

[3] Bertsekas, D.P.: A Distributed Algorithm for Assignment Problem, Unpublished LIDS Working Paper, MIT, 1979.

[4] Bertsekas, D.P.: A New Algorithm for the Assignment Problem, Math. Programming 21, pp. 152-171 (1981).

[5] Bertsekas, D.P.: The Auction Algorithm: A Distributed Relaxation Method for the Assignment Problem, Annals of Oper. Research 14, pp. 105-123 (1988).

[6] Bertsekas, D.P.-Eckstein, J.: Dual Coordinate Step Methods for Linear Network Flow Problems, Math. Programming 42, pp. 203-243 (1988).

[7] Derigs, U.: The Shortest Augmenting Path Method for Solving Assignment Problems-Motivation and Computational Experience, Annals of Oper. Research 4, pp. 57-102 (1985).

[8] Glover, F.-Glover, Glover, R.-Klingman, D.: Threshold Assignment Algorithm, Math. Programming Study 26, pp.12-37 (1986).

[9] Goldfarb, D.: Efficient Dual Simplex Method for the Assignment Problem, Math. Programming 33, pp. 187-203 (1985).

[10] Hung, M.: A Polynomial Simplex Method for the Assignment Problem, Oper. Research 31, pp. 595-600 (1983).

[11] Jonker, R.-Volgenant, A.: Improving the Hungarian Assignment Algorithm, Oper. Research Letters 5, pp. 171-175 (1986).

[12] Jonker, R..-Volgenant, A.: A Shortest Augmenting Path Algorithm for Dense and Sparse Linear Assignment Problem, Computing 38, pp. 325-340 (1987).

[13] Kennington, J.L.-Wang, Z.: An Empirical Analysis of the Dense Assignment Problem: Sequential and Parallel Implementations, ORSA J. on Comp. 3, pp. 299-306 (1991).

[14] Kuhn, H.: The Hungarian Method for the Assignment Problem, Naval Research Logistics Quarterly 2, pp.93-97 (1955).

[15] Wein, J.M.-Zenios, S.A.: On the Massively Parallel Solution of the Assignment Problem, J. of Parallel and Distr. Comp. 13, pp.228-236 (1991).

PART III

TOOLS FOR AUTOMATIC PARALLELIZATION

13

APPLICATIONS OF GRAPH SCHEDULING TECHNIQUES IN PARALLELIZING IRREGULAR SCIENTIFIC COMPUTATION

Apostolos Gerasoulis, Jia Jiao and Tao Yang*

Department of Computer Science
Rutgers University
New Brunswick, NJ 08903, USA

**Department of Computer Science*
University of California
Santa Barbara, CA 93106, USA

ABSTRACT

Parallelizing irregular scientific computation efficiently is a challenging open research area. In this paper we investigate the applicability of graph scheduling techniques to solving irregular problems in distributed memory machines. Our approach is to express irregular computation in terms of a macro-dataflow task model and use an automatic scheduling system to map task graphs and also generate parallel code based on the scheduling result. We study the performance of run-time execution of an irregular static schedule and report our solutions with the automatic scheduling system for sparse matrix computation and hierarchical 2D N-body simulations. We also examine the run-time performance of static mapping for the adaptive n-body simulation. Our preliminary experiments in nCUBE-2 have produced promising results with regard to the practicality of automatic scheduling for irregular problems.

1 INTRODUCTION

Solving irregular problems, where the patterns of computation and communication are unstructured and/or changing dynamically, is very important to many scientific computation applications, and the parallelization of such problems is extremely difficult and presents a great challenge. Although some automatic

A. Ferreira and J. D. P. Rolim (eds.), Parallel Algorithms for Irregular Problems: State of the Art, 245–267.

compiler systems such as PARAFRASE-2 [21] are able to conduct dependence analysis and partitioning for regular problems, they are generally unable to parallelize irregular problems efficiently. One approach in tackling irregular problems which involve DOALL parallelism and iterative computation is the CHAOS system [24], and recently there has been an attempt to extend HPF compilers to handle irregular computation [22]. In this paper, we will discuss how the graph scheduling techniques can guide the mapping of irregular computations on multiprocessors and effectively utilize parallelism hiden in complex data dependence structures.

Task scheduling theory and algorithms have been of long-standing research interest because of the promise that such approach could better utilize the power of multicomputers. Finding optimal scheduling solutions is difficult, especially in the presence of communication. Instead many scheduling heuristic algorithms have been proposed in the literature [9, 12, 18, 25, 26, 3, 20]. Also many attempts have been made to build scheduling systems that predict the performance of task graph computation so that better parallel programs could be written, [9, 26, 27]. Our work tries to demonstrate the practicality of a scheduling system on a large class of application programs. We will first briefly discuss task graph scheduling techniques and the execution of a schedule and then address the following issues: 1) The discrepancy between the static performance predicted by the scheduler at compile time and the actual run-time performance. Since there is always a discrepancy between the weight estimation and the actual run-time performance, it is of interest to know that small estimation errors will not dramatically alter the run-time performance behavior. 2) The automatic scheduling approach in solving irregular problems such as sparse matrix factorization and n-body simulation.

The paper is organized as follows. Section 2 discusses the macro-dataflow task graph model used to represent parallel computation. Section 3 analyzes run-time performance of static scheduling. Section 4 discusses how scheduling could be used for irregular sparse matrix and n-body computation.

2 TASK COMPUTATION MODEL AND SCHEDULING

A task graph is a directed acyclic task graph (DAG) defined by a tuple $G = (V, E, \mathcal{C}, \mathcal{T})$ where V is the set of task nodes and $v = |V|$ is the number of nodes, E is the set of communication edges and $e = |E|$ is the number of edges, $\mathcal{C}$ is the set of edge communication costs and $\mathcal{T}$ is the set of node computation costs.

The value $c_{i,j} \in \mathcal{C}$ is the communication cost incurred along the edge $e_{i,j} = (n_i, n_j) \in E$, which is zero if both nodes are mapped in the same processor. The value $\tau_i \in \mathcal{T}$ is the execution time of node $n_i \in V$. $PRED(n_x)$ is the set of immediate predecessors of n_x and $SUCC(n_x)$ is the set of immediate successors of n_x. We use the ***macro-dataflow task model*** of execution. A task receives all input in parallel before starting execution, executes to completion without interruption, and immediately sends the output to all successor tasks in parallel.

Scheduling parallel tasks with precedence relations over distributed memory multiprocessors has been found to be much more difficult than the classical scheduling problem [7]. This is because data transferring between processors requires substantial transmission delays. Without imposing a limitation on the number of processors, the scheduling problem that ignores communication overhead is solvable in polynomial time. When communication overhead is present, however, the problem becomes NP-complete [25, 6]. Several scheduling algorithms have been proposed in the literature, e.g. [3, 9, 19, 18, 25, 26]. The main optimization in task scheduling is to consider three goals in task assignment and ordering: eliminating communication overhead, overlapping communication with computation, and balancing load. There is a trade-off the algorithm needs to decide: placing two tasks in different processors *parallelizes* the computation process but also could induce higher communication cost. Placing tasks in one processor eliminates communication cost but also *sequentializes* the computation process. Also the current scalable architectures such as iPSC/860 and NCUBE-2, require coarser grain task partitioning because of a large overhead in the startup time for message transmission.

We have developed efficient scheduling algorithms and also a system called PYRROS that integrates scheduling with code generation techniques to produce scheduled code. The input of PYRROS scheduling system is a weighted task graph and the associated sequential C or Fortran code. The output is a static schedule and parallel code for a given architecture. While other algorithms could be used in PYRROS, we will briefly discuss the algorithms of PYRROS.

There are two classes of scheduling strategies. One is to travel through the given graph once, and produce a solution in one stage. Another is a multi-stage approach which divides the scheduling process into several stages [18, 25]. We have used the following multi-stage approach to design low-complexity scheduling algorithms:

a. **Clustering:** The goal of clustering is to identify "useful" parallelism and eliminate unnecessary communication. Sarkar [25] was one of the first to recognize the importance of clustering, which he calls internalization prepass. This problem is equivalent to scheduling on an unbounded number of processors. PYRROS uses the Dominant Sequence Algorithm (DSC) with a complexity of $O((v+e)\log v)$ to automatically determine the clustering for task graphs. It performs a sequence of clustering refinement steps and at each refinement step, it tries to zero an edge to reduce the parallel time (PT). In [29], we show that DSC performs well for general DAGs by examining a set of randomly generated DAGs and also produces the optimal solutions for fork, join and coarse grain tree DAGs.

b. **Mapping clusters to processors:** After finding the clusters, PYRROS uses a variation of *work profiling method* suggested by George et. al. [10] for cluster merging in order to balance the load of computation among the p given processors. This method is simple and has been shown to work well in practice. The complexity of this algorithm is $O(v \log v)$. Now we have p virtual processors (or clusters) and p physical processors. Since physical processors are not completely connected, we have to take the processor distance into account. We use a heuristic algorithm due to Bokhari [2]. His algorithm starts from an initial assignment, then performs a series of pairwise interchanges so that the cost function (the overall communication volume) is reduced monotonically.

c. **Task ordering:** After the physical mapping has been fixed, the interprocessor communication delay between tasks can be determined. We further order the execution of tasks within each processor with a goal of overlapping the communication with computation to hide communication latency so that the total parallel time is minimized. Finding a task ordering that minimizes the parallel time is NP-hard even for chains of tasks [15]. We have proposed a heuristic algorithm RCP based on the critical-task-first principle and compared its performance with others for randomly generated graphs, and we have shown that RCP is optimal for fork and join DAGs (see [28]).

The complexity of our algorithm is $O((v+e)\log v + p^3)$. In [13], we have compared this multi-stage method with the one-stage ETF method [16] using randomly generated graphs. We find that compared to ETF, our algorithm has a lower complexity but produces competitive or even better solutions.

Tasklist = ordered tasks assigned to processor i based on the static schedule.
While Tasklist is not empty **Do**
Remove the first task T_x in the list,
Examine the incoming edges of this task T_x. If data items it needs are not in the local memory, issue a message-receiving to fetch desired data items from the communication buffer.
Perform the task computation.
Send out the produced data items to the processors that need these data items.
ENDWHILE

Fig. 1. The run-time execution of a static schedule.

3 THE RUN TIME METHOD FOR EXECUTING A SCHEDULE

Once a schedule is determined by the PYRROS scheduling algorithms or other scheduling algorithm. This schedule must be executed on the real architecture. Let us assume that the target is a message-passing parallel machine, each processor has its own private memory and a communication buffer to hold the outgoing and incoming messages. The basic model of the run-time schedule executor for processor i is described in figure 1.

Parallel machines are complex systems and accurate estimation of communication and computation weight at compile time is very difficult. Therefore, it is important to know the sensitivity of the schedule executor on inaccurate weight estimation and also other run time weight variations.

For example, in solving the N-body problem as described in Section 4.2, the scheduling tool can be used to map the DAG onto parallel processors. Particles gradually move during each iteration, and the weights of computation and communication change between iterations. Because there is a cost of rescheduling at each iteration it is of interest to know how the weight variations will affect the run-time scheduler so that rescheduling is done only after a sufficiently large number of iterations.

3.1 An analysis of run-time performance

We address the sensitivity of the scheduling performance on the variation of task communication and computation weights. Let G be the graph for which the scheduling algorithm produces a schedule S. At run-time, the weights of this graph change, we call the new graph as G^r. We examine the performance of schedule S when used to execute G^r.

In [11], we define the grain of a task T_x as

$$g_x = min\left(\frac{\min_{T_k \in PRED(T_x)} \tau_k}{\max_{T_k \in PRED(T_x)} c_{k,x}}, \frac{\min_{T_k \in SUCC(T_x)} \tau_k}{\max_{T_k \in SUCC(T_x)} c_{x,k}}\right)$$

and the *granularity* of a DAG G as

$$g(G) = \min_{x=1:v} \{g_x\}.$$

We assume that at run-time, each communication weight $c_{i,j}$ is changed to $c^r_{i,j}$ and the computation weight to τ^r_i. They satisfy

$$(1 - f^c_1)c_{i,j} \le c^r_{i,j} \le (1 + f^c_2)c_{i,j}, \qquad (1 - f^\tau_1)\tau_i \le \tau^r_i \le (1 + f^\tau_2)\tau_i$$

where $f_1 = max(f^c_1, f^\tau_1)$ and $f_2 = max(f^c_2, f^\tau_2)$. The lower and upper bounds in the inequalities above are chosen to be sharp.

Let the length of the schedule S for G be $PT(G)$. The run-time method in Fig. 1 uses the mapping and task ordering of S to execute graph G^r, producing a parallel time $PT(G^r)$.

Let us assume that $PT(G) \le B \times PT_{opt}(G)$ where $PT_{opt}(G)$ is the optimal schedule length for G where the minimum is taken over all possible schedules S. Similarly, let $PT_{opt}(G^r)$ is the optimal schedule length for G^r. If B is close to 1 then the scheduler, say for example the PYRROS static scheduler, produces close to the optimum schedules. What happens to the actual run-time execution of this schedule is described by the following property:

Property 1 *Assume that* $PT(G, S) \le B \times PT_{opt}(G)$, *then*

$$\frac{PT(G^r)}{PT_{opt}(G^r)} \le min\left(\frac{1+f_2}{1-f_1}, \frac{1+f^\tau_2}{1-f^\tau_1}(1+\frac{1}{g(G^r)})(1+\frac{1}{g(G)})\right) \times B$$

Proof: Given a schedule S of G which defines the mapping and ordering of task to processors. By adding an edge from T_x to T_y if T_x is executed immediately before T_y in the same processor and T_x and T_y are independent, then we obtain the execution graph G_s for G under schedule S. The graph structure of G_s is identical to G_s^r, the graphs differing only in the values for communication and computation weights. We set $PT(G^r)$, the length of the run-time schedule for the method in Figure 1 when using schedule S. We have that

$$PT(G^r) = \sum_{CP(G_s^r)} (\tau_i^r + c_{i,i+1}^r) \leq (1 + f_2) \sum_{P(G_s)} (\tau_i + c_{i,i+1}) \leq (1 + f_2) PT(G).$$

where $CP(G_s^r)$ is the critical path in G_s^r and $P(G_s)$ is the same path in G_s which might not be a critical path in G_s because of the change in the weights. The summations are taken over all nodes in the corresponding paths in the graph.

Now, let $PT_{opt}(G^r)$ be the length of the optimal schedule $S = ropt$ for G^r. Let $PT^*(G)$, be the parallel time of this optimum schedule applied to G. We have that

$$(1 - f_1)PT_{opt}(G) \leq (1 - f_1)PT^*(G) = \sum_{CP(G_{ropt})} ((1 - f_1)\tau_i + (1 - f_1)c_{i,i+1}) \leq$$

$$\leq \sum_{P(G_{ropt}^r)} (\tau_i^r + c_{i,i+1}^r) \leq \sum_{CP(G_{ropt}^r)} (\tau_i^r + c_{i,i+1}^r) = PT_{opt}(G_{ropt}^r).$$

The proof of the first bound is obvious.

To prove the second bound we use the definition of granularity to derive another upper bound for run-time communication with respect to computation: $c_{i,i+1}^r \leq \frac{\tau_i^r}{g(G^r)}$. We have that

$$PT(G^r) = \sum_{CP(G_s^r)} (\tau_i^r + c_{i,i+1}^r) \leq (1 + \frac{1}{g(G^r)}) \sum_{CP(G_s^r)} \tau_i^r \leq$$

$$\leq (1 + f_2^r)(1 + \frac{1}{g(G^r)}) \sum_{P(G_s)} (\tau_i + c_{i,i+1}) \leq$$

$$\leq (1 + f_2^r)(1 + \frac{1}{g(G^r)}) \sum_{CP(G_s)} (\tau_i + c_{i,i+1}) = (1 + f_2^r)(1 + \frac{1}{g(G^r)}) PT(G).$$

For the lower bound assume again that the schedule $S = ropt$ is optimum for G^r and $PT^*(G)$ is the corresponding parallel time when $ropt$ is applied to G.

Then we use the facts that $PT_{opt}(G) \leq PT^*(G)$ and $c_{i,i+1} \leq \frac{\tau_i}{g(G)}$ to show that

$$(1 - f_1^r)\frac{PT_{opt}(G)}{1 + \frac{1}{g(G)}} \leq (1 - f_1^r) \sum_{CP(G_{ropt})} \frac{\tau_i + c_{i,i+1}}{1 + \frac{1}{g(G)}} \leq \sum_{P(G^r_{ropt})} \tau_i^r$$

$$\leq \sum_{P(G^r_{ropt})} \tau_i^r + c^r_{i,i+1} \leq \sum_{CP(G^r_{ropt})} \tau_i^r + c^r_{i,i+1} = PT_{opt}(G^r).$$

Combining the lower and upper bounds we prove the property. □

This property implies that the run-time performance depends on how good the compile time schedule is and how small the variations of the weights at run-time are. It should be mentioned that for coarse grain task graphs it is easier to produce good schedules. As a matter of fact for a completely connected architecture with unbounded number of processors we can always produce a schedule, see [11], such that

$$\frac{PT(G)}{PT_{opt}(G)} \leq 1 + \frac{1}{g(G)}.$$

This implies that $B = 1 + 1/g(G)$ tends to 1 as the grain of graph becomes coarser. Also for coarse grain task graphs we can estimate the computation and communication weights much more accurately as long as there is no resource contention.

4 SOLVING IRREGULAR PROBLEMS

We first briefly discuss a solution to sparse matrix computation using the automatic scheduling techniques Then we turn to the problem of n-body simulation.

4.1 Sparse matrix factorization

We first study sparse matrix systems where dependence graphs are unstructured. Finding an efficient scheduling solution for such kind of problem is important. An example is the area of circuit simulation and testing where the solution of a set of differential equations is often used. These problems are solved numerically by discretizing in time and space to reduce the problem to a large set of nonlinear equations. The solution is then derived by an iterative method such as Newton-Raphson which iterates over the same dataflow graph

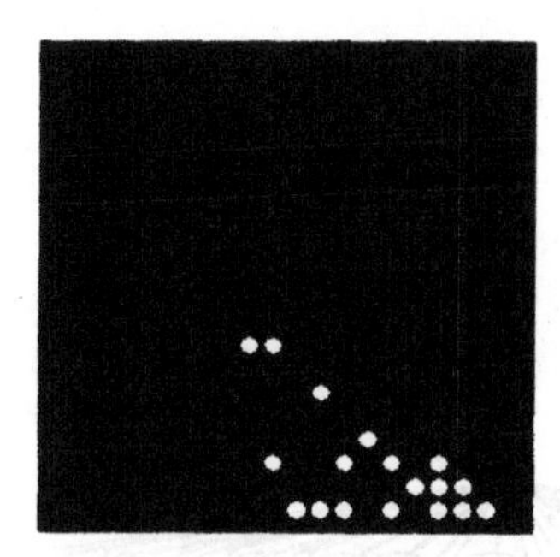

Fig. 2. A factorized sparse matrix with filled in elements.

in solving a sparse matrix system. The topology of the iteration matrix remains the same but the data change at each step [17].

Graph Structure for Sparse Cholesky factorization: For sparse matrices, the graph is irregular and can be derived using symbolic factorization algorithms. For a positive definite sparse matrix system $Ax = b$, we first perform a symbolic factorization to find the fill-in element in the resulting sparse matrix L where $A = L \times L^t$. Figure 2 illustrates a 20×20 factorized sparse matrix. Then we derive a task graph using this symbolic factorization. Figure 3 is the task graph for this 20×20 sparse matrix. This is equivalent to deleting all zero nodes of the regular dense Cholesky task graph. The complexity of this step is proportional to the number of the non-zero nodes and edges of the the irregular sparse graph.

Performance results: Figure 4 shows the speedup of PYRROS for test data matrix from Alvarado's sparse matrix manipulation system [1]. PYRROS achieves a maximum speedup of about 7 and the speedup upper bound is less than 11 for this matrix. (The upper bound is obtained by ignoring all the communication costs and assuming unlimited number of processors.) Since the speedup upper bound is loose, it is not clear if PYRROS performance can be improved further unless the communication cost is reduced. This example shows that PYRROS can be used for irregular sparse graph problems. Currently our partitioning method for the sparse matrix uses column block partitioning. We used PYRROS for several sparse matrices from the Harwell-Boeing test suite and found limited parallelism in these matrices. The results are preliminary and more extensive experiments are needed to determine when scheduling for sparse matrices is beneficial.

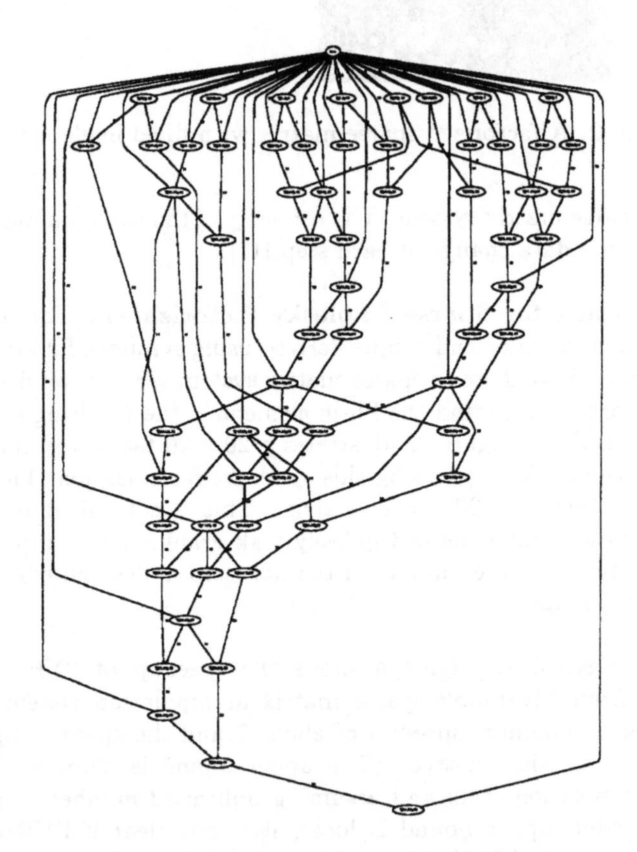

Fig. 3. A task graph for factorizing a 20 × 20 sparse matrix.

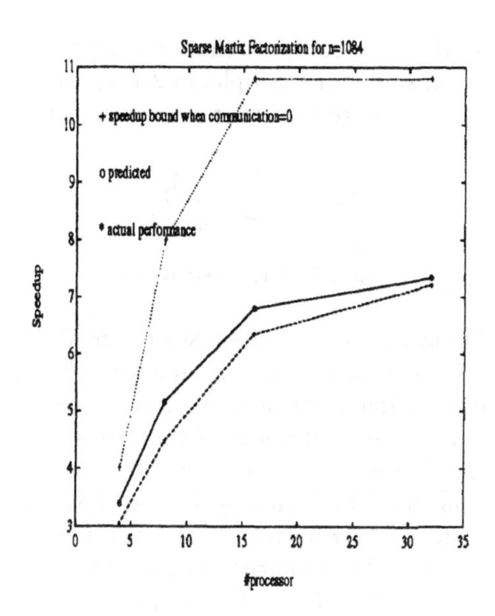

Fig. 4. The sparse graph performance for PYRROS.

Other researchers have used the PYRROS scheduling algorithms for sparse matrix computations[23, 5]. For example, in [5] the authors considered the iterative solution of linear systems for sparse matrix equations. They reported the highest speedups in the literature for these problems using the DSC algorithm, a clustering algorithm for PYRROS, and the CM5 scalable architecture. DSC achieved a speedup of 35 on 128 CM5 processors for small size sparse matrices, e.g. 5k size, as opposed to no higher than 4 speedup reported previously for similar size matrices [5]. The authors also report up to 75% improvements of DSC scheduling over several well known data driven mappings. The authors have used low level communication libraries in addition to scheduling to reduce communication and achieve what it appears to be the maximum possible performance for their matrices.

4.2 The N-body problem

The N-body problem is a common problem encountered in scientific computation applications. Given a distribution of n particles in space, and a rule governing their interactions, e.g. gravitational force, the problem is to compute the movement of the particles. This movement can be discretized in time steps, during each of which we compute the force applied on each particle, and update the particle positions. For example, in 2-D space force applied to a particle located at position z_j (note that we represent points in 2D using complex numbers) is :

$$F_j = \sum_{i=1, i \neq j}^{n} \frac{A_i}{z_i - z_j}$$

where A_i are the charge of particle i located at z_i.

Here we use PYRROS to map the 2D Fast Multipole Method (FMM). The basic idea of the algorithm is to subdivide the space into boxes. The evaluation of particle positions (interactions) between far-away boxes can be approximated by series of expansions. The coefficients of the series are computed once and used by all "far-away" boxes. Depending on the input distribution, the algorithm comes in two flavors: (1) Non-adaptive: all boxes are of the same size. This is applied to uniform distributions. (2) Adaptive: boxes are subdivided further in the regions of higher particle density. This is used for highly-non-uniform distributions. For details see Greengard's thesis ([14]. Here is some rough idea. The force on a particle located at z_j is estimated for an entire box of particles. Assume that the 2D space is recursively divided into 4^l boxes where l is the level of subdivision and that each box contains s particles(uniform distribution). We have $4^l s = n$ and the evaluation in the summation is re-written as follows:

$$F_j = \sum_{i=1, i \neq j}^{n} \frac{A_i}{z_i - z_j} = \sum_{box=1}^{4^l} \sum_{z_i \in ibox} \frac{A_i}{z_i - z_j}.$$

The inner summation is approximated by a multipole series expansion with p terms. Assume z_i in a box (call it b) centered at position z_0 is evaluated for particles z_j which is far away, i.e. $|z_i - z_0| \leq c \times |z_j - z_0|$ where $c < 1$ is a separation factor. Then

$$\sum_{z_i \in b} \frac{A_i}{z_i - z_j} = -\sum_{z_i \in b} \frac{A_i}{(z_j - z_0) - (z_i - z_0)} = -\frac{1}{(z_j - z_0)} \sum_{z_i \in b} \frac{A_i}{1 - \frac{z_i - z_0}{z_j - z_0}}$$

$$\approx -\frac{1}{(z_j - z_0)} \sum_{k=0}^{p} \frac{a_k}{(z_j - z_0)^k}$$

where $a_k = \sum_{z_i \in b} A_i (z_i - z_0)^k$ are the terms of the expansion. They only depend on the particles within the box so such terms can be computed once and used for ALL far away particles. In this way we reduce complexity from s^2 to $O(ps)$ when we evaluate the interaction between s far away particles. To reduce the total complexity further the computations must be performed hierarchical using the tree structure. By evaluating the interactions at a constant number of boxes at each level then the complexity reduces and the entire space evaluation is covered.

4.3 Scheduling for the non-adaptive FMM algorithm

We consider the non-adaptive algorithm first.

Graph Struture: Figure 5 is a 2-level ($l = 2$) regular subdivision and its corresponding quadtree. Although the tree has a regular structure, the task weights distribution is input-dependent and irregular.

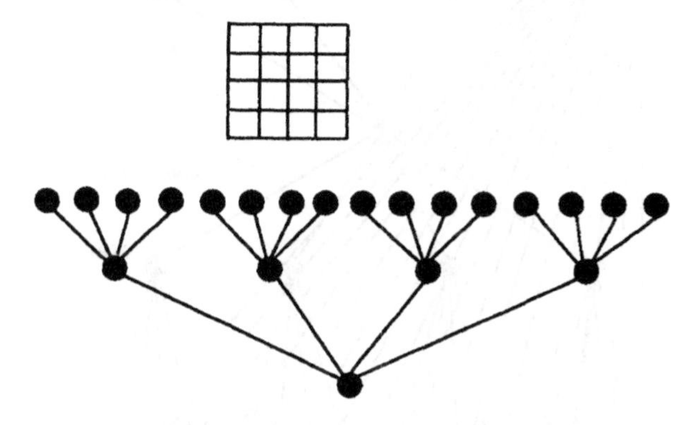

Fig. 5. A 2-level non-adaptive subdivision

The basic structure of the task graph is a concatenation of this in-tree with an out-tree which is its reverse. Because of the regular subdivisions, the tree is a complete quad-tree. The leaves of the tree correspond to the boxes at the finest level, and the internal nodes correspond to the boxes at higher levels. Following the thesis of Greengard, we call the in-tree "upward pass", and the out-tree "downward pass." There are additional edges between the two parts of the graph, which represent interactions between boxes. Figure 6 is an example of the task graph for a level 2 division. For clarity we only draw the communication edges for one leaf node.

Note that although in this graph, every leaf communicates with all the other leaves, this is not true for divisions of higher levels. A leaf communicates with it's neighbors and "interaction list" (which is essentially the list of all second near neighbors, See figure 7). The number of boxes in each of the lists is bounded by constant, so the graph has $O(4^l)$ edges. It's not easy to draw

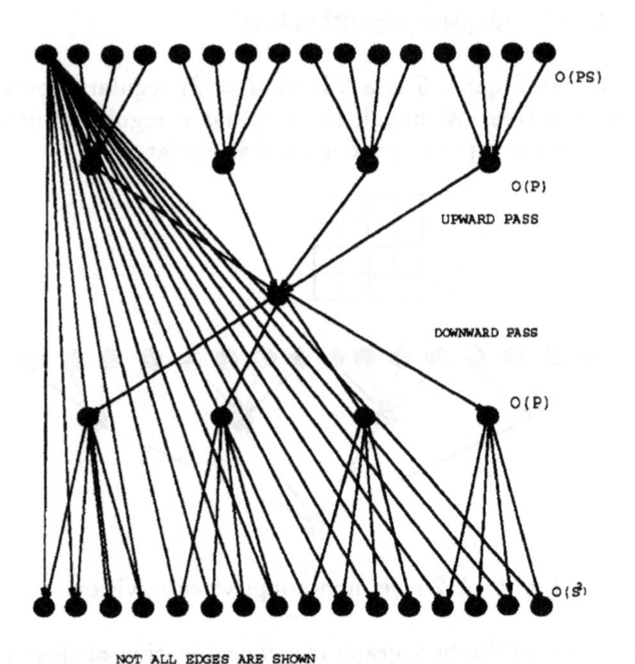

Fig. 6. Task graph for a 2-level subdivision. The order of task weights are also shown. Only communication from one leave of the upward pass tree to the leaves of the downward pass tree is shown. The others are deleted for simplicity of presentation.

these graphs clearly, but from the following description it should be easy to understand their structure.

Now we give some explanations of the functions of the nodes in the graph. These task definitions are derived directly from the algorithm description in Greengard's thesis, pp. 17-19. Since the task weights depends on the input and is irregular, we only describe the order of the weights in terms of number of particles in a leaf box s and number of terms in the expansion p. (For simplicity we use s as upper bound to the number of particles in each leaf box. In a uniform distribution the actual number would vary from box to box but not too much.)

1. A leaf node in the upward pass :
a. Generates multipole expansions from the particles in the box.
b. Sends the multipole expansion coefficients to its parent box in the upward pass. and also to those boxes in the interaction list corresponding to the leaf level of the downward pass.
c. Sends the particle positions to those neighboring leaf boxes (including itself) in the downward pass.
Such node has computational weight $O(p * s)$.

2. A non-leaf node in the upward pass:
a. Receives the multipole expansions from each child, and shifts the expansion to its own box center. Then sums them up to obtain the multipole expansion for the whole box.
b. Sends the multipole expansion to it's parent (if it's not the root), and also, to the boxes in its interaction list in the downward pass.

3. A non-leaf node in the downward pass :
a. Receives the multipole expansions from the boxes in its interaction list in the upward pass. Then converts these multipole expansions into local expansions. (Local expansion is a derived form of multipole expansion. It's used to reduce complexity, see Greengard thesis [14] for the definition of local expansion.)
b. Receives the local expansion from its parent, shifts it to its own center, then adds it to the local expansion computed in (a) above to obtain the total local expansion at this box.
c. Sends the local expansion to each of its children.

A non-leaf node of either kind has weights $O(p)$.

4. A leaf node in the downward pass:
It does the same as a non-leaf node, (Except that it does not send local expan-

proc	Time(ms)	PY-Run speedup	PY-compile speedup
p=4	320017	3.96	3.98
p=8	160320	7.90	7.95
p=16	81258	15.6	15.7
p=32	41638	30.4	31.0
p=64	21463	59.0	61.0

Table 1. The performance of the non-adaptive FMM on NCUBE-2s.

sion since it has no children) plus:
a. Receives the particle positions from neighboring boxes in the upward pass, and evaluates nearby interaction using direct, pairwise summation.
b. Computes the far away interaction using the local expansion, by evaluating the expansion at each particle in its box.
c. Sums up the "nearby" and "far away" interactions, obtaining the total interaction and updates particle positions.
Such node has weight $O(s^2)$.

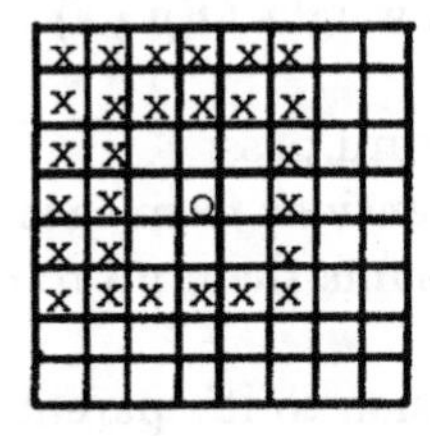

Fig. 7. A 3-level non-adaptive subdivision. X are the boxes in the interaction list of box O. Any box has most 27 boxes in its interaction list, and at most 8 neighbors

Granularity: The graph is fine grain near the root but coarse grain at the leaf level where high degree of parallelism exists. So the speedup is expected to be very high.

Performance results: We use an example of 80000 randomly distribution particles, with a 5-level (32*32) subdivision. This graph has about 2700 nodes and 30000 edges. PYRROS scheduler can handle graph with such size quite rapidly. Our result in the case is very encouraging. The speed up is almost linear. See table 1 . (Note this is the speedup for the first iteration. We will discuss the issue of the iterative execution later.)

The largest problem we were able to run on the NCUBE-2s machine is a problem with $8 * 10^5$ particles, with 128 processors, each having 16MB of memory. This shows that our system is capable to handle large real world problems.

4.4 Scheduling for the adaptive FMM algorithm

Although the non-adaptive algorithm has very good performance for uniform distributions of particles, it cannot handle highly non-uniform distributions encountered in some applications, such as galaxy simulations. Adaptive subdivision must be used for such distributions. Figure 8 illustrates an example of an adaptive subdivision and the corresponding unbalanced tree for this partition. The real task graph consists of the concatenation of this in-tree with the symmetric out-tree and edges between the two parts.

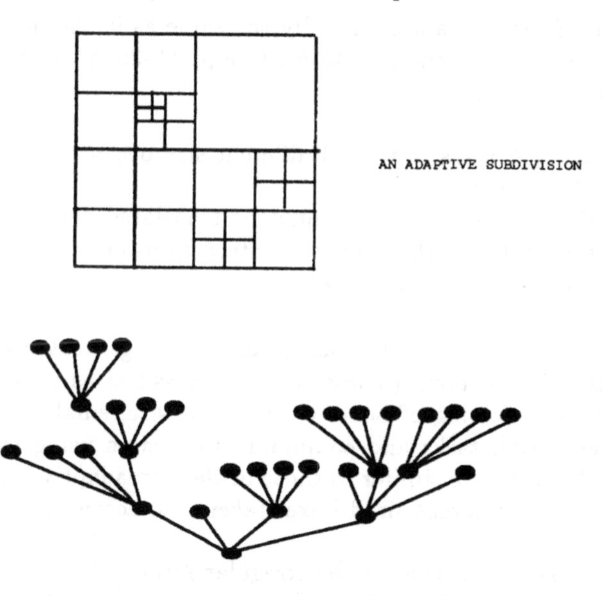

Fig. 8. An adaptive subdivision

Graph structure: The task graph structure is similar to that of the non-adaptive case in the sense that it still has an upward pass (in-tree) and a downward pass (out-tree). But the tree is unbalanced, and the height is larger for the places with higher particle density. So the tree structure varies, de-

#proc	Time(ms)	PY-Run speedup	PY-compile speedup
p=4	105542	3.94	3.97
p=8	53663	7.76	7.83
p=16	28270	14.7	15.1
p=32	16648	25.0	28.0
p=64	10458	40.0	45.4

Table 2. The performance of the adaptive FMM on NCUBE-2s. Time is given in milliseconds(ms).

pending on the input distribution. For galaxy distributions, the tree is highly unstructured.

The function of the nodes are basically the same as in the non-adaptive algorithm, with the following complications: (see pp.28-38 in Greengard's thesis for detailed explanations.)

a. Neighboring leaf boxes can be of different size because they may not be in the same level of the tree.
b. Because Multipole expansions cannot be translated between boxes of different sizes, we must deal with "far-away" interactions between boxes in different levels of the tree in some special way.

With these extra cases in mind, a task graph can be generated by first creating different lists for each box. (These lists are classified as neighbors, far-away interaction lists consisting of boxes of the same size, smaller size, and bigger size) then generating tasks and communication edges using these lists. The functions of the tasks are similar to those in the non-adaptive task graph, with the special cases of different sized boxes taken into account.

Performance results: Due to the irregular structure, it's necessary to use automatic scheduling tool to schedule such task graphs. PYRROS scheduling tool is well-suited for this purpose. Our example is a highly non-uniform galaxy distribution with 40000 particles. This distribution results in a highly unbalanced tree, with the maximum depth of the leaves being as large as 9, and the minimum depth of the leaves being only 1. Such kind of distribution is very difficult to parallelize. Nevertheless, PYRROS is able to achieve good performance, with speed up of 40 using 64 processors. See table 2.

processors	slow-down(%)
p=8	4.0
p=16	4.9
p=32	6.1
p=64	7.5

Table 3. The increase in parallel time after 100 iterations.

4.5 Static scheduling performance for iterative executions

The N-body problem is a dynamic problem, in the sense that particles are moving, their positions are updated during each time step. So they could move from one box to another, resulting in changes in the task graph. Notice that in the non-adaptive algorithm, only task weights are changed due to particle movement; however in the adaptive case, a task graph structure could also be affected. When the space is re-partitioned, if we choose to maintain the same partition before doing any rescheduling, only task weights are affected. Therefore, the schedule derived for the initial particle distribution is good for only initial several iterations and the performance deteriorates gradually when particles move in the space.

We find that a static scheduling tool still could be useful for such applications. This is because generally the movement of particles is small during each time step. Even if the particles have high velocity, the time step δt has to be chosen small enough to ensure stability. On the other hand, PYRROS scheduler for coarse grain graphs does not need precise weight information, and we expect that small perturbations should not lead to significant deterioration in performance, as we saw in section 3 . Thus a schedule can be re-used for many steps, until its performance deteriorates to the point where rescheduling becomes beneficial. The interesting question is how frequent we need to reschedule, since this will determine the overall performance of PYRROS for the N-body problem. We investigate this next.

Using a uniform distribution of 10000 particles, we chose a $\delta t = 0.001$ which for this case is large enough to get sizable movement of particles in each step, but small enough to ensure stability. Our experiment runs iteratively on 8-64 processors. The same initial schedule is used for all iterations, and we calculated the deterioration of the performance: i.e. $\frac{PT_{100}-PT_1}{PT_1}$, where PT_1 is the parallel time for the first iteration and PT_{100} is the parallel time for the 100th iteration. The result are shown in table 3.

Note that the slow-down in parallel time is only 7.5% for 64 processors (About 130 iterations are needed for the parallel time to increase 10%,in this particular case.) Of course, these data are only for this particular uniform distribution. But in general, if the change of the task graph happens slowly and gradually, we could say that an initial schedule can be reused for a substantial number of steps with reasonable performance. Combined with run-time reschedule to correct the imbalance when it gets too large, this approach could be useful in many real world applications, not just N-body. We plan to investigate this approach further.

5 CONCLUSIONS

We have presented our approach to the solutions of sparse matrix problem and n-body simulation using automatic mapping techniques. We have provided an analysis on the run-time execution performance of static schedule. Our experiments show that the automatically scheduled task computation could attain a good performance for irregular problems provided that sufficient task parallelism is available. We have not yet obtained good performance for sparse matrix factorization partially because the partitioning method we use does not yield sufficient parallelism for the scheduling algorithms to explore. Recently Chong and Schreiber [4] had a similar situation in solving a sparse triangular system and they proposed a new method which results in task graphs with more parallelism and leads to good speedups after applying the PYRROS algorithms.

6 ACKNOWLEDGEMENTS

The work presented here was in part supported by ARPA contract DABT-63-93-C-0064 under "Hypercomputing and Design" project, by the Office of Naval research under grant N000149310114, by NSF Research Initiation Award CCR-9409695. The content of the information herein does not necessarily reflect the position of the Government and official endorsement should not be inferred.

REFERENCES

[1] F. L. Alvarado, The sparse matrix manipulation system users manual,

University of Wisconsin, Oct. 1990.

[2] S. Bokhari, On the mapping problem, *IEEE Trans. Comput.* C-30 3(1981), 207-214.

[3] D. Y. Cheng and S. Ranka, An optimization approach for static scheduling of directed-acyclic graphs on distributed memory multiprocessors, Report, Syracuse Univ, 1992,

[4] Frederic T. Chong and Robert Schreiber, Parallel sparse triangular solution with partitioned inverses and prescheduled DAGs, Tech Report, 1994.

[5] Frederic T. Chong, Shamik D. Sharma, Eric A. Brewer and Joel Saltz, Multiprocessor Runtime Support for Fine-Grained Irregular DAGs, To appear in *Toward Teraflop Computing and New Grand Challenge Applications.* Eds. Rajiv K. Kalia and Priya Vashishta. Nova Science Publishers, Inc. New York. 1995.

[6] Ph. Chretienne, Task Scheduling over Distributed Memory Machines, *Proc. of Inter. Workshop on Parallel and Distributed Algorithms*, (North Holland, Ed.), 1989.

[7] E.G. Coffman and R.L. Graham, Optimal scheduling for two-processors systems, *Acta Informatica*, 3 (1972), 200-213.

[8] J.J. Dongarra, . and D. C. Sorensen, SCHEDULE: Tools for Developing and Analyzing Parallel Fortran Programs, in *The Characteristics of Parallel Algorithms,* D.B. Gannon, L.H. Jamieson and R.J. Douglass (Eds), MIT Press, 1987, pp363-394.

[9] H. El-Rewini and T.G. Lewis, Scheduling parallel program tasks onto arbitrary target machines, *J. of Parallel and Distributed Computing*, 9(1990), 138-153.

[10] A. George, , M.T. Heath, and J. Liu, Parallel Cholesky factorization on a shared memory processor, *Lin. Algebra Appl.*, 77(1986), pp. 165-187.

[11] A. Gerasoulis and T. Yang, On the granularity and clustering of directed acyclic task graphs, *IEEE Trans. on Parallel and Distributed Systems.*, Vol 4, No 6, June 1993 pp. 686-701.

[12] A. Gerasoulis and T. Yang, A comparison of clustering heuristics for scheduling DAGs on multiprocessors, *J. of Distributed and Parallel Computing,* Vol. 16, No. 4, pp. 276-291 (Dec. 1992).

[13] A. Gerasoulis, J. Jiao, and T. Yang, A multistage approach to scheduling task graphs. To appear in DIAMCS Book Series on Parallel Processing of Discrete Optimization Problems. AMS publisher. Edited by P.M. Pardalos, K.G. Ramakrishnan, and M.G.C. Resende.

[14] Leslie Greengard, The Rapid Evaluation of Potential Fields in Particle Systems Ph.D thesis, Yale University, 1987.

[15] J.A. Hoogeveen, S.L. van de Velde, and B. Veltman, Complexity of scheduling multiprocessor tasks with prespecified processor allocations, CWI, Report BS-R9211 June 1992, Netherlands.

[16] J. J. Hwang, Y. C. Chow, F. D. Anger, and C. Y. Lee, Scheduling precedence graphs in systems with interprocessor communication times, *SIAM J. Comput.*, pp. 244-257, 1989.

[17] N. Karmarkar, A new parallel architecture for sparse matrix computation based on finite project geometries, *Proc. of Supercomputing '91*, IEEE, pp. 358-369.

[18] S.J. Kim and J.C Browne, A general approach to mapping of parallel computation upon multiprocessor architectures, *Proc. of Int'l Conf. on Parallel Processing*, vol 3, pp. 1-8, 1988.

[19] C.L. McCreary and D.H. Gill, Automatic determination of grain size for efficient parallel processing, *Communications of ACM*, vol. 32, pp. 1073-1078, Sept., 1989.

[20] S. Pande, D. Agrawal and Jon Mauney A threshold Scheduling Strategy for Sisal on distributed Memory Machines *IEEE Parallel and Distributed Technology*, vol. 21, pp. 223-236, 1994.

[21] C. D. Polychronopoulos, M. Girkar, M. Haghighat,C. Lee, B. Leung, and D. Schouten, The Structure of Parafrase-2: An advanced parallelizing compiler for C and Fortran, in *Languages and Compilers for Parallel Computing*, D. Gelernter, A. Nicolau and D. Padua (Eds.), 1990.

[22] Ravi Ponnusamy, Yuan-Shin Hwang, Joel Saltz, Alok Choudhary, Geoffrey Fox. Supporting Irregular Distributions in FORTRAN 90D/HPF Compilers, University of Maryland, Department of Computer Science and UMIACS Technical Reports CS-TR-3268, UMIACS-TR-94-57

[23] R. Pozo, Performance modeling of sparse matrix methods for distributed memory architectures, in *Lecture Notes in Computer Science, No. 634, Parallel Processing: CONPAR 92 - VAPP V*, Springer-Varlag, 1992, pp. 677-688.

[24] Saltz, J., Crowley, K., Mirchandaney, R. and Berryman,H., Run-time scheduling and execution of loops on message passing machines, *J. of Parallel and Distributed Computing,* Vol. 8, 1990, pp. 303-312.

[25] V. Sarkar, *Partitioning and Scheduling Parallel Programs for Execution on Multiprocessors,* The MIT Press, 1989.

[26] M. Y. Wu and D. Gajski, Hypertool: A programming aid for message-passing systems, *IEEE Trans. on Parallel and Distributed Systems,* vol. 1, no. 3, pp.330-343, 1990.

[27] T. Yang and A. Gerasoulis, PYRROS: Static task scheduling and code generation for message-passing multiprocessors, *Proc. of 6th ACM Inter. Conf. on Supercomputing,* Washington D.C., July, 1992, pp. 428-437.

[28] T. Yang and A. Gerasoulis, List scheduling with and without communication delay, *Parallel Computing,* Vol 19, 1993, pp. 1321-1344.

[29] T. Yang and A. Gerasoulis, DSC: Scheduling parallel tasks on an unbounded number of processors, *IEEE Trans. on Parallel and Distributed Systems.*, Vol. 5, No. 9, 951-967, 1994.

14

ATREDIA: A MAPPING ENVIRONMENT FOR DYNAMIC TREE-STRUCTURED PROBLEMS

Angela Sodan

GMD Institute for Computer Architecture and Software Technology
Rudower Chaussee 5, D - 12489 Berlin, Germany
email: angela@first.gmd.de

ABSTRACT

Problems with dynamic tree-structured behavior are usually highly irregular with respect to the shape of potential processes. Such problems require a special mapping on to a parallel machine, with appropriate partitioning of the tree as well as dynamic load balancing. Atredia is an environment providing several tools for mapping dynamic tree-structured problems. These include a granularity controller (for partitioning), a load balancer, a scheduler, and a profiler. Innovative features of Atredia are its support of selection from a set of given granularity-control and load-balancing strategies and their parameterization according to the characteristics of the respective application, the fact that it uses explicit granularity control at all, and its use of a systematic and formalized approach. The latter is realized by performing classifications and calculations based on a model of the application and the machine, and by obtaining dynamic behavior characteristics of the specific application via profiling. One of Atredia's main aims is applicability to large real-life problems.

1 INTRODUCTION

Dynamic tree-structured behavior means that the shape of the problem's potential processes is tree-like, and that these trees are evolving dynamically at runtime. Such behavior can be found as search trees in discrete optimization problems [12] (e.g. in VLSI chip layout or in flight crew assignment) but also in theorem-proving, computer algebra, in AI planning and problem-solving, and also in some numeric problems. Thus, dynamic tree-like behavior forms an

A. Ferreira and J. D. P. Rolim (eds.), Parallel Algorithms for Irregular Problems: State of the Art, 269–296.

important and broad class of problems, and it is worth looking for appropriate solutions.

Besides being dynamic in nature, trees usually are not balanced, but highly irregular in their structure. Furthermore, the individual nodes of the tree, i.e. the potential processes, are usually too small to be cost-effective as a process, and their number can be extremely large. We have found examples with million of nodes, and with an execution time per node in the range of microseconds. This makes it obvious that appropriate partitioning of the tree as well as dynamic load balancing are necessary.

To meet our requirements, a mapping solution should, in addition to being able to handle highly unbalanced and dynamic trees:

- provide scalable strategies (for problems with a high degree of parallelism and for massively parallel machines)
- be able to map to distributed memory machines
- take into account typical properties of AI applications in strategies and cost calculations, like large arguments and garbage collection
- free the user from the need to consider hardware and application runtime behavior (allowing the user to concentrate on the semantics of the application)
- provide at least a semi-automatic approach for selection and parameterization of strategies
- take into account the characteristics of the respective application
- choose the most efficient and simplest strategy that is adequate to avoid unnecessary overhead (what has been promoted as the main philosophy of MANNA's operating system PEACE [8])
- work in practice for large real-life applications

The general approach of the solution offered by Atredia is to provide an environment with a set of tools, comprising a profiler, a granularity controller, a dynamic load balancer, and a scheduler. The granularity controller (partitioning) and load balancer both provide several strategies which are selected and parameterized according to user specifications and to the application's dynamic characteristics. So far, there are only a few environments available for mapping trees. Atredia is more flexible and more systematic than the others, being based on a system and application model, a cost function, and some decision rules.

The research presented in this paper was conducted as part of the development of a parallel Lisp system for the MANNA machine [8], which is a massively parallel machine with distributed memory. Since the interconnection network is a hierarchical crossbar, the communication costs for all pairs of nodes are approximately the same. The parallel Lisp system (parLisp) is an extension of AKCL (Austin Kyoto Common Lisp) with features for expressing parallelism explicitly. AKCL was chosen because it compiles to C. Thus, parLisp is only dependent on MANNA's operating system PEACE, and it currently runs on MANNA as well as on a SUN workstation network. In addition to the parallel runtime system, we have developed a static analyzer for finding data dependencies (thus supporting the user in parallelizing the sequential version of a program), a special visualization tool for dynamic trees, and the mapping system described here.

The mapping concepts presented in this paper are, however, independent of the development environment, i.e. of both the MANNA machine and Lisp, and apply, for example, to C or C++ as implementation languages as well.

In the sequel, we begin by looking at other research in this field, then go on to describe the Atredia environment in general. This is followed by a more detailed consideration of the main components of the environment. Finally, we present the results from our experiments and conclude with a summary and an outline of future work.

2 COMPARISON WITH OTHER RESEARCH

There are many approaches for partitioning and assigning processes in the case of fairly static program behavior by the compiler (in the symbolic area e.g. [19]). Recursions and hence dynamic trees are, however, excluded from such considerations. There exists much research on general load balancing, and the application side has also proposed and applied several load balancing strategies for dynamic search trees. Most approaches, however, fail to give proper consideration to the tree-like nature of the behavior and at most consider the specific characteristics of applications for explaining different qualities of results.

Hardly any research has been done on granularity control for dynamic behavior. Two main simple dynamic heuristics are suggested. [16] memorizes all potential tasks and assumes that the nearer the tasks are to the top of the tree, the larger they are. Tasks are assigned remotely from top to bottom on receiver request (or else executed locally). However, task memorization involves overhead even if finally no process is created, and is thus not acceptable for large numbers of small tasks. Moreover, the heuristic does not work for problems where there is a large number of small tasks, even near the top of the tree. The idea of limiting the number of processes by memorization is, however, useful and we have integrated it. [15] also memorizes spawned tasks in the form of a stack, splitting this stack in the case of a remote request for work assignment. This strategy is, however, limited to receiver-initiated balancing and still needs a criterion for determining up to which size stacks should be split (or resulting task portions are no longer cost-effective). Another drawback is that stack splitting can result in multiple movement of tasks, which is unacceptable in the case of large arguments, since this would increase the already high communication costs. Both heuristics can be subsumed under our priority and merging mechanisms as a special case and can be applied together with other strategies.

In general, only single strategies for mapping are provided, and within the few existing environments like CHARM [13] the user has to choose and parameterize a strategy manually. Only [5] presents a systematic approach for parameterizing its load-balancing strategy in accordance with application behavior parameters, but this is for a single strategy and for the very restricted situation of programs with limited irregularity only. An environment with automatic selection and parameterization for a set of strategies still remains to be developed, although also elsewhere this deficit has been recognized and research is currently under way to at least evaluate parallel performance information with a view to reconsidering and improving mapping decisions [22]. There would still appear to be a gap between the very abstract models for complexity analysis [10], and the practical results from implementations. More detailed models for system and application behavior are needed, although they can easily become too complex to handle.

3 GENERAL FEATURES OF THE ENVIRONMENT

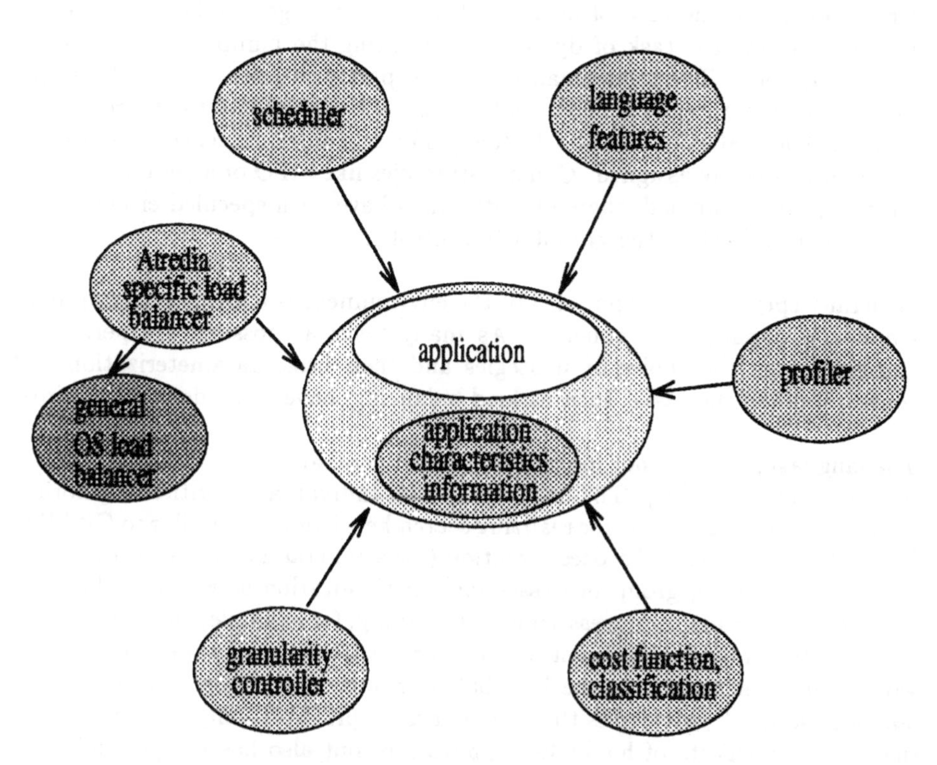

Fig. 1. Components of the Atredia mapping environment

The environment provides several tools and subcomponents which together perform the task of mapping. A granularity controller partitions the dynamic tree and forms processes of cost-effective granularity by considering computation and communication times as well as the number of processes. Good possibilities for balancing the load (requiring a high-enough number and a low-enough granularity of processes) have to be weighed against the maintenance and communication overheads involved. The granularity controller takes decisions on the selection of strategies according to the general characteristics of the application's tree and uses a cost function based on a machine and application model for configuring the strategies appropriately. The information about the general characteristics and the detailed behavior parameters of the application are obtained by collecting data with a profiler and by statistically evaluating and classifying the collected data.

A dynamic load balancer assigns the processes formed by the granularity controller to nodes. In the case of an unpredictable but large number of processes, it also performs the task of dynamically limiting the number of processes to the useful ones. The load balancer can be partly realized by standard operating system strategies for load balancing. A component strongly related to dynamic load balancing is the scheduler. The scheduler influences the order in which processes are assigned. General strategies like FIFO or a detailed priority handling can be applied where priorities may have been specified either by the user or determined by the granularity controller.

Generally speaking, the approach of the environment is hybrid, featuring both static and dynamic components. As many tasks as possible are performed statically, e.g. the choice of strategies and their basic parameterization. Of course, some of the tasks, such as load balancing, have to be done at runtime.

The language, which is not looked at in detail here, also provides some features related to mapping [23]. The user can specify individual priorities or priority-assignment strategies, e.g. for realizing search heuristics [14] similar to CHARM [13], delay the release of process creation (for semi-static assignment of depth-first-guided search), group processes for synchronization as well as balancing, and form hierarchies of process groups. Grouping, for example, allows processes to be created individually, but still handled collectively, using the same strategy, within the same request to the load balancer, etc. The philosophy behind the language support provided is that the user is, in principle, only concerned with the semantic aspects of his or her application, but also has the possibility of influencing the mapping with some specifications clearly separated from the algorithm.

4 GRANULARITY CONTROL

Granularity control has the two main functions of creating processes of effective granularity and, closely related to that, of limiting the number of processes to the useful ones. The ideal result would be a partitioning, where the possibilities for parallel execution and overhead are perfectly weighed for optimum overall performance. However, even the partitioning of static trees is NP-hard, and our trees evolve dynamically, which means, for example, that often even the number of processes cannot be foreseen. Hence, we can only expect an approximate solution, and a main prerequisite is predicting the characteristics of the tree as well as possible.

Thus, at each point at which a process may potentially be split, we have to predict the granularity, or at least the minimum granularity, of that process before it starts executing, which usually means predicting the subtree evolving from that process. Since trees are created by recursion (more precisely by a nesting of recursion and iteration) at the same point within the code, we have to take the decision dynamically in accordance with some criterion. Atredia currently provides the following strategies for predicting and controlling the granularity:

- subtree creation by argument size in the case of correlation of argument size and granularity
- subtree creation by level in the case of a strong correlation of level and granularity (which is equivalent to a limited irregularity of the tree)
- horizontal plus vertical fragmentation of the tree
- merging of steps or subtrees

The last two strategies address the situation in which no real dependence criterion can be found. Fragmentation is often adequate in cases where argument sizes shrink sub-linearly to the number of subtasks on descent. Merging is of advantage, especially if arguments can be shared or are very small so that communication costs for the merger are smaller than the sum of the individual ones would be (for small arguments, the fixed basic communication costs are significant and have to be paid only once then). Both strategies frequently benefit from reducing the load balancing costs as well. The user can also specify his or her own granularity-control criteria, which then usually relate to the semantics of the application (and so cannot be extracted by the system). They should, however, be easy to compute and be integrable into cost estimations.

Since the number of processes may not be predictable at compile time, we provide the following strategies for dynamic adaptation of the process number:

- delayed task creation, and
- dynamic adaptation of granularity

Delayed task creation memorizes tasks and creates processes for remote execution only in the case of low load on other nodes. Otherwise, tasks are executed locally. This is based on the idea of lazy task creation presented in [16], but it is used here only as an additional strategy and not as a basic granularity-control heuristic. Moreover, in our system, it can work with receiver- as well as sender-initiated strategies. Dynamic adaptation of granularity assumes that the current behavior will stay the same for a significant period of time; feedback on current behavior is then given to granularity control for adjusting control parameters.

In addition, granularity control can

- assign internal priority strategies or individual internal priorities.

Individual priorities, for example, are meaningful if processes are created at different levels with a different prognosis of the granularity. In this case, higher priority should be given to tasks with higher granularities.

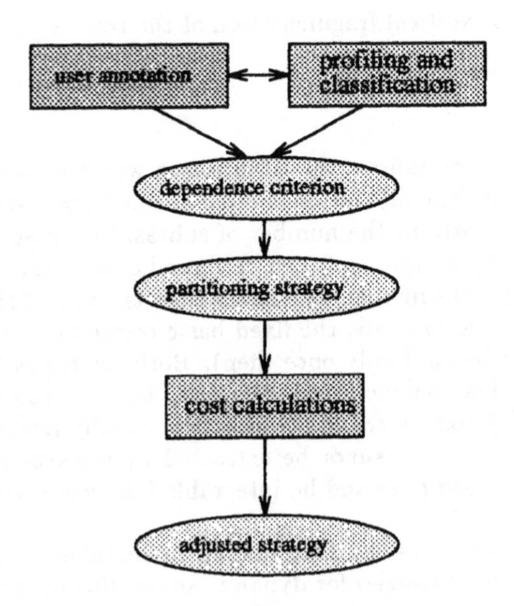

Fig. 2. The steps opf static granularity control

Granularity control applies one of the strategies or a combination of them. The dynamic part of granularity control consists in the dynamic limitation of the process number and the checking of the control criterion. The rest of the work is done at compile time. Static granularity control proceeds in the following way: it evaluates and classifies the application behavior, and checks the critical path, i.e. whether the application is parallelizable at all (see example given in Section 10). This is an extremely important step which is often omitted in direct approaches using manual parallelization. Granularity control then looks for a dependence criterion and chooses one strategy or a set of strategies. Next, costs and speed-up are estimated on the basis of an internal model of system and application and the concrete behavior parameters obtained. The chosen strategies then are parameterized accordingly.

In general, an explicit granularity control as opposed to heuristics has the advantages of being independent of a specific load balancing strategy, of assigning remotely only tasks that are cost-effective (no special position of task granularities within the tree is necessary to make the strategies work), and of imposing overhead only for the cost-effective tasks (otherwise maintenance overhead would occur for all potential tasks, and this is usually greater than the time needed for checking control criteria).

5 PROFILING AND EVALUATION OF DATA

Information about dynamic program behavior is currently obtained in Atredia merely by profiling. Atredia's profiler can be used for the sequential program version, thus being able to supply information even before the parallelization is finished (see example in Section 10). The profiler operates on a trace basis, which enables it to differentiate between individual calls. GPROF-like profilers (such as those provided by UNIX and by Allegro Common Lisp [6]) statistically inspect the running program and only deliver approximated mean values per function (or at least mean values per call path if the stack configuration is considered during inspection). This, however, is not sufficient for highly dynamic behavior.

In Atredia, the application under consideration is profiled with different inputs, which, of course, have to be representative. In our experiments, we had no problems choosing inputs and always found the same typical behavior for different runs of the same application. On the other hand, profiling does, of course, involve extra effort and time, and more importantly: it is partly machine-dependent. Static weight analysis (e.g. [17]) cannot, however, handle recursions, given to the current state of the art. Nor is weight analysis suitable for application to complex programs, and it may be too inaccurate, even for small programs. Thus, profiling is the only option. In the future, we plan to use a combination of static weight analysis and profiling to make our approach more machine-independent (and possibly also more time-efficient). Most of the data listed below does not depend on the machine on which profiling is executed - the only critical factor is time information:

- tree structure (all processes and the dependence relations between them)
- computation time per step/subtree
- dynamically allocated data per step/subtree
- argument sizes (input and output) per step
- branching factor

After collection, the data is statistically evaluated for abstraction from the concrete run. We compute

- the means and variation coefficients (standard deviation normalized in relation to mean) for computation time, argument sizes, etc.

In the case of high variation coefficients, we also consider

- the percentage of small values (far below the mean) and the maximum value (influencing the worst case)

Before carrying out these evaluations, however, the data must normally be classified, and the distribution parameters are then computed for each class. For example, we may be interested in the distribution of the computation time per level or per argument size (or a range of sizes). This is closely related to finding control criteria and to checking

- the correlation between computation time and level or argument size, respectively

Examples of other evaluations carried out are the computation of

- the critical path
- the relation of the starting phase to the problem size
- the degree of irregularity

Since the amount of data collected (it may easily be in the range of several Mbytes of memory) can become a problem, the profiler provides the possibilities of only collecting selectively specified data (e.g. only computation times) and of dynamic, incremental evaluation. The latter, however, means fixing all evaluations in advance, i.e. it prevents us from reconsidering the data from different views (for example, for distribution functions over ranges of values, the ranges cannot be refined if they have been chosen too coarse). Also, some evaluations cannot be carried out incrementally.

6 SYSTEM AND APPLICATION MODELS

The models form the basis for the cost estimations and may be divided into a system and an application model. The system model describes machine and runtime system behavior. Completely general application models are not realistic for the resultant cost estimations. Here, however, we confine our attention to tree-like behavior anyway, so that the problem becomes manageable. First, we look at the system model and then the application model.

The machine model makes the following basic assumptions:

- memory is distributed and communication is by message-passing,
- communication time is the same for all pairs of nodes (diameter is 1)
- communication costs are described by

$$C_{write/read,\, call/return}\,(arguments)\;\; for\ the\ communication\ runtime$$

$$CL_{call/return}\,(arguments, LD)\;\; for\ the\ communication\ latency$$

  Both include process creation as well as coding and decoding overhead (for passing linked data structures) and depend on argument sizes (and types). C_{write} is the communication time at the sending side, and C_{read}

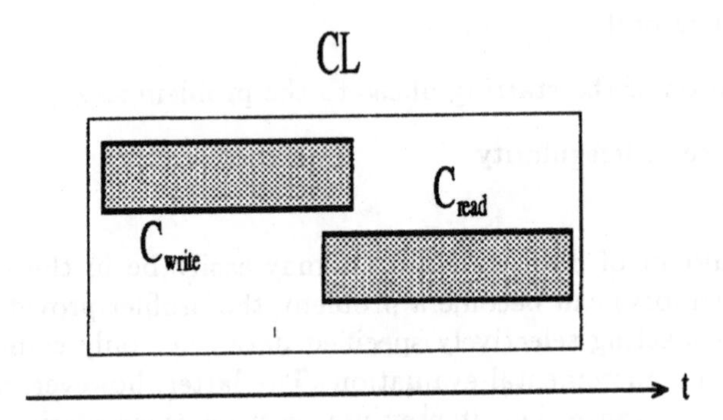

Fig. 3. Communication model

the communication time at the receiving side. Both may overlap. CL also depends on delays because of high system load LD.

There may be a separate communication processor node, which then drives the basic communication protocol, thus partly relieving the application processor of that task, thereby reducing C (but increasing usually CL). For the sort of applications considered by us, however, coding and decoding costs usually dominate, so that the difference is not very significant. And the same applies to the overlapping of C_{write} and C_{read}.

- garbage collection costs are described by

$$GC\,(allocated\_data,\, survival\_rate)$$

GC depends on the amount of allocated data as well as on the percentage that does not become garbage (to be traversed and, potentially, to be copied); GC is usually super- linear because of repeated garbage collection phases and accumulating survived data.

- there exists a function per load-balancing strategy, which describes its costs (per task) in the form:

$$LBAL_{parent/child}\,(strategy, N_{nodes}, N_{local\_tasks}, N_{group}, LD)$$

$$later\ abbreviated\ as:\quad LBAL_{parent/child,\, N_{group}}$$

$LBAL$ depends on the number of machine nodes N_{nodes}, the current number of local tasks N_{local_tasks}, on the number of tasks which may be assigned within a single balancer request (N_{group}), and on the overall system load LD.

The application model assumes the following behavior:

- process structure is tree-like (hence, containing partial dependencies), and trees evolve dynamically
- processes are short-term processes
- communication takes place only at call and return points between the spawning process and the spawned subprocess
- tasks are usually split within a loop (which can form a group) where the iteration number mostly corresponds to the branching factor
- data operated upon is allocated dynamically and passed in arguments for access (i.e. long-term data and its partitioning and distribution do not need to be considered)
- several parallelization centers (simple iterations, recursion iteration trees) may be nested in the case of large applications

Note, that garbage collection may influence execution time significantly, as can be seen from the following function (with $P(task)$ being the pure computation time and SV the survival rate):

$$T_{par\_runtime}(task) = P(task) + GC\left(\left(ALLOC(task) + ARG_{in}(task) + \sum_{i=1}^{subtasks\_of\_task} ARG_{ret}(subtask_i) \right), \; SV \right)$$

Since we assume that each task has its own local heap, GC may on the one hand be lower than in a sequential context, but on the other hand, additional space for input and return arguments has to be collected. Execution time for a task may thus differ in a sequential and a parallel execution context for applications where garbage collection costs dominate.

7 COST FUNCTIONS

For our cost functions, we give some consideration to the context of the computation which can influence the costs in the following ways:

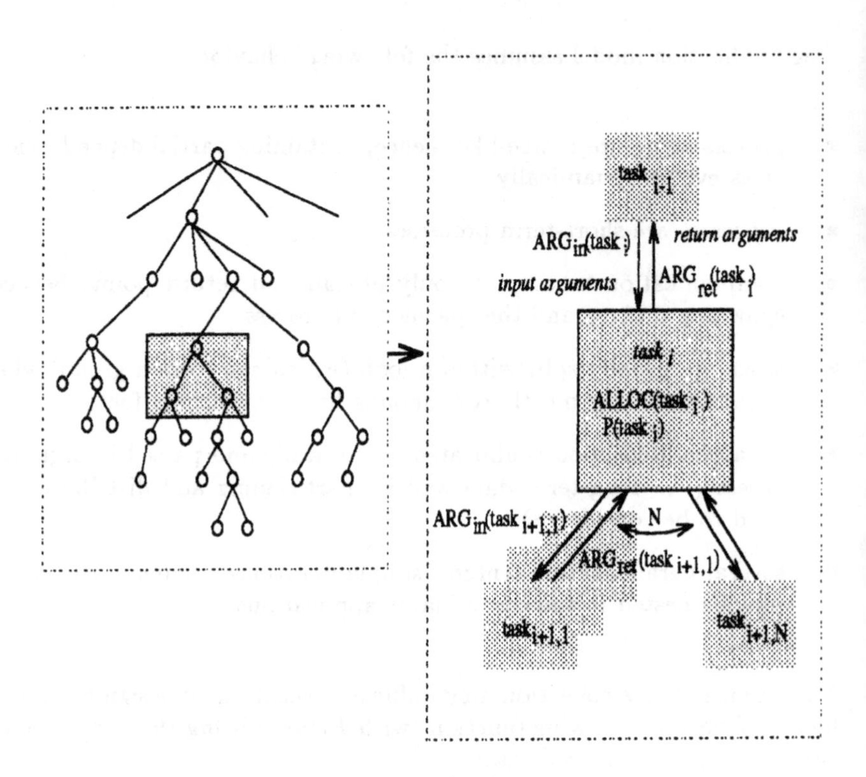

Fig. 4. Model of task

- executed in parallel (and for how many tasks overhead has to be paid that is not outweighed by parallel execution)
- computation and communication can be overlapped to some degree (locally and remotely)
- the costs for load balancing depend on the number of nodes and tasks

A minimum requirement for a cost-effective task granularity is that the remote-execution overhead on the spawning side is significantly lower than its sequential execution time would be:

$$O_{parent}(task) = C_{call,parent}(task) + C_{return,parent}(task) + LBAL_{parent,1}$$

$$C_{parent}(task) + LBAL_{parent,1}$$

$$\begin{aligned} S_{task}^{max} &= T_{seq\_runtime}(task) \,/\, O_{parent}(task) \\ &\leq T_{seq\_runtime}(task) \,/\, C_{parent}(task) \end{aligned}$$

S_{task}^{max} is the maximum relative speed-up (at the spawning side) for an individual task and is obtained provided that the parent can execute other tasks during remote subtask processing (requiring approximately the time $C_{call,child} + T_{par\_runtime}(task) + C_{return,child} + LBAL_{child}$), and that the receiver node is free for execution immediately, i.e. does not sequentialize the subtask execution. It should apply $S_{task}^{max} \gg 1$, i.e. $T_{seq\_runtime} \gg C_{parent}$, and all iterative calculations for finding best granularities only consider granularities which are at least high enough for fulfilling this minimum requirement.

We currently provide cost functions for the two main situations described below:

1) Overall current number of tasks $\ll$ number of nodes, and tasks are created in a group (e.g. in the same loop).

Then, the speed-up for a group of tasks being split can be determined by

$S_{task\_group} = T_{seq\_runtime}(task\_group) \,/\, T_{par\_real\_time}(task\_group)$ *and*

$$\begin{aligned} &T_{par\_real\_time}(task\_group) = \\ &\quad \sum_{i=1}^{task\_group} \left(C_{call,parent}(task_i) + C_{return,parent}(task_i)\right) + \\ &\quad LBAL_{parent,task\_group} + T\_C\_rest(task\_group) \quad with \end{aligned}$$

$$T\_C\_rest(task\_group) \leq max\,\{T_{par\_runtime}(task) \mid task \in task\_group\}$$

T_C_rest is the idle time (or time to be filled with other tasks' computations) which depends on the children's communication and computation times as well as their accidental order of creation.

Theorem 1 *Assume $T(task_i)$ and $C(task_i)$ are the same for all tasks in the group. Then the maximum speed-up is $\frac{T_{seq\_runtime}(n*task)}{n*C_{parent}(task)}$ or, if garbage collection time is not relevant, then $\frac{T_{seq\_runtime}(task)}{C_{parent}(task)}$. Note, however, that $\frac{T_{seq\_runtime}(n*task)}{n*C_{parent}(task)} \geq \frac{n*T_{seq\_runtime}(task)}{n*C_{parent}(task)} = \frac{T_{seq\_runtime}(task)}{C_{parent}(task)}$.*

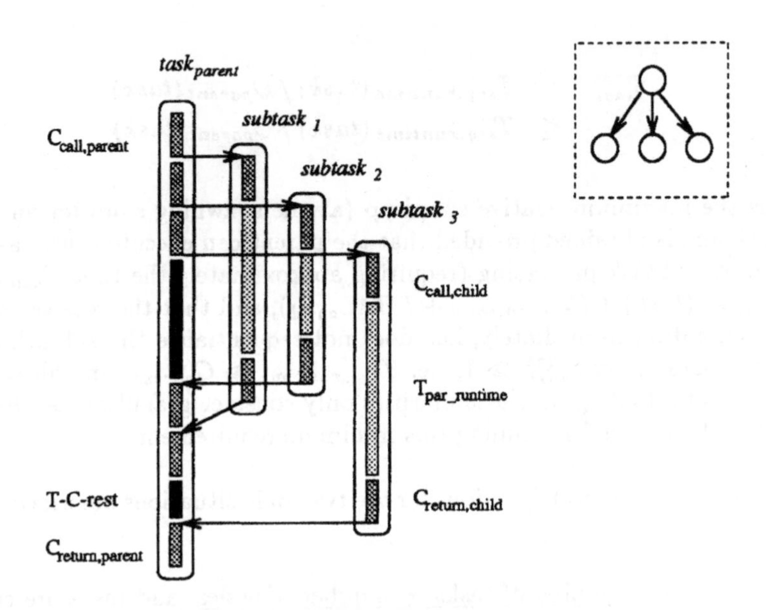

Fig. 5. Time behavior for splitting a group of tasks

This can easily be seen from:

$$S_{task\_group} = \frac{T_{seq\_runtime}(n*task)}{n*C_{parent}(task)+LBAL_{parent,task\_group}+T\_C\_rest}$$

$$\leq \frac{T_{seq\_runtime}(n*task)}{n*C_{parent}(task)}$$

The theorem shows the significance of the communication time, but also that further advantage can be taken from reduced garbage collection costs. The speed-up becomes close to the upper limit if the number of tasks within the group is large.

2) Number of tasks $\gg$ number of nodes, and tasks are created at arbitrary times during computation.

Then, a different view is more appropriate. In most cases, the whole system is then busy, and the aim is to keep it busy by shifting tasks from high-loaded to low-loaded nodes. We use the following rough estimation for the parallel execution time:

$$
\begin{aligned}
&T_{par\_runtime}(program) = \\
&\quad N_{lctask} * MAINT + \\
&\quad N_{lctask} * (1 - MOVE\_FRAC) * T_{seq\_runtime}(task)) + \\
&\quad N_{lctask} * MOVE\_FRAC * \\
&\quad (C_{call,parent}(task) + C_{return,parent}(task) + \quad ; from \\
&\quad LBAL_{parent}(strategy, N_{nodes}, N_{local\_tasks}, N_{group}, LD) + \\
&\quad C_{call,child}(task) + C_{return,child}(task) + T_{par\_runtime}(task) + \quad ; to \\
&\quad LBAL_{child}(strategy, N_{nodes}, N_{local\_tasks}, N_{group}, LD))
\end{aligned}
$$

This estimation uses mean values for all basic variables and assumes the same behavior per node, i.e. that 1) $N_{lctask} = N_{all\_tasks}/N_{nodes}$ of tasks are executed per node, that 2) for each a maintenance overhead $MAINT$ must be paid, that 3) there is only a single predicted task granularity (an extension to different ones is possible in quite a straightforward manner), and that 4) there is a proportion of tasks being exchanged between nodes, and a proportion of tasks being executed where they are created. We further assume that, during the overall computation with its changing load situations, the same number $N_{lctask} * MOVE\_FRAC$ of tasks is moved from and to the node ($0 \leq MOVE\_FRAC < 1$), i.e. assigned remotely in both directions (*to* and *from* in the formula). For tasks that are exchanged, communication and load-balancing costs have to be added, and for tasks that are moved to the node, runtime is $T_{par\_runtime}$. Here, idle times of nodes are assumed to be part of the $LBAL$ overhead. The most difficult thing is to determine $MOVE\_FRAC$, which we estimate mainly by the degree of irregularity in the tree. $MOVE\_FRAC$ depends, however, also on the number of machine nodes, on the number of tasks, and on the quality of the load balancing strategy, i.e. on how much tasks are moved, but sequentialized and not filling gaps. Of those, the critical factor is irregularity, currently determined by variation of recursion depth and computation time per granularity, related to the problem size. We are, however, still experimenting with finding best irregularity measures.

This formula still assumes knowledge of the number of tasks. It can, however, easily be transformed into one based upon creation rates (number of processes created per time unit).

Example 1 *Assume to have the choice between two different granularities with parameters as shown below. 1) describes the lower granularity, and 2) the larger one.*

1) $T_{par\_runtime}(task) = T_{seq\_runtime}(task) = T_1, N_{local\_tasks} = NLT_1, C = C_1,\ LBAL = L_1, MAINT = M_1, MOVE\_FRAC = 1/MVF_1$

2) $T_{par\_runtime}(task) = T_{seq\_runtime}(task) = T_2, N_{local\_tasks} = NLT_2 = 0.5\ NLT_1, C = C_2 = 1.5\,C_1,\ LBAL = L_1, MAINT = M_1, MOVE\_FRAC = 1/MVF_1$

$$S_{partitioning\_1}(program) = \frac{T_{seq\_runtime}(program)}{T_{par\_runtime}(program)} =$$
$$NLT_1 * M_1 + NLT_1 * (1 - 1/MVF_1) * T_1 + \frac{NLT_1}{MVF_1} * (C_1 + T_1 + L_1)$$

$$S_{partitioning\_2}(program) = \frac{T_{seq\_runtime}(program)}{T_{par\_runtime}(program)} =$$
$$0.5 * NLT_1 * M_1 + NLT_1 * (1 - 1/MVF_1) * T_1 +$$
$$0.5 * \frac{NLT_1}{MVF_1} * (1.5\,C_1 + 2\,T_1 + L_1)$$

That means, for both the computation time parts are the same, but for 1) the maintenance part and especially the communication part are larger.

Cost estimations are applied by the granularity controller while looking for the best granularity. It then iterates over all appropriate granularities and estimates the corresponding parallel runtime. Finally, the granularity providing the best, i.e. the smallest parallel runtime is chosen. Calculations are currently based on mean values only. Variation coefficients, however, not only influence the choice of strategies but are also considered in respect to granularities. If maximum task granularities are too high, then a smaller granularity than that indicated as optimum by the estimations is chosen.

With our estimation formulas, we have taken a first step towards considering the characteristics of the tree, e.g. the branching factor, the overall degree of parallelism, and the degree of irregularity. However, a great deal remains to be done in order to find the best estimation formulas.

8 LOAD BALANCING AND SCHEDULING

Load-balancing strategies in Atredia have general properties to the extent that they are made hierarchical ([11], [21]) in the case of a large number of tasks and machine nodes, and that they can be central or decentral. An internal interface is established which allows either standard OS strategies to be used or special ones to be implemented (which is currently the case). Atredia does, in any case, select and configure the strategies and perform all tasks that are not standard like queue handling.

Load balancing in Atredia has special aspects in that

- we are faced with an assignment problem only (migration does not make sense for short-term processes)
- creation, too, can be central or decentral, since trees fold up dynamically
- internal queueing (memorization) must be provided for dynamically limiting the number of processes, and it is applied independently of whether the load-balancing strategy used is central or decentral
- node assignment (via the load balancer) and argument passing usually have to be done in two separate communications because of the large size of the arguments (multiple movements of arguments must be avoided); for the same reason, tasks should be moved for final assignment only and not passed around several times
- a fast reaction to local load changes is necessary because of the potentially high irregularity of the tree (that means purely receiver-initiated strategies are not normally adequate)
- load assignment can often be optimized by groupwise handling (e.g. tasks created in a loop can be announced to the load balancer all at once - if not, lower and upper bounds with respect to task numbers control communication with the load balancer anyway)
- a special strategy, like duplication of work [18] or pre-assignment for the top nodes of the tree, may be necessary for the starting phase where the tree folds up, if the amount of time taken by that phase is large; otherwise, latency, which also includes load-balancing overhead, may significantly delay the point in time at which all nodes are busy and contribute to the solution of the problem
- potentially, hierarchies of different strategies may also be applied, namely, in the case of hierarchies of parallelization centers

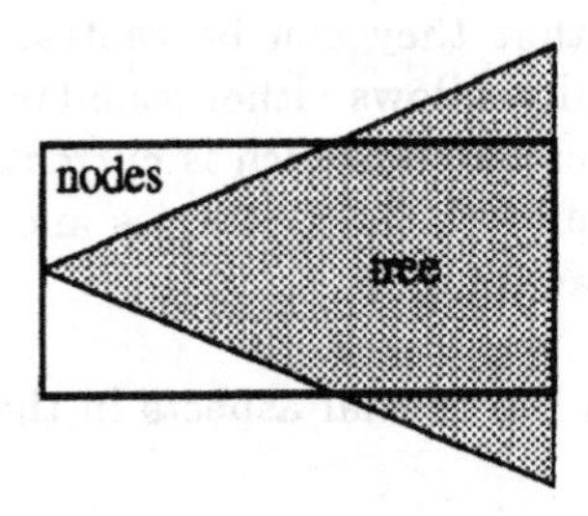

Fig. 6. Relation of starting phase and number of nodes

Scheduling determines the order of task assignment. General heuristics like FIFO (used as part of lazy task creation in [16]) or individually assigned priorities can be applied, chosen either by the user or by the system. If chosen by the user, scheduling strategies reflect (or should reflect) the semantics of the application, like e.g. search heuristics. Individual application-assigned priorities (e.g. in best-first search for pruning the search space) have to be globally ensured as far as possible, which in turn involves restrictions in permissible load-balancing strategies and causes overhead. System-assigned priorities, on the other hand, are merely suggestions and may just be considered locally. For example, the best dynamic-scheduling heuristic is to assign the critical path first, which can be approximated by remotely assigning the largest granularities first (if task creation is multi-level, then priority has highest-level/largest-granularity) [4].

9 DECISION RULES FOR SELECTION AND CONFIGURATION

Several rules for selection and configuration have already been mentioned above. The most important of these are:

- correlation of granularity and level or argument size → apply corresponding granularity control strategy
- high variation of granularities (and no concentration on small values) or high maxima → choose lower granularity than optimums with respect to mean value suggest

- high variation of granularities and high ratio of small values $\rightarrow$ try to apply merging
- massive parallelism $\rightarrow$ hierarchical load balancing
- starting phase of tree is significant in time-consumption $\rightarrow$ apply special task assignment strategy in that phase
- large arguments $\rightarrow$ assign and move tasks once only

Some other important rules are given below. The list is, however, by no means exhaustive. There are many rules of minor importance, and we are still working on deriving more rules.

- priorities to be balanced globally $\rightarrow$ apply a central load-balancing strategy
- several synchronization points in tree $\rightarrow$ use more subtle organization in multiprocessing and load balancing to avoid deadlocks
- very high irregularity $\rightarrow$ apply distributed creation

10 EXPERIMENTAL RESULTS

Since our load-balancing implementation is not yet finished, we present below results from simulations. Data is collected on the MANNA machine (application traces and system communication parameters, which are: $C_{call,parent} = C_{call,child} = 0.85\,msec + CODING\,(args) + TRANS\,(args)$, $C_{return,parent} = C_{return,child} = 0.57\,msec + CODING\,(args) + TRANS\,(args)$, $CL^{min}_{call} = 1.2$ $msec$, $CL^{min}_{return} = 0.64\,msec$, $CODING\,(args) = 3.6\,\mu sec * number\_list\_cells$, $TRANS\ (args) = 0.0227\,\mu sec * bytes\_in\_message$). The simulation then assumes an unlimited number of nodes and assigns each process its own node. The results presented here, then, mainly demonstrate the appropriateness of the profiling approach and the usefulness of granularity control.

The first example is the Triangle program of the Gabriel Lisp benchmark set [7]. Triangle is a board game which searches for all move sequences leading to a desired final position. With maximum exploitation of parallelism, the program would create about 6 million processes, with a step processing time in the microseconds range. Table 1 gives some statistics on the dynamic behavior of Triangle as obtained by the profiler. Rows correspond to recursion depths,

N is the number of tasks, $vcff$ is the variation coefficient (standard deviation related to mean), $mean\,T$ is the mean subtree execution time, $max\,T$ the maximum time, and *trivs* are small subtrees below a certain threshold. All time information is in msec. Triangle is of limited irregularity (depth is bounded) and the average subtree size depends strongly on the recursion depth. Hence, level becomes the control criterion. The variation coefficient, however, is quite high, which is due to the high percentage (about 90 %) of trivial tasks per level. We additionally applied merging (which, here, is similar to loop partitioning) with 18 tasks per merger. This increased the probability of a reasonable process granularity to $1 - 0.9^{18} = 0.85$. By choosing level 5 for process spawning, we got 118 processes only, an average processing time of 0.23 sec, and a maximum processing time of 1.15 sec. The overall speed-up is good, as can be seen from Table 2 (*critical path* is without communication overhead, *distributed* means that overhead for all communications is included).

The next example is the rewriting phase of the Boyer-Moore theorem-prover in a simplified version used for Lisp benchmarking [7]. This is an example with high irregularity and unbounded depth. Subtree sizes, however, depend on the input argument size (here being the number of list elements). T_{minsub} in Table 3 does not denote the whole time spent in the subtree computation, but merely the time for recursive descent and trying and applying substitutions (rewriting the term once at each recursion step). The time subsequently taken for trying a rewrite on the resulting term again is, however, excluded. Substitution rules are semantic, and the substituted term and all computations on it cannot be expected to be predictable in terms of size and time. *ntasks larger args* is the number of tasks with larger arguments (the sum of the tasks in the rows below), and *%tasks* the percentage of tasks. The Boyer-Moore example creates 53711 potential processes (Table 3, however, also includes sequential rewriting

level	N	mean T	vcff	max T	trivs N/ %
1	1	27582.1	0	27582.1	0/0
2	36	775.8	380	14470.3	33/92
3	108	252.5	349	4496.2	95/88
4	468	58.3	357	1310.4	409/87
5	2124	12.8	358	345.9	1858/87
6	9576	2.9	366	91.5	8380/88
7	43056	0.6	385	36.1	38402/89

Table 1. Excerpt from profiling results for Triangle

	run time	speed-up
sequential	27.58 sec	
critical path	1.29 sec	21.3
distributed	1.44 sec	19.1

Table 2. Performance results for Triangle

tasks), very many of which are below the cost-effective size. With a minimum argument size of 30 list elements, we get a reasonable task granularity and 2028 processes only.

Nevertheless, the speed-up is only slight (see Table 4). The problem size, however, is small, and so the communication cost is relatively high. Additional costs for checking the control criterion are also involved (we do not compute the complete argument size for large arguments, but stop when the chosen threshold is reached; the mean checking time is 18.6 μsec). Checking is, however, mostly done in parallel, too. Under these conditions, the distributed runtime is relatively good, and not too far from ideal time (critical path). More elaborate tests with other inputs are necessary, ideally conducted using the real Boyer-Moore theorem-prover. This we are currently working on.

argument size	number tasks	ntasks larger args	T_{minsub} (msec)	vcff	%tasks < 0.5 msec	max T_{minsub} (msec)
1	11282	79742	0.056	19	100	0.33
2	43391	36351	0.088	3845	99.9	540.05
3	2269	34082	0.810	1963	97.6	543.35
4	1257	32825	0.562	17	10.7	1.12
$1 \leq n < 10$	76784	14240	0.375	1892	74.3	563.58
$10 \leq n < 20$	6246	7994	3.087	737	0.2	533.13
$20 \leq n < 30$	1614	6380	6.465	571	0.2	566.44
$30 \leq n < 40$	39	6341	9.444	27	0.0	16.04
$40 \leq n < 50$	994	5347	9.352	466	0.3	536.35

Table 3. Excerpt from profiling results for Mini-Boyer-Moore, rewrite

	run time	speed-up
sequential	3.51 s	
critical path	0.32 s	10.9
distributed	0.79 s	4.4

Table 4. Performance Results for Mini-Boyer-Moore, rewrite

Both examples show that there can be a high ratio of small processes, even near the root. Our granularity control still works for such situations and avoids sending trivial processes to remote nodes.

The next example is the real Boyer-Moore theorem prover [2]. Boyer-Moore is not an ideal candidate for parallelization, since it is written mainly following a sequential thought model, but it does have the important advantage of being available as public domain software. Also, despite its non-optimal general construction, it has three processing centers which are interesting for parallelization, located at different abstraction levels [9]. Originally, we favored the medium center, at which we measured a large number of processes with cost-effective mean granularities. However, a full trace and a critical path estimation [3] showed that for different inputs the critical path always took nearly as much time as the sequential execution, even without taking into account communication overheads. In other words, there was hardly any speed-up (see Table 5). This can be explained by the fact that the trees are ill-formed and practically restricted to a single path. Although these results were disappointing, they did prove the usefulness of our tools. We obtained these results merely by marking the points for splitting processes and without restructuring the program with respect to data handling. Thus, the mapper saved us a lot of unnecessary work, and we were able to focus our attention on another level more promising for parallelization. This shows how important it is to first check the maximum potential parallelism of an application by computing the critical path. Such a step should form part of any systematic parallelization approach.

Measurements for a combination of parallelization on level 1 (prove-lemma) as well as on level 3 (rewrite) are given in Table 6. The results show that, for this approach, at least the basic requirement is fulfilled, namely, that the inherent potential parallelism is sufficient. Work on the realization of a parallelization for Boyer-Moore at these levels is in progress, but we are not yet at a point where we can present performance results. Even if the final speed-up, with all overhead included, were only 5-10, this would be a significant improvement,

input file	T-seq	T-step	crit path	speed-up
RSA	152.7 sec	0.31 sec	137.4 sec	1.1
Goedel	25116.1 sec	2.33 sec	20311.1 sec	1.2
Tautology	163.2 sec	0.13 sec	110.6 sec	1.4

Table 5. Critical path of Boyer-Moore, clause simplification

input file	T-seq	crit path	speed-up
Wilson	213.5 sec	2.5 sec	86.5
TMI	44.6 sec	6.0 sec	7.4
Tautology	163.2 sec	7.5 sec	21.7

Table 6. Critical path of Boyer-Moore, prove-lemma and rewrite

since there are inputs which run for up to several hours sequentially (Goedel, for example, takes 6 h 59 min and a factor of 10 faster would mean a runtime of only 42 min).

The Boyer-Moore example also demonstrates that the same typical behavior can be observed for different inputs (here we have shown only a small excerpt, but we obtained the same behavior for all inputs), and that profiling on samples of input is adequate. Also, the mapper assumption about a typical behavior characteristic per application is proved to be right.

11 SUMMARY AND DISCUSSION

We here presented an environment for mapping dynamic tree-structured problems. Unlike other approaches, the environment is open for integration of a set of strategies for granularity control and load balancing, and it constitutes a step towards a systematic and formally based solution. This is accomplished by incorporating modelling and cost estimation and featuring automatic selection and parameterization of strategies by the system. The tree-structure is partly taken into consideration in the cost estimation formulas (e.g. the degree of irregularity, branching factor), and in strategies (e.g. no pure receiver-initiated approach for load balancing, relating the starting phase to the problem size).

The approach is hybrid, comprising static and dynamic components. Several tools are provided, such as a profiler for obtaining information about the dynamic behavior characteristics of the specific application. This information is then exploited for granularity control and load-balancing configuration. The results obtained demonstrate the usefulness of the approach with respect to profiling and granularity control, and also its applicability to large real-life problems.

Load balancing is at present still under implementation. Besides expecting to get more expressive results once the load-balancing tool is available, we plan to experiment with more and larger applications. We are, currently, in the process of parallelizing a large theorem-prover, and results will soon be available. This, we hope, will give us more insight into which classifications are appropriate, what cost estimations are best, what other decision rules can be derived, and how much can really be fully automated or what sort of user interaction is necessary. For example, it may sometimes be impossible to establish granularity control criteria by the system, user hints being necessary (e.g. the number of variables within a polynomial may be a criterion for estimating granularities in the computer algebra problem of polynomial decomposition [15]). In addition, we wish to check whether general strategies can work efficiently. The theorem-prover has already brought to light many subtle problems which, at worst, will require very specific handling. Nevertheless, we still believe that we can address these problems using the general approach and an appropriate configuration.

We also plan to integrate weight analysis and to obtain some of the behavior information by this method instead of by profiling. In addition, we will look into whether our strategies and cost estimations need extensions for machines with non-uniform communication diameters. Hierarchical strategies create localities in communication anyway, so that if an equal communication diameter in the neighborhood can always be assumed, there may be no great difference (as already investigated elsewhere, cf. [1]).

Acknowledgements

The work on which this paper is based was conducted in the context of the MANNA project and funded by the German Federal Ministry of Research and Technology under grant 413-4001-01 IR 201 C. I am indebted to Heiko Bock and Peter Kabat for their work on the profiler and the Boyer-Moore theorem-prover, respectively. My thanks also go to Phil Bacon for polishing up my English.

REFERENCES

[1] Ishfaq Ahmad. A Semi Distributed Task Allocation Strategy for Large Hybercube Supercomputers, IEEE Supercomputing 1990.

[2] Robert S. Boyer and J. Strother Moore: A Computational Logic. Academic Press, New York, 1988.

[3] Heiko Bock. Konzeption und Implementierung eines Profilers zur Gewinnung von symbolischen Anwendungen. Diploma thesis, Technical University Berlin, 1994.

[4] Christophe Coroyer and Zhen Liu. Effectiveness of Heuristics and Simulated Annealing for the Scheduling of Concurrent Tasks: An Empirical Comparison. Proc. PARLE'93, Parallel Architectures and Languages Europe, Springer-Verlag, 1993.

[5] Masakazu Furuichi, Kazuo Taki, Nobuyuki Ichiyoshi. A Multi-Level Load Balancing Scheme for OR-Parallel Exhaustive Search Programs on the Multi-PSI. PPOPP 1990.

[6] Franz Incorporated. Allegro Composer, Franz Inc., 1990.

[7] Richard P. Gabriel. Performance and Evaluation of Lisp Systems. MIT Press, 1985.

[8] Wolfgang K. Giloi. From SUPRENUM to MANNA and META - Parallel Computer Development at GMD FIRST. Proc. 1994 Mannheim Supercomputing Seminar, Sauer-Verlag, Munich 1994.

[9] Peter Kabat. Parallelisierung des Boyer-Moore Theorembeweisers. Bachelor thesis, Technical University Berlin, 1994.

[10] Vipin Kumar and Anshul Gupta. Analyzing Scalability of Parallel Algorithms and Architectures, Journal of Parallel and Distr. Computing, 1994.

[11] Vipin Kumar and Anshul Gupta. Scalable Load Balancing Techniques for Parallel Computers, Journal of Parallel and Distributed Computing, 1994.

[12] V. Kumar, A. Grama, A. Gupta, and G. Karypis. Introduction to Parallel Computing - Design and Analysis of Algorithms. Benjamin/Cummings Publ. Company, 1994.

[13] L.V. Kale and S. Krishnan. CHARM++: A Portable Concurrent Object Oriented System Based on C++. OOPSLA93.

[14] L.V. Kale, B. Ramkumar, V. Saletore, and A.B. Sinha. Prioritization in Parallel Symbolic Computing. In Halstead and Ito (eds.), Proc. US/Japan Workshop on Parallel Symbolic Computing: Languages, Systems, and Applications, Oct. 1992, Springer-Verlag, 1993.

[15] Wolfgang Küchlin, Universität Tübingen, private communication, Sept. 1994.

[16] Eric Mohr, David A. Kranz, and Robert H. Halstead. Lazy Task Creation: A Technique for Increasing the Granularity of Parallel Programs. Proceedings ACM Conference on Lisp and Functional Programming, 1990.

[17] Brian Reistad and David K. Gifford. Static Dependent Costs for Estimating Execution Time. ACM Conf. on Lisp and Functional Programming, June 1994.

[18] A. Reinefeld and V. Schnecke. Work-Load Balancing in Highly Parallel Depth-First Search. Proc. SHPCC'94, Knoxville/Tennessee, May 1994.

[19] Vivek Sarkar. Partitioning and Scheduling Parallel Programs for Multiprocessors. MIT Press, 1989.

[20] Angela Sodan and Hua Bi. A Semi-Automatic Approach for Parallelizing Symbolic Processing Programs. First Int. Symp. on Parallel Symbolic Comp., Linz/ Austria, Sept. 1994.

[21] Amitabh B. Sinha and Laxmikant Kale. A Load Balancing Strategy for Prioritized Execution of Tasks. Internat. Parallel Processing Symposium, Los Angeles/CA, April 1993.

[22] Amitabh B. Sinha and Kaxmikant V. Kale. A framework for intelligent performance feedback. ICPP-94.

[23] Angela Sodan. Parallelisierung von Lisp - Inwieweit bieten deklarative Sprachmittel Vorteile? Workshop on "Entwicklung, Test und Wartung deklarativer KI-Programme" at 18th German Ann. Conf. on AI, Saarbrücken, Sept. 1994, Springer Press (short version), GMD-Berichte (long version).

15

REGULARISING TRANSFORMATIONS FOR INTEGRAL DEPENDENCIES

Graham M. Megson and Lucia Rapanotti

Department of Computing Science, The University, Newcastle upon Tyne, NE1 7RU, UK

e-mail: *Graham.Megson@newcastle.ac.uk*
Lucia.Rapanotti@newcastle.ac.uk

ABSTRACT

The high-level synthesis of parallel algorithms characterised by irregular data dependencies is a challenging task for algorithm designers. In this paper we show that for problems specified as systems of integral recurrence equations, a systematic derivation of regular arrays is possible. The approach combines established techniques based on the space-time mapping of regular (typically linear) data dependencies onto processor arrays, with systematic ways of regularising the data dependencies through powerful forms of localisation. The methods we present are applicable to a range of combinatorial problems, including the Knapsack problem, which we adopt as a case study.

1 INTRODUCTION

The problem of automatically detecting and extracting parallelism from algorithms is a significant challenge to researchers (see, e.g., [9, 4, 3, 5]). In particular it is the degree of regularity (or lack of it) in a problem that often determines the success (or more often) the failure of systematic methods. The most successful parallel compilers rely mainly on clever mechanisms for detecting regularity either explicitly or implicitly through the application of transformations to rewrite the computation and task dependencies ([1, 3, 4, 9, 5, 8, 11, 13, 14, 16, 17]). In essence generating parallel programs can be reduced to determination of a, mapping or, pair of functions, the timing and allocation functions respectively. The timing function determines the exact time (or when) a task

A. Ferreira and J. D. P. Rolim (eds.), Parallel Algorithms for Irregular Problems: State of the Art, 297–322.

or computation is to be evaluated while the allocation determines the location (or where) the evaluation is to be executed. The timing and allocation must be made in manner that observes constraints imposed by the dependency of the variables in the computation and the (partial) ordering of tasks. For example we cannot evaluate two computations at the same place at the same time, and we cannot perform a computation before all its operands (arguments) are known. Mapping problems which satisfy all the constraints are termed conflict-free.

It is an interesting endeavour to determine the type of mappings required for different types of computational problem. For example if we confine our attention to so-called regular problems where the tasks are known apriori and the data dependencies have recurrent structures in the computation space it is well-known that the timing and allocation functions can be determined explicitly by solution of linear programming problems ([11, 13, 14]). Such problems arise in the mapping of many nested DO-loop structures and form a significant class of practical problems, mainly in numerical linear algebra ([12, 8]). The existence of explicit functions is useful because the parallelism is static and can be determined at compile time. Unfortunately the task of identifying explicit functions is not possible in general and becomes more difficult as the regularity of the problem is lost. In such cases the data dependencies become non-trivial and the associated tasks dynamic so that compile-time mapping is impossible.

Faced with such difficulties it is easy to see why automatically parallelising compilers cannot offer a general solution to mapping problem. Nevertheless it is intriguing and practically important to explore the relationships between regular and irregular problems in both static and dynamic settings. In this paper we consider the problem of systematically regularising computations with statically defined task structures but whose data dependencies can be very irregular. In particular we are interested in dependencies which can be non-linear functions of index variables. Such problems arise naturally in a range of dynamic programming problems such as the Knapsack problem.

The paper is organised as follows. Section 2 introduces some basic concepts regarding recurrence equations and data dependencies in high-level synthesis. Section 3 introduces integral recurrence equations, while Section 4 discusses their localisation through decomposition and uniformisation. Section 5 presents the Knapsack problem as an illustration of the techniques. Finally Section 6 draws some conclusions.

2 BASIC CONCEPTS

In this section we recall some basic definitions concerning recurrence equations and data dependencies and their geometric and operational interpretation. We will also discuss some basic requirements and assumptions behind the equation transformations of the following sections.

2.1 Recurrence Equations and Data Dependencies

We consider algorithms specified as *systems of recurrence equations.* The equations express recurrence relations among the algorithm variables and the variables are indexed in an n-dimensional Euclidean lattice space, called the *index space.* The generic form of a recurrence equation is

$$z \in D: \; U(z) = f(\ldots, V(\mathcal{I}(z)), \ldots),$$

such that, for each point z of the index space, in the set D, the value of the indexed variable U at z is equal to the evaluation of a certain function f on an arbitrary but finite number of arguments, each argument being the value of an indexed variable V at the index point $\mathcal{I}(z)$. D is the *equation domain* and is a convex polyhedral set in the index space. In order to guarantee the existence of a finite design, the equation domain is either finite or has at most one infinite direction[1]. f is a known function of constant complexity, strictly dependent on its arguments. $\mathcal{I}$ is called an *index function* and maps index points to index points. The evaluation of a variable at a point of the index space is called an *instance* of that variable. Variable instances are *functional* in that they assume a unique value. Therefore the domains of equations defining instances of the same variable are disjoint.

$\mathcal{I}$ defines a *data dependence relation* (or more simply *data dependence*) between variables U and V over the domain D. In particular, for each z in the domain D we say that the value of U at z is *data dependent* on the value of V at $\mathcal{I}(z)$. The data dependence itself is represented by the n-dimensional vector $z - \mathcal{I}(z)$. Of the data dependence defined by an index function $\mathcal{I}$ we provide a geometric representation as an n-dimensional *data dependence graph*[2] in the index space, whose nodes are the points z and $\mathcal{I}(z)$ and whose arcs are the n-dimensional vectors $z - \mathcal{I}(z)$. We denote such a graph by $\mathcal{DDG}_{\mathcal{I}}$.

[1] Domains with one infinite direction arise naturally, for instance, in the synthesis of digital signal processing applications in which the signals are modeled as streams of data in time.

[2] An n-dimensional graph is an ordinary directed graph [2]

We may interpret the $\mathcal{DDG}_\mathcal{I}$ by considering the nodes as computations in space and time and the arcs as sequentiality constraints among computations, so that in order to perform the computation of U at the point z we need to have performed the computation of V at the point $\mathcal{I}(z)$ and transferred the corresponding value from $\mathcal{I}(z)$ to z along the arc $z - \mathcal{I}(z)$. Therefore any array design correctly derived from this data dependence graph performs the two computations at distinct time cycles, with $V(\mathcal{I}(z))$ computed before $U(z)$. With this operational interpretation we obtain a computational model based on data flow in which the arcs of the $\mathcal{DDG}_\mathcal{I}$ are seen as ***abstract direct data channels*** between pairs of points $(z, \mathcal{I}(z))$ (they are abstract in that they do not imply any particular implementation and direct in that they do not involve intermediate points).

The definitions and interpretation above naturally extend to: the data dependence graph of an equation, being the union of the data dependence graphs of its index mappings; and the data dependence graph of a system of equations, being the union of the data dependence graphs of its equations.

A particular type of dependencies which play an important rôle in high-level synthesis is represented by ***uniform data dependencies***. A uniform data dependence is defined through a linear index function which is just a translation according to a certain vector, i.e., for each $z \in D$, $\mathcal{I}$ is defined as $\mathcal{I}(z) = z+d$, for some vector d in $\mathbf{Z}^n$. Therefore the data dependence graph consists of a replication of the vector $-d$ on each point of the domain. Because of their regularity (see also the discussion in the following section), uniform data dependencies allow a direct derivation of regular array designs. Most of the techniques which we discuss in this paper aim at establishing such a regularity starting from generic integral data dependencies.

2.2 Data Broadcasts, Data Conflicts and Localisation

In terms of architecture, a ***data broadcast*** is a situation in which the same data is transferred from one cell directly to several cells. In the data dependence graph $\mathcal{DDG}_\mathcal{I}$, this corresponds to a situation in which more then one node is a direct descendent of the same node, or, equivalently, the index mapping $\mathcal{I}$ is not injective.

A ***data conflict*** is a situation in which two or more distinct pieces of data require the use of the same communication channel during the same time cycle. In the abstract model, a data conflict is related to the ***overloading*** of the nodes of a data dependence graph. Intuitively, a node of the graph is overloaded if several instances of the same variable have to be transferred through the node. Overloading is not a problem *per se*, but it becomes a problem when the number of instances to be transferred through the node depends on the problem's size, as this fact violates the axiomatic requirement that cells' memory and bandwidth of a regular array design have to be independent from the size parameters of the problem. (The reason for this requirement is that it guarantees the *extensibility* of the resulting designs.) Indeed, if we could assume unbounded memory and bandwidth, the axiom would always be satisfied and overloading would not be a problem.

Overloading may arise because of transformations of the data dependence graph such as *localisation*. Localisation aims at enforcing some *regularity properties* on the data dependencies. These properties are *locality* (i.e., data dependence relations involve only pairs of neighbour index points) and *uniformity* (i.e., all the data dependence vectors of a certain data dependence relation "look" the same for each point of the domain). Localisation replaces global non-uniform data dependencies with local uniform data dependencies, by replacing the direct arc between each pair of points with a pipelining/routing path which involves a number of intermediate nodes. The shape of these paths is determined by a choice of *localisation direction vectors*. If the intermediate nodes introduced by localisation belong to several of paths, then overloading may appear. An illustration of localisation and consequent overloading is given in Fig. 1. In part a) of the figure a generic data dependence is given. In part b) the data dependence is decomposed into two constituent components (solid and dashed arrows, respectively): note that overloading appears on the first component. Finally in c), the second component of the data dependence is made uniform, and this yields more overloading. Note that in the figure, overloading develops through localisation from data broadcasts.

In the definition of localisation techniques, precautions should be taken to avoid the introduction of overloading. As we will see, the appearance of overloading is related both to the presence of data broadcasts and to the choice of the localisation direction vectors. In Sections 4.2 and 4.3, we introduce two forms of localisation for integral data dependencies which guarantee that overloading is not generated. We call them: *decomposition*, which replaces a "complex" data dependence with a set of simpler data dependencies, and *uniformisation*, which replaces such simpler data dependencies with uniform pipelinings or routings of the data.

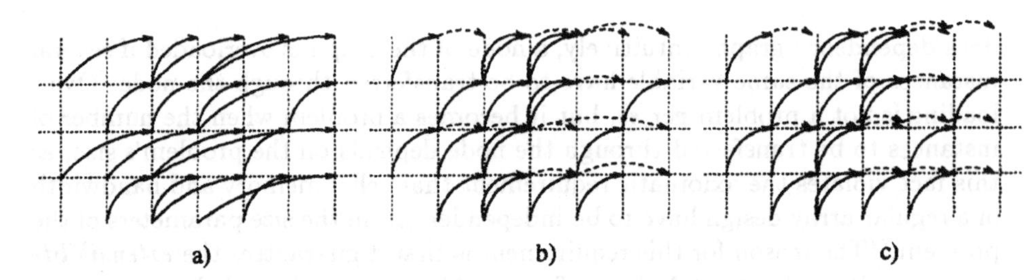

Fig. 1. Localisation and overloading: a) a generic data dependence; b) a possible decomposition; c) a possible uniformisation.

2.3 Space-Time Mappings

Intuitively, a recurrence equation is computable if and only if none of its variable instances depends on itself. When this is the case, there exists an integral function, called the *timing function*, which associates a computation time with each variable instance of the equation in such a way that its data dependencies are reflected in a partial order of the computation times of the variable instances. For realistic implementation of the algorithms, only a finite number of computations must be allowed at each instant of time and the first set of computations must occur at a precise point in time, which represents a lower bound for all computation times of the variable instances. Therefore, given a recurrence equation over a domain D, with index functions $\mathcal{I}_j$, for j in an index set J, a timing function for the equation is a mapping $t : \mathbf{Z}^n \rightarrow \mathbf{Z}$, such that[3]:

i) for all $z \in D$ and for all $\mathcal{I}_j$ with $j \in J$, $t(z) > t(\mathcal{I}_j(z))$ whenever $z \neq \mathcal{I}_j(z)$;

ii) for all $\tau \in range_D(t)$, there exists only a finite set of points z in D such that $t(z) = \tau$ (*finiteness*);

iii) there exists $\bar{\tau} \in range_D(t)$, such that for all $\tau \in range_D(t)$, $\bar{\tau} \leq \tau$ (*existence of a lower bound or boundedness*).

When the domain D is bounded, conditions ii) and iii) are trivially satisfied. When a system of equations is computable, among all possible schedules *affine timing functions* are of particular interest because an optimal scheduling can be calculated by solving a linear programming problem built on the geometry of the system of equations (see, e.g., [13, 15, 8]). A timing function t over D is an affine timing function if and only if there exist $\lambda \in \mathbf{Z}^n$ and $\alpha \in \mathbf{Z}$ such that

[3] Where $range_D(t)$ denotes the range of t over D.

for all $z \in D$, $t(z) = \lambda \cdot z + \alpha$. λ may be regarded as the normal vector of a family of parallel hyperplanes intersecting D, each hyperplane characterised by a particular instant of time (so called *isochronous hyperplanes*). It is possible to show that given an affine timing function t for a recurrence equation over a domain D, an infinite family of affine timing functions can be derived from t by the introduction of integral delays, i.e., given an affine timing function t such that, for all points $z \in D$, $t(z) = \lambda \cdot z + \alpha$, then for all $\beta \in \mathbf{Z}$, $t_\beta(z) = t(z) + \beta = \lambda \cdot z + \alpha + \beta$ is also an affine timing function for the equation. It is usual to consider timing functions ranging over the natural numbers[4]. This is not a restriction because any affine timing function with negative lower bound $\bar{\tau}$ can be transformed into a non-negative affine timing function by adding a delay equal to $\bar{\tau}$.

An *allocation function* for a recurrence equation determines the distribution of the computations among a set of processing elements, together with the interconnections of these processors. An allocation function can be seen as a mapping from the n-dimensional space-time of the recurrence equation to an r-dimensional space of processing elements, where, in general, $r < n$. It is usual to assume $r = n - 1$. Given a timing function t, an allocation function needs to be *consistent* with t, in that the partial order defined by t on the computations must be preserved. More precisely, given a recurrence equation over a domain D and a timing function t for the equation, an allocation function for the equation is a mapping $a : \mathbf{Z}^n \rightarrow \mathbf{Z}^{n-1}$ such that for all $z, z' \in D$, $t(z) = t(z')$ implies $a(z) \neq a(z')$ (*consistency*). As for timing functions, we are mainly interested in affine allocation functions, i.e., allocation functions a over D such that there exists $\sigma \in \mathbf{Z}^{(n-1)\times n}$ and for all $z \in D$, $a(z) = \sigma \cdot z$. Such an allocation function defines a projection from $\mathbf{Z}^n$ to $\mathbf{Z}^{n-1}$ and can be characterised by a *projection direction vector* u, orthogonal to all the rows of σ.

The pair $[t, a]$ constitutes a space-time mapping and uniquely determines an array design for the recurrence equation. The image of D under the allocation function a defines the processor space of the design. For each point $z \in D$, $[t, a](z)$ is an n-dimensional vector whose first component represents the time when computation z occurs and the remaining $n - 1$ components are the coordinates of the computation in the processor space. Given a data dependence vector $z - \mathcal{I}(z)$, then its image under the allocation function a is a corresponding communication channel in the processor space, while its image under the timing function t decreased by one, represents the delay associated with the

[4]Considering non-negative affine timing functions is necessary in order to generate affine schedulings for the equations automatically.

channel (communication is never instantaneous and a channel with zero delay, requires one time cycle to transfer a piece of data).

2.4 Linearity and Convexity

Synthesis techniques for regular array designs rely on a number of assumptions on the equations, their domains and index functions. In particular they assume that: all index expressions (i.e., equation domain definitions and index mappings) are affine expressions of the indices; all data dependencies are *static*, i.e., the index functions need to be completely specified; all equation domains are convex regions.

In the rest of the paper we will introduce a class of recurrences characterised by generic integral (i.e., not necessarily linear) index functions. Their systematic treatment will be achieved through a generalisation of well-known transformational synthesis techniques (see [13]). One of the consequences of relaxing the linearity constraints on the data dependencies is that non-convexity may appear in the definition of the new equation domains which are usually introduced by localisation. Convexity needs then to be enforced. This is achieved by extending the domains to their convex closure with the addition of "dummy" computation points. Control variables (see, e.g., [16, 17, 18]) are then introduced to suitably identify the computation points of interest in these enlarged domains.

Because of the non-linear nature of some of the transformations involved and in order to guarantee that all variable instances are well-defined, we make the assumption that a variable instance assumes a particular *undefined value*, denoted by $\perp$, on each index point outside its definition domain[5].

3 INTEGRAL RECURRENCE EQUATIONS

In this section we discuss a class of recurrences, called *integral recurrence equations*, characterised by data dependencies which can be expressed as integral combinations of a finite number of direction vectors in the index space. In their generic form, their index functions can be expressed as summations of a finite

[5] By *definition domain* of a variable we mean the union of the domains of the equations having that variable on their left-hand side.

number of basic components, each component being a direction vector d_j in $\mathbf{Z}^n$ multiplied by a non-negative integral function g_j. More formally:

Definition 3.1 *[Integral Data Dependence]* Let D be a convex polyhedron in $\mathbf{Z}^n$. An index function $\mathcal{I} : \mathbf{Z}^n \to \mathbf{Z}^n$ defines an integral data dependence over D if and only if, for all $z \in D$, $\mathcal{I}(z) = z + \sum_{j=1}^{m} g_j(z)d_j$, where, for $j = 1, \ldots, m$, $g_j : \mathbf{Z}^n \to \mathbf{Z}$ are functions non-negative and bounded over D and $d_j \in \mathbf{Z}^n$.
■ 3.1

The boundedness of the functions g_j over D means that the computation of a variable instance cannot depend on variable instances at an arbitrary long distance in the direction of d_j in the lattice space (the intended notion of distance is the usual Euclidean distance). This property is necessary for realistic implementation of the algorithms. Indeed if the equation domain is bounded, any integral function defined on it is also bounded. The requirement for the functions g_j to be non-negative over D is not restrictive as any integral function can be rewritten as the sum of its positive and negative parts. This definition of index function is general enough to include any integral function from $\mathbf{Z}^n$ to $\mathbf{Z}^n$ which is component-wise bounded over the equation domain[6]. When the number of addenda defining an integral data dependence reduces to one, we regard the corresponding equation as belonging to a special subclass of integral recurrences, so-called *atomic integral recurrences.* Atomicity in this context stands for the simplest form of integral data dependencies which we want to consider. Intuitively, an atomic integral data dependence is characterised by vectors aligned according to some direction vector d, and with lengths varying according to the values of an integral function g, which is a multiplicative coefficient of d. More formally:

Definition 3.2 *[Atomic Integral Data Dependence]* Let D be a convex polyhedron in $\mathbf{Z}^n$. An index function $\mathcal{I} : \mathbf{Z}^n \to \mathbf{Z}^n$ defines an atomic integral data dependence over D if and only if, for all $z \in D$, $\mathcal{I}(z) = z + g(z)d$, where $g : \mathbf{Z}^n \to \mathbf{Z}$ is a function non-negative and bounded over D and $d \in \mathbf{Z}^n$.
■ 3.2

In the following, given a data dependence defined by an atomic integral index mapping $\mathcal{I}$, we will sometimes refer to the vector d as the *direction of the data*

[6] This fact can be easily proved by considering, for instance, the standard basis of $\mathbf{Z}^n$, and rewriting the function as an integral combination of the vectors of this basis, possibly separating the positive and negative part of each resulting component integral function.

dependence and to the function g as its *length function*. Therefore by *length of the data dependence vector* $z - \mathcal{I}(z)$ we will mean the value $g(z)$. Note that this value corresponds to the modulo of $z - \mathcal{I}(z)$ if only if d is a unit vector.

Atomic integral recurrences represent a special subclass of integral recurrences as: uniformisation techniques of more traditional classes of recurrence equations naturally extend to them; and any integral recurrence can always be replaced by an equivalent system of atomic integral recurrences. This is explained in the following section.

4 LOCALISATION OF INTEGRAL RECURRENCE EQUATIONS

In this section we discuss localisation techniques for integral recurrences and the conditions which guarantee that by applying the transformations no overloading of the data dependence graph occur.

4.1 Choice of the Localisation Direction Vectors

The appearance of overloading through localisation is related both to the presence of data broadcasts and to the choice of the localisation direction vectors. Note that as any index mapping $\mathcal{I}$ is a function, then $\mathcal{DDG_I}$ is not overloaded; and that if $\mathcal{I}$ is also injective then $\mathcal{DDG_I}$ does not contain any data broadcasts. Therefore, we want to define localisation techniques which guarantee that the derived data dependence graphs correspond to injective index functions. In doing so we will rely on the important result stated in the following proposition, in which $lin(D)$ denotes the *direction of the domain* D[7]:

Proposition 4.1 Let us consider a domain $D \subseteq \mathbf{Z}^n$ and an atomic integral index function $\mathcal{I}$ defined as $\mathcal{I}(z) = z + g(z)d$ for each z in D. If $d \notin lin(D)$ then $\mathcal{I}$ is injective over D.

Proof: From linear algebra, if $z, z' \in D$ then $z - z' \in lin(D)$. Assume that there exist $z, z' \in D$, such that $z \neq z'$ and $\mathcal{I}(z) = \mathcal{I}(z')$. We want to prove that

[7] The direction of an affine space is the unique linear space parallel to that affine space. The affine closure of a set is the unique affine space which contains the set. The direction of a set is the direction of its affine closure. See [10] for an introduction to linear algebra.

this assumption always implies a contradiction with respect to the proposition's hypotheses and therefore for all $z, z' \in D$, $z \neq z'$ implies $\mathcal{I}(z) \neq \mathcal{I}(z')$. There are only two possibilities, both leading to a contradiction. If $g(z) = g(z') = c \geq 0$, then $\mathcal{I}(z) = \mathcal{I}(z')$ implies $z + cd = z' + cd$, i.e., $z = z'$. Otherwise, if $g(z) \neq g(z')$, then $g(z') - g(z) = c \neq 0$ and $\mathcal{I}(z) = \mathcal{I}(z')$ implies $z - z' = cd$, i.e., $d \in lin(D)$. ■ 4.1

Note that the converse is not true as $\mathcal{I}$ injective only implies that no data broadcasts are defined by $\mathcal{I}$, independently for the direction of d.

4.2 Decomposition

The decomposition of an integral data dependence is realised by replacing its index function $\mathcal{I}$ by a system of composable index functions, each function defining an atomic integral data dependence, and such that their composition assumes the same values as $\mathcal{I}$ for each point of the domain. The technique is defined so that an atomic component of an integral data dependence is "extracted" at each application. The decomposition has to guarantee that such an extracted component is defined by an injective index mapping. Therefore we have two possibility. Either we can recognise injectivity of such a component and hence exploit it directly, or we cannot decide about its injectivity, hence we can exploit only geometric properties and Proposition 4.1 comes in hand. When injectivity is decidable we may apply:

Proposition 4.2 Let us consider an integral recurrence equation $z \in D : U(z) = f(V(\mathcal{I}(z)))$, where $D \subseteq \mathbf{Z}^n$ and $\mathcal{I}(z) = z + g(z)d + \sum_{j=1}^{m} g_j(z)d_j$. If $\mathcal{I}_1$, defined as $\mathcal{I}_1(z) = z + g(z)d$ for each $z \in D$, is injective over D then the equation is equivalent, over D, to the following system of integral recurrence equations:

$$\begin{aligned} z \in D: \quad U(z) = &\quad f(V_1(\mathcal{I}_1(z))) \\ z \in D_1: \quad V_1(z) = &\quad V(\mathcal{I}_2(z)), \end{aligned}$$

where V_1 is an auxiliary variable, $\mathcal{I}_2(z) = z + \sum_{j=1}^{m} g_j(\mathcal{I}_1^{-1}(z))d_j$, $\mathcal{I}_1^{-1}$ is a left inverse of $\mathcal{I}_1$ over D_1, and D_1 is the convex closure of $\mathcal{I}_1(D)$.

Proof: *[sketched]* It amounts to showing that it is possible to define a left-inverse of $\mathcal{I}_1$ over D_1 (which comes from the injectivity of $\mathcal{I}_1$) and that the composition of the new index functions $\mathcal{I}_2 \circ \mathcal{I}_1$ is equivalent, over D, to the original index function $\mathcal{I}$. ■ 4.2

When function invertibility cannot be exploited, the following two forms of decomposition are possible:

Proposition 4.3 Let us consider an integral recurrence equation $z \in D : U(z) = f(V(\mathcal{I}(z)))$, where $D \subseteq \mathbf{Z}^n$ and $\mathcal{I}(z) = z + g(z)d + \sum_{j=1}^{m} g_j(z)d_j$. If $d \notin lin(D)$ then the equation is equivalent, over D, to the system of integral recurrence equations:

$$\begin{aligned} z \in D: \quad U(z) = & \quad f(V_1(\mathcal{I}_1(z))) \\ z \in D_1: \quad V_1(z) = & \quad V(\mathcal{I}_2(z)), \end{aligned}$$

where V_1 is an auxiliary variable and the new domain D_1 is the convex closure of $\mathcal{I}_1(D)$. The index functions are $\mathcal{I}_1(z) = z+g(z)d$ and $\mathcal{I}_2(z) = z+\sum_{j=1}^{m} g'_j(z)d_j$, where each g'_j is defined as $g'_j(z) = g_j(z + l(z)(-d))$, with l the linear mapping $l(z) = (\pi \cdot z - \theta)/\eta$, $\pi \in (lin(D) + \langle d \rangle) \cap lin(D)^{\perp}$[8], $\pi \cdot z = \theta$ a hyperplane containing D and $\eta = \pi \cdot d$.

Proof: *[sketched]* For each $z \in D$, each g'_j is equal to $g_j(z)$ on all the points of D_1 on the half-line $\{z + ld \mid l \geq 0\}$ (i.e., the half-line "starting" at z, having the direction of d). Therefore each g' is well-defined over D_1. To prove the result we need to prove that the new system of equations is equivalent to the original equation over D, i.e., that for all $z \in D$, $\mathcal{I}_2 \circ \mathcal{I}_1(z) = \mathcal{I}(z)$. The proof makes use of Proposition 4.1 and consists of a simple algebraic rewriting of the expressions. ■ 4.3

Proposition 4.4 Let us consider an integral recurrence equation $z \in D : U(z) = f(V(\mathcal{I}(z)))$, where $D \subseteq \mathbf{Z}^n$ and $\mathcal{I}(z) = z + g(z)d + \sum_{j=1}^{m} g_j(z)d_j$. If $d \in lin(D)$ and $dim(D \cup \mathcal{I}(D)) < n$, then the equation is equivalent, over D, to the following system of integral recurrence equations:

$$\begin{aligned} z \in D: \quad U(z) = & \quad f(V_1(\mathcal{I}_1(z))) \\ z \in D_1: \quad V_1(z) = & \quad V_2(\mathcal{I}_2(z)) \\ z \in D_2: \quad V_2(z) = & \quad V(\mathcal{I}_3(z)), \end{aligned}$$

where V_1, V_2 are auxiliary variables, the new domain D_1 is the convex closure of $\mathcal{I}_1(D)$, and D_2 is the convex closure of $\mathcal{I}_2(D_1)$. The index functions are: $\mathcal{I}_1(z) = z + g(z)\hat{d}$, $\mathcal{I}_2(z) = z + \sum_{j=1}^{m} g'_j(z)d_j$ and $\mathcal{I}_3(z) = z + g'(z)(-\pi)$, where $\hat{d} = d + \pi$, $\pi \cdot z = \theta$ is a hyperplane containing D, $\pi \in lin(D \cup \mathcal{I}(D))^{\perp}$, and

[8] Where $\langle d \rangle$ denotes the space spanned by d and $lin(D)^{\perp}$ the space orthogonal to $lin(D)$.

g'_j, g' are defined as $g'_j(z) = g_j(z + l(z)(-\hat{d}))$, $g'(z) = l(z)$, with l the linear mapping $l(z) = (\pi \cdot z - \theta)/\eta$ and $\eta = \pi \cdot \hat{d}$.

Proof: *[sketched]* For reasons similar to the proof of Proposition 4.3, g'_j and g' are well-defined functions on their respective domains. To prove the result we need to prove that the system of equations above is equivalent to the original equation over D, i.e., that for all $z \in D$, $\mathcal{I}_3 \circ \mathcal{I}_2 \circ \mathcal{I}_1(z) = \mathcal{I}(z)$. This proof exploits the particular choice of vectors π and $\hat{d}$, makes use of Proposition 4.1 and consists of a simple algebraic rewriting of the expressions. ■ 4.4

Note that the application of Proposition 4.4 may require an increase in the space dimensionality to satisfy condition $dim(D \cup \mathcal{I}(D)) < n$. Also note that, if condition $d \notin lin(D)$ of Proposition 4.3 holds, then, because of Proposition 4.1, $\mathcal{I}_1$ is injective and Proposition 4.2 is also applicable. However, it may be the case that a left inverse of $\mathcal{I}_1$ is difficult to define, hence Proposition 4.3 is preferable. Moreover, Proposition 4.3 is more general in that it is based on geometric properties only, hence is more suitable for the development of automatic tools.

From the previous propositions we can derive the following *decomposition property*:

Theorem 4.5 Any integral recurrence equation $z \in D : U(z) = f(V(\mathcal{I}(z)))$, where $D \subseteq \mathbf{Z}^n$ and $\mathcal{I}(z) = z + \sum_{j=1}^{m} g_j(z)d_j$, can be rewritten as a system of atomic integral recurrence equations in $\mathbf{Z}^r$, with $r \leq (n + m)$.

Proof: *[sketched]* The result follows from the application of the decomposition techniques to each of the components of $\mathcal{I}$. As there are m components in the definition of $\mathcal{I}$ at most m increases of the space dimensionality are required. ■ 4.5

4.3 Parametric Uniformisation

After decomposition, uniformisation allows one to derive systems of uniform recurrences to which space-time mappings can be applied directly to obtain regular array designs. Uniformisation replaces a non-uniform atomic integral data dependence with an equivalent uniform pipelining/routing scheme of the

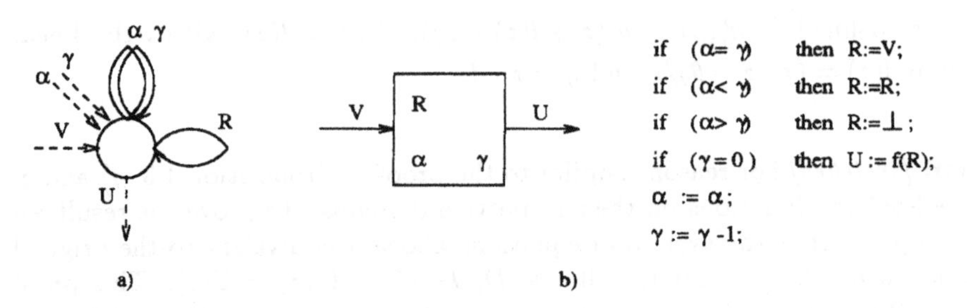

Fig. 2. For $u = d, \bar{g} > 0$: a) projection of the line $z + ld$, for $z \in D$; b) corresponding processing element.

data. As uniformisation is a form of localisation, the choice of the uniformisation direction vectors is related to the same type of geometric considerations that we have made in the previous section for decomposition.

Because of the convexity requirement on the equation domains (see the discussion in Section 2.4), control variables are introduced to identify the subsets of index points which constitute the routing paths replacing the original data dependence vectors. Like other variables, a control variable is defined by a subsystem of (in our case, uniform) recurrence equations. The control variables we introduce are initialised so that the length of each routing path corresponds to the length of the replaced data dependence vector (this length being the value of the integer function g on the corresponding point of the domain (see Section 3)).

Theorem 4.6 Let us consider an atomic integral recurrence equation $z \in D : U(z) = f(V(\mathcal{I}(z)))$, where $D \subseteq \mathbf{Z}^n$ and $\mathcal{I}(z) = z + g(z)d$. If $d \notin lin(D)$, there exists a system of conditional uniform recurrence equations in $\mathbf{Z}^n$ which is equivalent, over D, to the equation.

Proof: *[sketched]* The proof amounts to defining a system of uniform equations from the given equation and proving that they are equivalent. The system of equations may be defined as follows. Let $\bar{g}$ be the maximum value of g over D. If $\bar{g} = 0$ the equation is uniform as the index function trivially reduces to $I(z) = z$ for all $z \in D$. If $\bar{g} > 0$, consider: $\pi \cdot z = \theta$ a hyperplane containing the domain D such that π is a vector in the space $(lin(D) + \langle d \rangle) \cap lin(D)^{\perp}$ (where $\langle d \rangle$ denotes the space spanned by d and $lin(D)^{\perp}$ the space orthogonal to $lin(D)$); $\eta = \pi \cdot d$ and $D_1 = \{z + ld \mid z \in D, 0 \leq l \leq \bar{g}\}$. The system of equations (in which R, α and γ are new variables) is:

$$\begin{aligned}
z \in D_1, \gamma(z) = 0 : \quad & U(z) = && f(R(z)) \\
z \in D_1, \alpha(z) > \gamma(z) : \quad & R(z) = && \perp \\
z \in D_1, \alpha(z) < \gamma(z) : \quad & R(z) = && R(z+d) \\
z \in D_1, \alpha(z) = \gamma(z) : \quad & R(z) = && V(z) \\
z \in D_1, \pi \cdot z < \eta\bar{g} + \theta : \quad & \alpha(z) = && \alpha(z+d) \\
z \in D_1, \pi \cdot z = \eta\bar{g} + \theta : \quad & \alpha(z) = && g(z - \bar{g}d) \\
z \in D_1, \pi \cdot z < \eta\bar{g} + \theta : \quad & \gamma(z) = && \gamma(z+d) - 1 \\
z \in D_1, \pi \cdot z = \eta\bar{g} + \theta : \quad & \gamma(z) = && \bar{g}.
\end{aligned}$$

The proof that this system of equations is equivalent to the initial equation, makes use of Proposition 4.1 and consists of a simple algebraic rewriting of the expressions. ■ 4.6

The system of equations in the proof of Theorem 4.6 defines the following routing scheme, which uses a routing variable R and control variables α and γ. For each point z in the domain D, the value $V(\mathcal{I}(z))$ is collected by R (when α and γ assume the same value) and pipelined along the direction of d to the point z, where $U(z)$ is computed. The control variable α carries the value $g(z)$, while γ acts as a counter, initially set to the maximum value $\bar{g}$ of g over D, and decremented by one at each step.

As such a system of equations is uniform, given an affine timing function t, a regular design can be derived by projecting the system according to any direction u compatible with t. In particular, let us consider the case in which vector d is chosen as the projection direction. In this case, for each $z \in D$, all computation points on the line $z + ld$, with l ranging over $\mathbf{Z}$, are mapped to a single node in the processor space, with their data dependencies mapped to channels as illustrated in Fig. 2.a. In the figure, each channel is characterised by a zero delay (we have assumed $|\lambda \cdot d| = 1$) and dashed arrows indicate input/output channels (we have assumed that V is an input and U an output). The corresponding processing element and its behaviour are given in Fig. 2.b. Note that the behaviour of the cell is dictated by the control variables α and γ, while R functions as a buffer of capacity one.

The second uniformisation technique is given in the following theorem, which, instead of presenting the basic case, introduces its parametrized form. The idea behind parameterisation is to allow the designer to specify the maximum admissible amount of overloading per node (i.e., the maximum acceptable number of abstract channels per node, this value being the parameter). In turn,

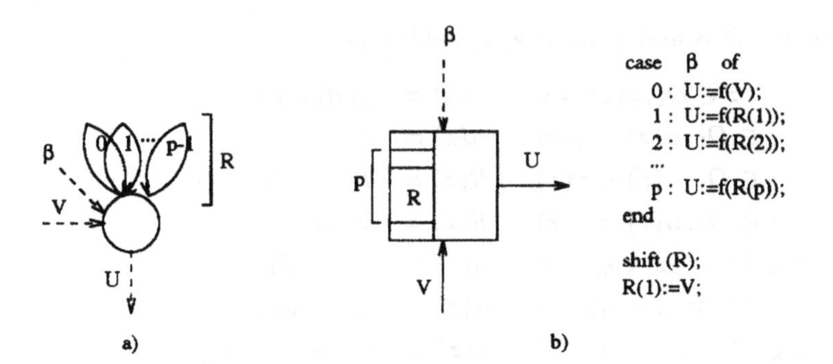

Fig. 3. For $u = d, \bar{g} = 0$: a) projection of the line $z + ld$, for $z \in D$; b) corresponding processing element.

this parameter translates onto some memory/bandwidth requirement for the cells of the resulting regular array designs. Indeed a unit parameter indicates that no overloading is admissible.

Theorem 4.7 Let us consider an atomic integral recurrence equation $z \in D : U(z) = f(V(\mathcal{I}(z)))$, where $D \subseteq \mathbf{Z}^n$, $\mathcal{I}(z) = z + g(z)d$, $d \in lin(D)$ and $dim(D) < n$. Let $p \in \mathbf{P}$, where $\mathbf{P} \subseteq \mathbf{N}^+$. There exists a system of conditional uniform recurrence equations in $\mathbf{Z}^n$ parameterised with respect to p, which is equivalent, over D, to the equation.

Proof: *[sketched]* The proof amounts to defining a system of uniform equations from the given equation and proving that they are equivalent. The system of equations may be defined as follows. Let $g_{max} = \max_{z \in D} g(z)$ and $\bar{g} = \lfloor g_{max}/(p+1) \rfloor$.

If $\bar{g} = 0$, define the following system of equations (where R and β are new variables):

$$\begin{aligned} z \in D: \quad & U(z) = && f(R(z)) \\ z \in D, \beta(z) = 0: \quad & R(z) = && V(z) \\ z \in D, \beta(z) = 1: \quad & R(z) = && V(z + d) \\ & \dots \\ z \in D, \beta(z) = p: \quad & R(z) = && V(z + pd) \\ z \in D: \quad & \beta(z) = && g(z) mod_{p+1}. \end{aligned}$$

If $\bar{g} > 0$, consider: $\pi \cdot z = \theta$ a hyperplane containing the domain D such that π is a vector in the space $lin(D)^{\perp}$ (where $lin(D)^{\perp}$ the space orthogonal to $lin(D)$); $\eta = \pi \cdot \hat{e}$; $\hat{e} = d + \pi$ and $\check{e} = d - \pi$; $D_1 = \{z + l\hat{e} \mid z \in D, 0 \leq l \leq \bar{g}\}$; and $D_2 = \{z + (l_1 + l_2)d + l_3\check{e} \mid z \in D_1, 0 \leq l_1 \leq p, 0 \leq l_2 \leq (p-1)\bar{g}, 0 \leq l_3 \leq \bar{g}\} \cap \{z \in \mathbf{Z}^n \mid \pi \cdot z \geq \theta\}$. The system of equations (in which $R^1, R^2, R^3, \alpha, \beta$ and γ are new variables) is:

$$\begin{array}{rrl}
z \in D: & U(z) = & f(R^1(z)) \\
\\
z \in D_1, \gamma(z) < \alpha(z): & R^1(z) = & R^1(z + \hat{e}) \\
z \in D_1, \gamma(z) > \alpha(z): & R^1(z) = & \perp \\
z \in D_1, \gamma(z) = \alpha(z), \beta(z) = 0: & R^1(z) = & R^2(z) \\
z \in D_1, \gamma(z) = \alpha(z), \beta(z) = 1: & R^1(z) = & R^2(z + d) \\
& \cdots & \\
z \in D_1, \gamma(z) = \alpha(z), \beta(z) = p: & R^1(z) = & R^2(z + pd) \\
z \in D_2, \pi \cdot z > \theta: & R^2(z) = & R^3(z + \check{e}) \\
z \in D_2, \pi \cdot z = \theta: & R^2(z) = & V(z) \\
z \in D_2, \pi \cdot z \geq \theta: & R^3(z) = & R^2(z + (p-1)d) \\
\\
z \in D_1, \pi \cdot z < \eta\bar{g} + \theta: & \alpha(z) = & \alpha(z + \hat{e}) \\
z \in D_1, \pi \cdot z = \eta\bar{g} + \theta: & \alpha(z) = & \lfloor g(z - \bar{g}\hat{e})/(p+1) \rfloor \\
z \in D_1, \pi \cdot z < \eta\bar{g} + \theta: & \beta(z) = & \beta(z + \hat{e}) \\
z \in D_1, \pi \cdot z = \eta\bar{g} + \theta: & \beta(z) = & g(z - \bar{g}\hat{e}) mod_{p+1} \\
z \in D_1, \pi \cdot z < \eta\bar{g} + \theta: & \gamma(z) = & \gamma(z + \hat{e}) - 1 \\
z \in D_1, \pi \cdot z = \eta\bar{g} + \theta: & \gamma(z) = & \bar{g}.
\end{array}$$

In both cases above, the proof that the system of equations is equivalent to the initial equation, exploits the particular choice of vectors π, $\hat{e}$ and $\check{e}$, makes use of Proposition 4.1 and consists of a simple algebraic rewriting of the expressions. ■ 4.7

When $\bar{g} = 0$, the "parameterised" overloading is expressed by the control variable β, which assumes values in the range $[0..p]$, each value corresponding to a distinct abstract channel. Note that, as g is a function, only one of these channels is devoted to each point of the domain. When $\bar{g} > 0$, a routing scheme is defined, which uses routing variables R^1, R^2 and R^3, and control variables α, β and γ. For each point z in the domain D, the value $V(\mathcal{I}(z))$ is collected by R^2, subsequently transferred (possibly several times) between variables R^3

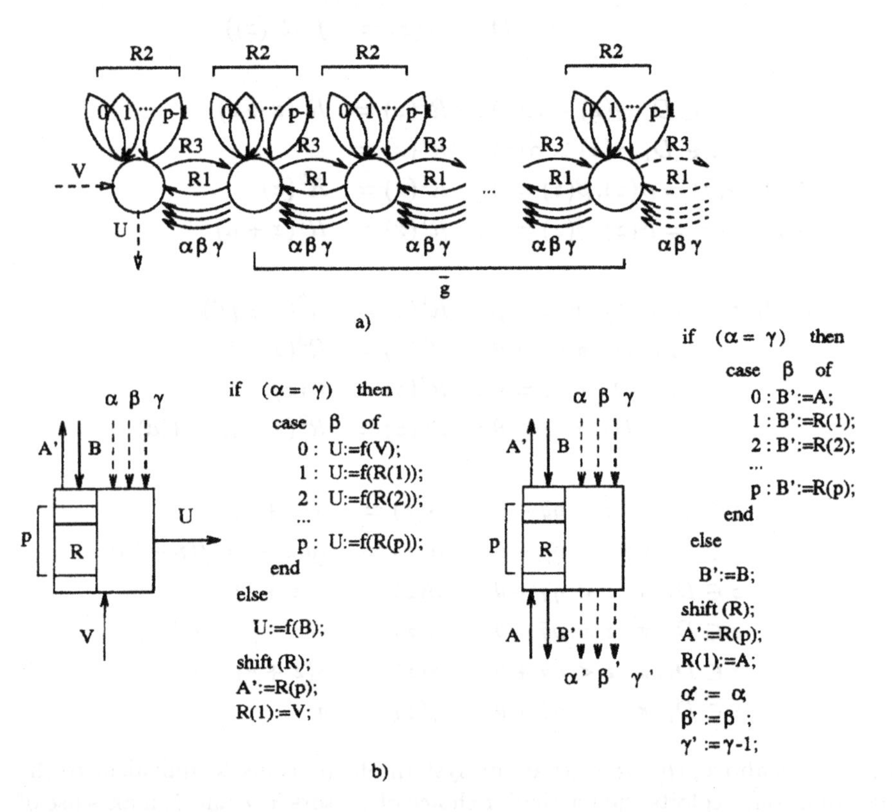

Fig. 4. For $u = d, \bar{g} > 0$: a) projection of the plane $z + l_1\hat{e} + l_2\check{e}$, for $z \in D$; b) corresponding processing elements.

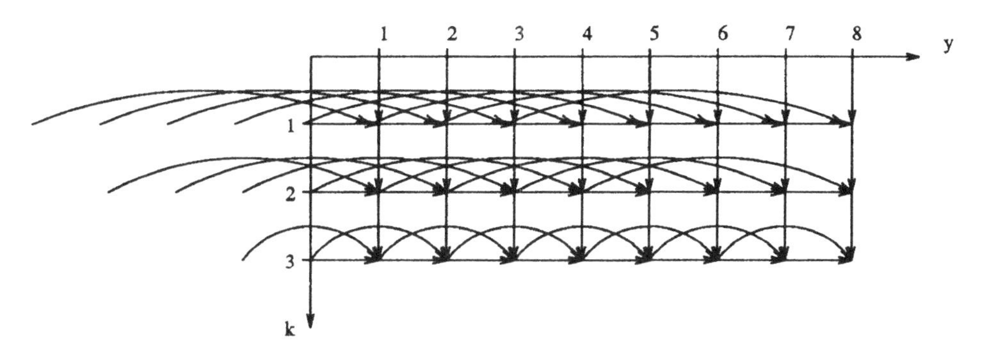

Fig. 5. Initial data dependence graph for the Knapsack problem.

and R^2, and finally transferred to R^1, which pipelines it along the direction of $\hat{e}$ to the point z where $U(z)$ is computed. The actual shape of the routing path varies according to the function g and the value of the parameter p, and is determined by the control variables α, β and γ.

Let us consider a projection of the systems of equations in the proof of Theorem 4.7 according to the vector d. When $\bar{g} > 0$, for each $z \in D$, all computation points on the plane $z + l_1\hat{e} + l_2\check{e}$, with l_1, l_2 ranging over $\mathbf{Z}$, are mapped to a set of $\bar{g} + 1$ nodes in the processor space as illustrated in Fig. 4.a. The leftmost of these nodes performs the actual computation of U, while the remaining $\bar{g}$ elements route the data, i.e., they function as delay elements according to the value of the control signals α, β and γ (which, in turn, depend on the value of the function g). The delay is achieved both by transferring the data between adjacent cells and using the channels at each cell, characterised by delays ranging between 0 and $p - 1$. These channels may be realised by the local memory of each cell: a RAM memory of size p allows one to simulate the p channels, with the values in the memory locations "shifted" of one position at each clock cycle. The behaviour of the cells is given in Fig. 4.b. When $\bar{g} = 0$, by projecting the system according to d, for each $z \in D$, the line $z + ld$ is mapped onto a processing element as illustrated in Fig. 3.

5 THE KNAPSACK PROBLEM

As an example of integral recurrence equations and their localisation, we consider the knapsack problem ([7, 6]), a classic combinatorial optimisation problem, which consists[9] of determining the optimal (i.e., the most valuable) selection of objects of given weight and value to carry in a knapsack of finite weight capacity. If c is a non-negative integer denoting the capacity of the knapsack, n the number of object types available, w_k and v_k, respectively, the weight and value of an object of type k, for $1 \leq k \leq n$ and $w_k > 0$ and integral, a dynamic programming formulation of the problem is represented by the following system of recurrences[10]:

$$\begin{aligned}
(k,y) \in D_1 : \quad F(k,y) &= \quad 0 \\
(k,y) \in D_2 : \quad F(k,y) &= \quad 0 \\
(k,y) \in D_3 : \quad F(k,y) &= \quad -\infty \\
(k,y) \in D_4 : \quad F(k,y) &= \quad f(F(k-1,y), F(k,y-w_k), V(k,y)) \\
(k,y) \in D_2 : \quad V(k,y) &= \quad v_k \\
(k,y) \in D_4 : \quad V(k,y) &= \quad V(k,y-1),
\end{aligned}$$

with f defined as $f(a,b,c) = max(a, b+c)$, for all a, b, c, and equation domains

$$\begin{aligned}
D_1 &= \{(k,y) \mid k = 0, 1 \leq y \leq c\} \quad & D_2 &= \{(k,y) \mid 1 \leq k \leq n, y = 0\} \\
D_3 &= \{(k,y) \mid 1 \leq k \leq n, y < 0\} \quad & D_4 &= \{(k,y) \mid 1 \leq k \leq n, 1 \leq y \leq c\}.
\end{aligned}$$

In this system, $F(k,y)$ represents the optimal solution of a knapsack sub-problem with k types of objects and knapsack capacity y. Hence, for k and y in the ranges $[1..n]$ and $[1..c]$, respectively, the optimal solution is given by $F(n,c)$. As boundary conditions, F assumes value 0 for $y = 0$ or $k = 0$, and value $-\infty$ for $y < 0$. The variable V carries the values v_k of the objects through the computation domain D_4, while the weights w_k of the objects define an atomic integral data dependence given by the index function $\mathcal{I}(k,y) = (k, y - w_k) = (k,y) + w_k(0,-1) = (k,y) + g(k,y)d$. All other data dependencies of the system are uniform. The data dependence graph of the system is given in Fig. 5, where we assume $c = 8, n = 3, w_1 = 5, w_2 = 4$ and $w_1 = 2$.

Parametric uniformisation, after re-indexing the equations in $\mathbf{Z}^3$ and by choosing $\pi = (0,0,1)$, delivers the system:

$$(k,y,z) \in D_1' : \quad F(k,y,z) = \quad 0$$

[9] This is one of the several variants of the knapsack problem. A complete presentation together with a number of applications is given in [7].

[10] This system of recurrences corresponds to the so-called forward phase of the knapsack problem in which the optimal carried value is computed. The corresponding combination of objects can be determined from the optimal solution in a second phase, known as the backward phase of the algorithm, which is essentially sequential and mainly consists of a backward substitution process on sub-optimal values of F. An efficient algorithm for the backward phase is given in [6].

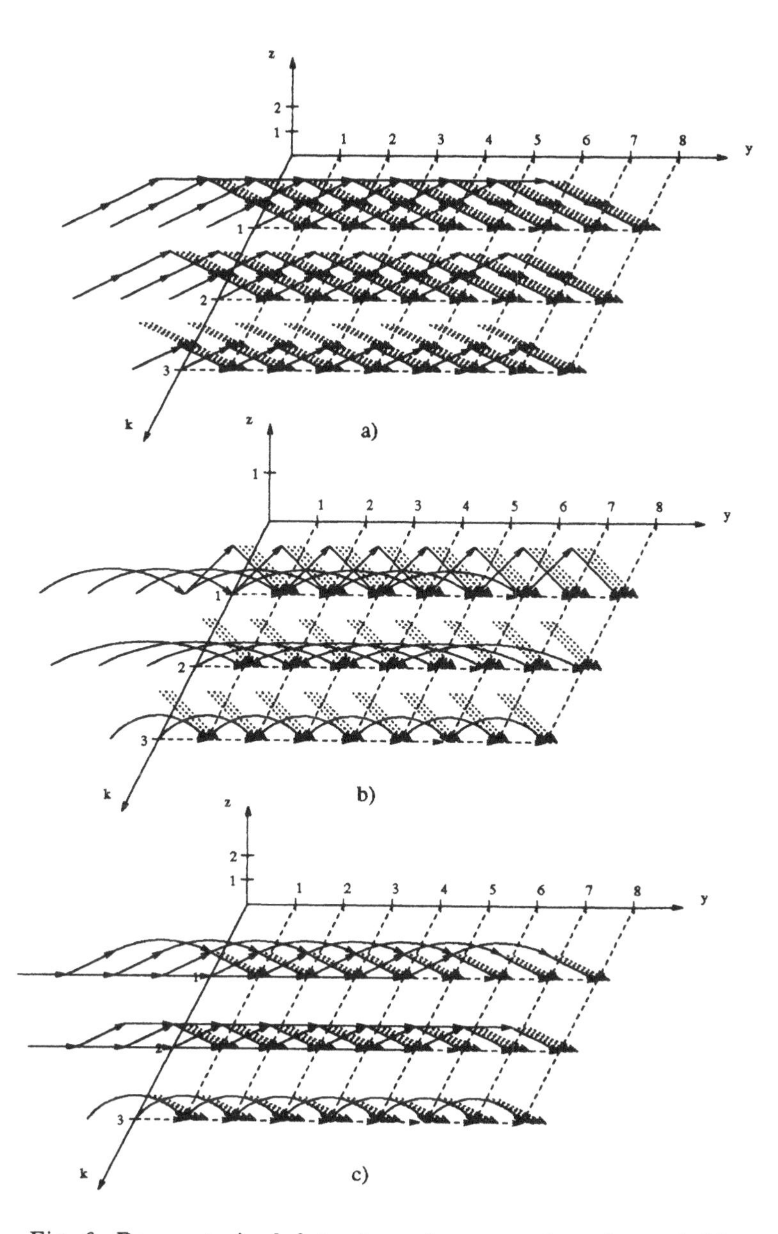

Fig. 6. Parameterised data dependence graphs: a) $p = 1$; b) $p = 2$; c) $p = 4$.

$$
\begin{aligned}
(k,y,z) \in D'_2: \quad F(k,y,z) &= 0 \\
(k,y,z) \in D'_3: \quad F(k,y,z) &= -\infty \\
(k,y,z) \in D'_4: \quad F(k,y,z) &= f(F(k-1,y,z), R^1(k,y,z), V(k,y,z)) \\
(k,y,z) \in D'_2: \quad V(k,y,z) &= v_k \\
(k,y,z) \in D'_4: \quad V(k,y,z) &= V(k,y-1,z)
\end{aligned}
$$

$$
\begin{aligned}
(k,y,z) \in D'_{4,1}, & \\
\gamma(k,y,z) < \alpha(k,y,z): \quad R^1(k,y,z) &= R^1(k,y-1,z+1) \\
(k,y,z) \in D'_{4,1}, & \\
\gamma(k,y,z) > \alpha(k,y,z): \quad R^1(k,y,z) &= \infty \\
(k,y,z) \in D'_{4,1}, & \\
\gamma(k,y,z) = \alpha(k,y,z), \beta(k,y,z) = 0: \quad R^1(k,y,z) &= R^2(k,y,z) \\
(k,y,z) \in D'_{4,1}, & \\
\gamma(k,y,z) = \alpha(k,y,z), \beta(k,y,z) = 1: \quad R^1(k,y,z) &= R^2(k,y-1,z) \\
\ldots & \\
(k,y,z) \in D'_{4,1}, & \\
\gamma(k,y,z) = \alpha(k,y,z), \beta(k,y,z) = p: \quad R^1(k,y,z) &= R^2(k,y-p,z) \\
(k,y,z) \in D'_{4,2}, z > 0: \quad R^2(k,y,z) &= R^3(k,y-1,z-1) \\
(k,y,z) \in D'_{4,2}, z = 0: \quad R^2(k,y,z) &= V(k,y,z) \\
(k,y,z) \in D'_{4,2}, z \geq 0: \quad R^3(k,y,z) &= R^2(k,y-(p-1),z)
\end{aligned}
$$

$$
\begin{aligned}
(k,y,z) \in D'_{4,1}, z < \bar{g}: \quad \alpha(k,y,z) &= \alpha(k,y-1,z+1) \\
(k,y,z) \in D'_{4,1}, z = \bar{g}: \quad \alpha(k,y,z) &= \lfloor g(k,y+\bar{g},z-\bar{g})/(p+1) \rfloor \\
(k,y,z) \in D'_{4,1}, z < \bar{g}: \quad \beta(k,y,z) &= \beta(k,y-1,z+1) \\
(k,y,z) \in D'_{4,1}, z = \bar{g}: \quad \beta(k,y,z) &= g(k,y+\bar{g},z-\bar{g}) mod_{p+1} \\
(k,y,z) \in D'_{4,1}, z < \bar{g}: \quad \gamma(k,y,z) &= \gamma(k,y-1,z+1) - 1 \\
(k,y,z) \in D'_{4,1}, z = \bar{g}: \quad \gamma(k,y,z) &= \bar{g},
\end{aligned}
$$

where $\bar{g} = \lfloor w_{max}/(p+1) \rfloor$ and w_{max} is the maximum weight of the objects. The new domains are defined as:

$$
\begin{aligned}
D'_1 &= \{(k,y,z) \mid k = 0, 1 \leq y \leq c, z = 0\} \\
D'_2 &= \{(k,y,z) \mid 1 \leq k \leq n, y = 0, z = 0\} \\
D'_3 &= \{(k,y,z) \mid 1 \leq k \leq n, y < 0, z = 0\} \\
D'_4 &= \{(k,y,z) \mid 1 \leq k \leq n, 1 \leq y \leq c, z = 0\}
\end{aligned}
$$

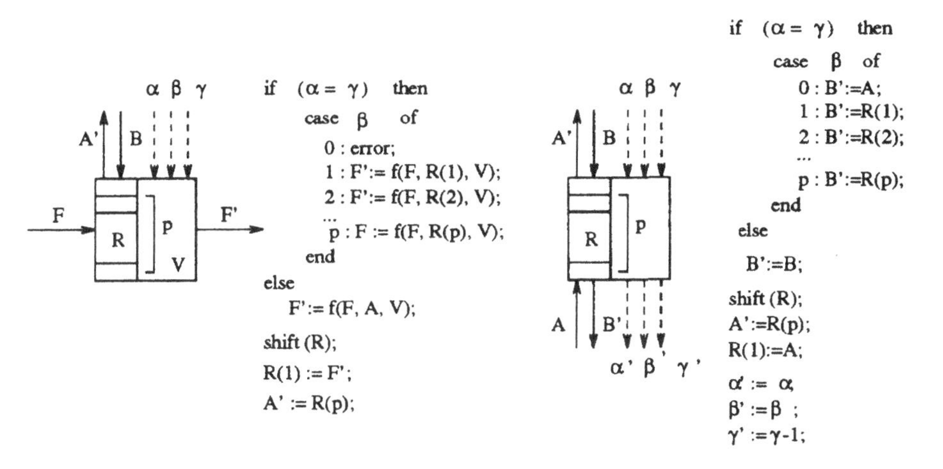

Fig. 7. Parameterised processing elements.

$$D'_{4,1} = \{(k, y, z) \mid 1 \le k \le n, 1 - z \le y \le c - z, 0 \le z \le \bar{g}\}$$
$$D'_{4,2} = \{(k, y, z) \mid 1 \le k \le n, pz + (1 - p) - (p + 1)\bar{g} \le y \le c - z, 0 \le z \le \bar{g}\}$$

For values of the parameter $p = 1, 2, 4$, the corresponding data dependence graphs are illustrated in Fig. 6, where the dotted arrows correspond to control variables (as before, we assume $c = 8, n = 3, w_1 = 5, w_2 = 4$ and $w_1 = 2$). An affine scheduling for the system of equations is $t(k, y, z) = k + y - 1 + \beta$, where $\beta = (p+1)\bar{g}+p$, with $\bar{g} = 2, 1, 1$ and $\beta = 5, 5, 9$ for $p = 1, 2, 4$, respectively. This scheduling allows us to project the whole system according to $u = (0, -1, 0)$ and generate the regular array designs of Fig. 8, which are characterised by processing elements of the two types depicted in Fig. 7. The first type of cell both computes the value of F and routes the result, while the second type of cell performs only routing functions. In Fig. 8, corresponding snapshots of the data and control flow are also given ("-" denotes $-\infty$). Note that the array designs are extensible with respect to all problem parameters c, n and $\bar{g}$, for any choice of p.

6 CONCLUSIONS

In this paper we have shown that algorithms specified as systems of integral recurrence equations can be systematically mapped onto regular arrays. This is realised by combining established techniques for the space-time mapping of uniform and affine data dependencies onto processor arrays, with systematic ways

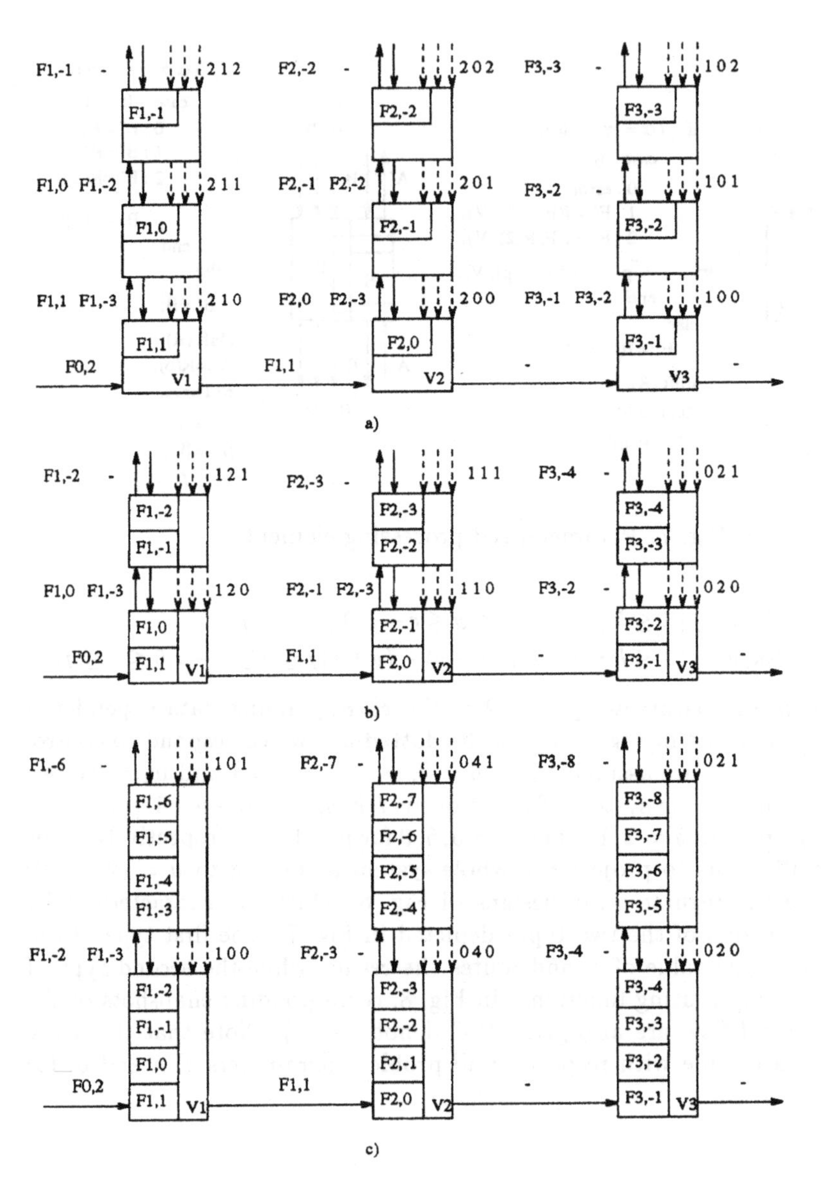

Fig. 8. Regular array designs: a) $p = 1$; b) $p = 2$; c) $p = 4$.

of regularising the data dependencies through powerful forms of localisation. In particular, through decomposition and uniformisation complex non-regular data dependencies can be replaced by a uniform routing of the data. Moreover, parameterisation adds flexibility to the design process allowing a better match between design features and architectural requirements. A practical advantage of the approach is that it consistently generalises established synthesis techniques for regular classes of problems, hence facilitates the upgrading of existing software environments for the synthesis of regular array designs to the systematic treatment of integral recurrences.

The method we have proposed cannot guarantee an optimal rewriting of the recurrences which minimises the space-time complexity of the arrays. Instead it provides a systematic framework for exploring the design space. The location of optimal solutions necessarily deals with combinatorial issues and demands approximate solutions or designer intervention.

Acknowledgements

This work was supported by REFLEX, Science and Engineering Research Council (SERC) grant no. GR/H46725.

REFERENCES

[1] P.R.Cappello, K.Steiglitz, *Unifying VLSI array design with linear transformations of space-time.* Advances in Computing Research, vol. 2, pp. 23-65, 1984.

[2] B.Carré, *Graphs and Networks.* Oxford Applied Mathematics and Computing Science Series, Oxford University Press, 1979.

[3] M.C.Chen, *A design methodology for synthesizing parallel algorithms and architectures.* Journal of Parallel and Distributed Computing, vol. 3, no. 4, pp. 461-491, 1986.

[4] J.A.B.Fortes, D.I.Moldovan, *Parallelism detection and transformation techniques useful for VLSI algorithms.* Journal of Parallel and Distributed Computing, vol. 2, pp. 277-301, 1985.

[5] D.Gelernter, A.Nicolau, D.Padua (eds.), *Languages and compilers for parallel computing.* Research Monograph in Parallel and Distributed Computing, Pitman, The MIT Press, 1990.

[6] T.C.Hu, *Combinatorial Algorithms.* Addison-Wesley Publishing Company, 1982.

[7] S.Martello, P.Toth, *Knapsack Problems: Algorithms and Computer Implementation.* John Wiley and Sons, 1990.

[8] G.M.Megson, *An Introduction to Systolic Algorithm Design.* Oxford Science Publications, 1992.

[9] W.L.Miranker, A.Winkler, *Spacetime representations of computational structures.* Computing, vol. 32, pp. 93-114, 1984.

[10] E.D.Nerig, *Linear algebra and matrix theory.* J.Wiley & Sons Inc., 1963.

[11] P.Quinton, "Automatic synthesis of systolic arrays from uniform recurrent equations", in *IEEE/ACM Proc. 11th Annual International Symposium on Computer Architecture*, 1984.

[12] P.Quinton, and Y.Robert, *Systolic algorithms and architectures.* Masson and Prentice Hall International, 1991.

[13] P.Quinton, and V.Van Dongen, "The mapping of linear recurrence equations on regular arrays", *J. of VLSI Signal Processing*, vol. 1, pp. 95-113, 1989.

[14] S.V.Rajopadhye, and R.M.Fujimoto, "Systolic array synthesis by static analysis of program dependencies", *PARLE - Parallel Architecture and Languages Europe*, Lecture Notes in Computer Science, vol. 258, pp. 295-310, Springer Verlag, 1987.

[15] S.V.Rajopadhye, and R.M.Fujimoto, "Automating systolic array design", *Integration - The VLSI Journal*, vol. 9, pp. 225-242, 1990.

[16] S.K.Rao, *Regular Iterative Algorithms and their Implementation on Processor Arrays.* PhD Thesis, Stanford University, 1985.

[17] J.Teich, and L.Thiele, "Control generation in the design of processor arrays", *J. of VLSI Signal Processing*, vol. 3, no. 1/2, pp. 77-92, 1991.

[18] J.Xue, *The formal synthesis of control signals for systolic arrays.* The University of Edinburgh, PhD Thesis, CTS-90-92, April 1992.

16

SOME STRATEGIES FOR LOAD BALANCING

Claude G. Diderich and Marc Gengler

Swiss Federal Institute of Technology - Lausanne
Computer Science Department
CH-1015 Lausanne, Switzerland

ABSTRACT

In this paper we discuss the needs for load balancing, also called scheduling. We exhibit different reasons that render static (compile-time) scheduling impossible and that determine the dynamic (run-time) load balancing schemes needed in order to get efficient parallel algorithms. One distinguishes between local load balancing policies where processors base their decisions on information about the load in some neighborhood and global load balancing policies where processors base their decisions on the load of the entire machine. Depending on the static information available and on the dependencies between the different tasks, some parallel algorithms accommodate with simple load balancing or load sharing mechanisms while others need more sophisticated solutions. The former are typically local while the later are global load balancing schemes. In particular, we analyze the branch and bound algorithm and show that it needs smart load balancing mechanisms ideally founded on global knowledge. We argue that for this algorithm a global load balancing policy may be interesting. Indeed, the best-first branch and bound algorithm can be defined as a sequence of independent computations allowing the design of a parallel algorithm that alternates between coarse grained parallel computation phases and so-called synchronization phases which provide perfect global load balancing.

1 INTRODUCTION

In sequential processing based on the Von Neumann model of computation [25], the sequence of computation steps is entirely determined by the program to be executed or is defined by the compiler or interpreter, in the case a so-

A. Ferreira and J. D. P. Rolim (eds.), Parallel Algorithms for Irregular Problems: State of the Art, 323–338.

urce program does not entirely specify the order of execution like for instance functional or logical programs. There is a unique flow of control. In parallel processing the situation is fundamentally different as there exist in general many flows of control or at least operation instances computed concurrently. The computational entities may execute many operations at the same time. Depending on the parallel execution model, the set of concurrent operations allowed to execute at a same moment is more or less restricted. In the SIMD model [8] all processing elements apply the same operation synchronously to different data items. In the MIMD model, processors are allowed to execute operations independently one from the other.

Thus, there comes an additional problem into play which consists in deciding what operations can be performed in parallel or what data can be computed in parallel. This is where load balancing or task scheduling problems come from. In the SIMD model only the data have to be distributed on the processors. In addition, in the MIMD model, tasks have to be allocated to processors. In the rest of this paper we will not distinguish between task scheduling and data distribution. As a task is a piece of executable code together with its local data, one can assume that all the code chunks are known to all processors and that only data need to be distributed on the processors. Consequently, we will interchangeably speak of load balancing and load sharing or of task scheduling. The model we consider is thus implicitly SPMD.

In this paper we address the problem of load balancing on distributed memory machines from a general point of view. Load balancing may be defined at compile-time if there is enough information available to decide before the actual execution what processor will be responsible for what computation. In this case, we say that the load balancing is static. Very often, it is impossible or only partly possible to allocate computations to processors at compile-time and we need to decide at run-time what computations have to be done on a given processor. The load balancing is said to be dynamic in this case. We will first uncover the fundamental reasons that imply a need for dynamic load balancing mechanisms. In a second step we will present and discuss some solutions for solving the load balancing problems for parallel branch and bound algorithms [11, 24, 31]. Thus, the solutions we will propose are more specifically designed for a particular problem. Coming up with a general solution for all kinds of problems is not very interesting because such a general solution will be too vague and must be reworked anyway to take into account the particularities of the given problem. A discussion that starts from the general problem settings in order to propose possible solutions for a given problem will be more valuable as it shows, hopefully, a rather general approach for designing load balancing schemes.

2 ANALYZING THE IMBALANCE

A static allocation of computations to processors is only possible when enough compile-time information is available. Basically, we need to know the following quantities:

- the set of all the tasks or computations to be done,
- their respective execution times,
- the dependencies between the tasks.

With these information, it is possible to decide on which processors tasks should be executed, what message passing is needed when a distributed memory machine is used and even, ideally, compute the execution time of the parallel algorithm for a given number of processors and a given size of the input data. For certain classes of programs like most of the numerical codes such knowledge on the program is readily available and the parallelization of the program can statically determine an optimal allocation. Such allocation are also possible when the exact execution times are not known but can be approximated sufficiently well so as to be considered exact times. Even when the set of tasks is unknown at compile-time, static allocation remains possible if the principle by which tasks are created is known at compile-time allowing to generate code that will realize the perfect allocation at run-time.

2.1 The Static Load Balancing Problem

Static allocation is possible in a pretty large number of situations. This is feasible at least in theory as the general task scheduling problem is known to be NP-hard [26]. The following decision problem known as the *precedence constrained scheduling* found in [9] was proved NP-complete by Ullman in [35].

Instance: Set T of tasks each having unit length, number $m \in Z^+$ of processors, partial order $\prec$ on T and a deadline $D \in Z^+$.

Question: Is there and m-processor schedule σ for T that meets the overall deadline D and obeys the precedence constraints, i.e. a function $\sigma : T \rightarrow Z^+$ such that 1) for all $u \geq 0$, the number of tasks $t \in T$ for which $\sigma(t) \leq u < \sigma(t) + 1$ is no more than m, 2) for all $t \in T$, $\sigma(t) + 1 \leq D$ and 3) $t \prec t'$ implies $\sigma(t') > \sigma(t)$.

Many problems aren't however sufficiently regular to allow static allocation of their tasks or data. This is the case whenever one of the three conditions given previously does not hold. First, the set of tasks or the principle by which tasks are created may not be known. In this case, obviously we cannot decide where to put the tasks. Even when the tasks are known, static allocation is impossible whenever the execution times cannot be determined at compile-time. And finally, static allocation remains unfeasible even if the two first conditions are met, but the dependencies between the tasks are unknown. Such a situation does not really seem to correspond to any reasonable programs, but there are many irregular problems for which the dependencies must be computed at run-time. Let us illustrate this point, which in our opinion was too often ignored, by comparing two irregular problems, namely the N-queens problem and the best-first branch and bound algorithm.

2.2 Examples

The N-queens problem

The N-queens problem is derived from the game of chess and consists in putting N queens on an $N \times N$ chess board in such a way that no queen is attacked by any other. Clearly, there will be one queen per row and per column placed in such a way that no diagonal is occupied by more than one queen. Searching the solutions to this problem yields irregular programs that are very time-consuming (on an 8×8 chess board there are already 92 solutions if we count the symmetric solutions and those obtained by rotations). A straight forward algorithm to solve this problem consists in exploring a search tree that is constructed according to the following principles: Any node node in the tree represents a (partial) configuration. The root node corresponds to the empty chess board and nodes at depth m represent a configuration where the m first rows contain each one queen, no queen attacking another queen. The tree is constructed in the following way. We select a partial configuration and, supposing that it is at depth m, we generate all its possible descendant nodes, which will be at depth $m + 1$, by constructing all configurations that contain a queen in row $m + 1$ and do not yield any attack. This set of descendants may be empty corresponding to an initial configuration that cannot be extended to a complete configuration. Problems are expanded until they don't have any descendants or their descendants correspond to complete configurations or solutions obtained at level N in the tree.

The algorithm we obtain is very simple but also rather irregular. There is certainly no way to know the tasks or partial configurations in advance. The execution times for exploring a sub-tree are also random as we do not know the size of the sub-tree rooted at a given configuration. Concerning the dependencies between configurations, things are much simpler. There are simply no dependencies between partial configurations that exist at any moment. Hence, the nodes of the tree can be expanded in just any order. This last point makes this algorithm easily parallelizable. It is indeed sufficient to allocate different sub-trees to different processors and let processors evolve independently one from the other. This parallelization will do exactly the same amount of work as the sequential algorithm. Nevertheless, it will probably not yield good speedups as the work load, i.e. the number of configurations, inspected by the different processors will greatly vary. In order to solve this load imbalance problem we can add a simple load balancing mechanism by means of which idle processors query work from busy processors. Other more sophisticated schemes such as load balancing policies which try to give roughly the same amount of work to each processor or, more generally, which try to equilibrate the load of the processors according to some given criterion, can also be used. There are many possible approaches [2, 7, 21] and we will later discuss such policies in detail.

The branch and bound problem

The branch and bound algorithm is much more difficult to parallelize efficiently. The additional complexity comes from the dependencies that exist between the different tasks or data items. The branch and bound algorithm is an algorithm for solving combinatorial optimization problems. It is mostly applied to NP-hard problems [9] in order to compute the optimal solution for such problems. Many research efforts try to find good approximate solutions for NP-hard problems using approaches like simulated annealing [15], neural networks [22] or genetic algorithms [33]. We will only give an informal description of the branch and bound algorithm and refer the reader to [11, 24, 31] for more detailed descriptions as well as to [10, 32] for parallel versions of the algorithm.

The branch and bound algorithm searches a state space S in order to find one state $s \in S$ which is optimal according to some given cost function f. Stated formally and more generally, it computes a non empty subset of the set $S^* = \{s \in S \mid f(s) \prec f(s'); \forall s' \in S\}$ where $\prec$ is a total order. Without loss of generality, $\prec$ can be assumed to be $\leq$ yielding minimization problems. The most classic example is the traveling salesman problem [17] which consists in finding a shortest roundtrip through a number of cities represented by a weighted complete directed graph in which nodes correspond to cities and the weighted arcs to distances [12, 13].

The branch and bound algorithm works similarly to the N-queens algorithm. It develops a tree in which each node represents a partial solution. The root node represents all possible roundtrips through the cities. Nodes are *branched* into new nodes which means that partial solutions are split into disjoint sub-solutions. This process is iterated until the solutions are complete, i.e. elements of the state space S. For the traveling salesman problem the branching may for example consist in considering a partial solution, choose two towns A and B which aren't connected already and create three new problems corresponding to the choices to go from A to B, from B to A or to forbid both of these two arcs. One at least of these three sub-problems will lead to the optimal roundtrip.

The branch and bound algorithm uses two functions called lower bound and upper bound which allow it to discard part of the search tree. A lower bound for a given problem P is any value that does not exceed the optimal solution that could be found when exploring the entire tree rooted at P. The upper bound function associates to a problem P the value of a feasible solution one can reach when completing the partial solution P. The notion of upper bound is naturally extended to the notion of best solution found so far. Any partial solution can be discarded whenever its lower bound exceeds the best current solution. This allows to cut off parts of the tree, i.e. avoid exploring some sub-trees, and reduces the complexity of the computation.

The lower bounds also allow to define a tree traversal strategy known as best first search. This strategy, or more precisely family of strategies, always selects one of the sub-problems with the best lower bound as a next candidate for branching. One may show that a best first strategy explores less nodes than any other strategy in the case one searches all best solutions [11]. When searching only for one optimal solution, this cannot be guaranteed anymore because of the non deterministic choice among equivalent candidates. Bad choices may lead to the well known speedup anomalies [16].

The parallelization of the best first branch and bound algorithm is difficult because this strategy introduces an explicit order in which nodes are visited. Apart for this point, the branch and bound algorithm and the algorithm for solving the N-queens problem are very similar. They have both an a priori unknown set of tasks with unforeseeable execution times, but the dependencies between tasks they generate are different. The N-queens algorithm has no dependencies while the branch and bound algorithm has dependencies which are clearly dynamic as the lower bound values that are used as dependencies are only computed during run-time.

For this reason, the parallelization we used for the N-queens does not give equally good results for the branch and bound algorithm. Indeed, if processors explore simultaneously different parts of the branch and bound tree, each one using a local best first search, one cannot guarantee that all processors work on best candidates, i.e. candidates that the sequential algorithm would also have visited. As a matter of fact, the best problems local to some processors may all turn out to be useless computations when compared to the problems held by another processor. But, using only our simple parallelization, processors cannot be aware of this fact.

2.3 Classes of irregular problems

From the previous example, one can state that an irregular algorithm which generates dynamically tasks which are related by some dependencies (most of the time these dependencies fix an order in which tasks must be considered) cannot be parallelized by simply distributing the tasks on the processors and let processors locally apply the algorithm. The reason for this is the fact that locally correct behaviors do not necessarily yield a globally correct behavior. In the case of the best-first branch and bound algorithm such an "incorrect" behavior, compared to the sequential best-first algorithm, finds nevertheless a solution with the correct optimal value although it might do useless work. But, for other algorithms, the solution found could be a different one or the parallelization could even produce wrong results, showing the parallelization scheme to be wrong.

As a matter of fact, the dependencies between tasks define a partial order over the entire set of tasks. They thus need a global knowledge rendering any strictly local approach wrong or probably inefficient. The load balancing schemes must thus either construct a global state using a global approach or, depending on the nature of the dependencies, provide at least all processors with information that are close to the global state of the system allowing them to take nearly optimal local decisions.

In the branch and bound algorithm there are two pieces of information that are global. First, the set of tasks and their dependencies (lower bounds) as already mentioned and, second, the value of the best known solution or upper bound.

Indeed, this value must be known to all processors in order to allow them to use the cut condition whenever possible.

The main conclusion of this discussion is thus the fact that the existence of dependencies between dynamically created tasks or data rely on a global knowledge and complicate the problem of load balancing and the design of smart parallel algorithms. The two next sections show how this problem can be solved for the parallel best first branch and bound algorithm using, on one hand, local policies and, on the other hand, global policies. In terms of the classification defined in [1] these approaches can be characterized as global, dynamic, physically distributed, cooperative. The local approaches are furthermore sub-optimal and approximate while the global approaches are classified optimal.

3 LOCAL POLICIES

Local load balancing policies have been studied in a large number of papers. See for instance [1, 2, 3, 7, 19] as well as references therein. Many researchers have developed models of load balancing policies in order to compare different approaches [2, 3, 7]. The objectives were to exhibit the parameters that mostly influence the system, to determine the steady states or to study how *processor trashing* can be avoided. Processor trashing designs the situation where processors spend all their time transferring data or tasks. Our interest in these studies is relatively limited because we want to investigate the situation where tasks interdepend while all the models cited above assume that the tasks are independent.

Before considering more precisely the load balancing problem for the branch and bound algorithm let us briefly describe the approach used in [19] which is well adapted to the N-queens problem. Le Sergent and Berthomieu consider a graph of interconnected processors that maintain each two treshold values, an upper and a lower value, which represent the minimum respective maximum number of tasks the processor considered wants to have. If the number of tasks changes, due to task accomplishment or tasks creation, and reaches a treshold, then it triggers the load balancing mechanism. A processor that reaches a treshold will thus, depending on whether it touched the upper or lower bound, offer work or ask for work. Such a mechanism works fine as long as the average processor load – which is not known explicitly – falls into the interval given by the two treshold values. Problems that may generate large and fast load changes cannot be treated this way. Indeed, if the load on all processors increase quickly, all processors will exceed the upper bound. They will all become bidders while no one is willing to accept problems. [19] include a means by which the treshold values can dynamically be adapted. Processors with a work load outside of their load interval try to send or receive problems to or from other processors. If this does not lead to a situation where their load is again within their interval then this interval is adapted according to given functions. These functions must be chosen in an adequate manner in order to avoid processor trashing (see also [7]).

For the parallel best-first branch and bound algorithm all processors would ideally know the value of the best current solution and the dependencies between their problems and all the problems held by other processors. With such information, they could locally compute an optimal behavior, applying the cut condition based on a correct upper bound value and knowing when their local problems are less interesting than problems on other processors. A local load balancing approach in which processors interact only within some neighborhood can, of course, not provide the information needed. All that is possible is an information propagation from neighboring processors to neighboring processors that gives every processor a view of the system that comes close to the actual global state of the machine. Such an approach is used by Lüling and Monien and described in [20, 21].

Lüling and Monien [20, 21] define the load of a processor P_i as the result of a weight function w computed over the set of lower bound of the problems $\{x_1, \ldots, x_k\}$ of P_i and the value b of the best solution found so far or, more precisely, the view P_i has of this value. As possible weight functions they consider the following functions $w(P) = k$ which simply counts the number of problems, $w(P) = \min_j\{x_j\}$ which takes only the lower bound of the best problem into account or $w(P) = \sum_{j=1}^{k}(c - x_j)^2$ and $w(P) = \sum_{j=1}^{k} e^{x-x_j}$ which consider all the lower bound values weighted differently.

The load balancing algorithm tries to achieve a situation where the biggest difference in the load of a processor P and its neighbor processors $\{P_1, \ldots, P_n\}$ is less than a given constant Δ, i.e.

$$max_{j \in \{P_1, \ldots, P_n\}} \mid w(P) - w(P_j) \mid \leq \Delta$$

If the constant Δ is selected to be equal to the unit and the weight function w is chosen so as to consider only the value of the best lower bound, this algorithm tries to provide equally good problems to all processors. As the neighborhood of one processor intersects the neighborhoods of other processors the weights are propagated through the entire network and the biggest possible difference between any two processors is determined by Δ and the diameter of the network. The values of the best current solutions or upper bounds are of course also propagated in this fashion.

In [20] Lüling and Monien show very good results of their parallel best first algorithm of the symmetric traveling salesman problem [17] using the lower and upper bounds computation functions of [12, 13, 14, 36]. They used machines with up to several hundreds of processors and solved large problems of the TSPLIB [30] with very good efficiencies. The interconnection graph was chosen as a DeBruijn graph [18] offering a very low diameter and insuring thus a that the differences of the loads are quite small.

The following properties of the branch and bound algorithm are important for yielding these excellent results. Wah and Yu in [37] and Gengler and Coray in [10] developed modelizations, compared in [6], of the dynamic behavior of the branch and bound algorithm which show that the number of problems having a given lower bound increases exponentially with the lower bound. This means that a small number of lower bound values corresponds to a large number of problems. This is especially true when using the bounding function that compute good bounds. Many problems will thus have the same lower bounds reducing considerably the frequency at which the load balancing mechanism is called.

As an intermediate conclusion one could say that local load balancing strategies which keep processors at nearly the same load level achieve a sufficiently good global equilibrium whenever the diameter is not too large. Moreover, such load balancing mechanisms are not too expensive if there are many equivalent tasks providing good candidates to all processors and not introducing too fast changes of the loads.

As we will see in the next section these same properties are also required to obtain efficient parallelizations of irregular problems based on global load balancing strategies.

4 GLOBAL POLICIES

The fundamental justification for using a load balancing scheme based on a global state is the fact that the parallelization of an algorithm which dynamically generates dependent tasks can in principle not be scheduled efficiently (or even not at all) by having the processors take local decisions. Any processor can schedule a local task only if it has checked that the task is available and allowed to be executed. On shared memory machines this can easily be done by maintaining a unique list of tasks and have all processors fetch tasks from it or put tasks into it. This principle may be kept on distributed memory machines under the form of a master/slave parallelization. The master processor is in charge of the list of tasks, hands out tasks to the slaves and inserts new tasks coming from the slaves into the list. But, it is well known that the master/slave principle is not well adapted to massive parallelism, the master becoming quickly a bottleneck. Things are even worsened by the fact that the entire list of tasks has to be kept in the master's memory.

If we want to weaken the centralized control and go towards a distributed control we need to make some hypotheses on the problem. One hypothesis that is often verified is the property that the set of dynamic tasks may be split into classes of independent tasks which can be executed in parallel, the classes being considered one after the other. In this case, a possible parallelization consists in having all processors work independently on local tasks that belong to the class of tasks currently treated. Once this is done, all processors synchronize, build up the global state, compute their dependencies and decide what set of tasks can be executed next. The parallel algorithm becomes thus an alternation of computation steps and synchronizations which schedule the tasks. Gengler and Coray in [11] showed that the best-first branch and bound algorithm can be formulated in this way.

This does not necessarily lead to a good parallelization as the efficiency one can obtain will clearly be determined by the number of synchronizations needed and the volume of computations corresponding to the treatment of one set of independent tasks. The parallel computations must be large and the number of synchronizations small. If these conditions are met, one gets a sufficient parallelism to keep all processors busy. Moreover, synchronizations happen rarely making their expense neglectable. For the branch and bound algorithm, modelizations due to Wah and Yu [37] or Gengler and Coray [10] show that such a behavior is indeed the case for a large class of problems and problem instances. Experiments on massively parallel computers given in a companion paper [4] confirm the predictions obtained in the models for realistic problems.

In our computation/synchronization parallelization paradigm, the actual load balancing is realized during the synchronization step. At the beginning of a synchronization, all processors first compute the set of tasks that can be executed in parallel as the next step. Then they distribute these tasks equally on all processors. This redistribution is in fact close to a static load balancing scheme as the set of tasks is now known. Moreover, all tasks that are considered for parallel execution are independent allowing a load balancing algorithm of the N-queens kind.

Moreover, if the execution times of the tasks are known, we can compute an optimal schedule applying any of the available solutions. But, for most irregular problems the execution times are unknown. The load balancing can thus be only approximate and depends on the particular problem considered. In our implementations of the branch and bound algorithm, for instance, the load balancing policy consists in giving to all processors the same number of problems, taking also the problems' estimated complexity into account. The later is chosen to be equal to the number of unsolved constraints.

If the work loads given to the processors for one computation step cannot approximately be scheduled statically with enough precision to yield well balanced execution times, we may add other load balancing mechanisms to improve the situation. These can either be local schemes like the algorithms proposed for the N-queens problem or global schemes which could be based on interrupt driven message passing and provide a means to idle processors to force additional load balancing steps. In our branch and bound algorithms we use this latter approach when needed.

A last important point in a global load balancing policy is the information and task exchanges occurring during the synchronization steps. The computation of an optimal load balancing for a given set of tasks spread over all processors must be done with a reasonable complexity. Three questions arise. What information on the number of tasks owned by the different processors are needed? What is the complexity of the routing computation? How can tasks be exchanged by the processors in order to obtain the perfect load balancing but without generating too much contention in the network?

As all tasks are identical in the branch and bound algorithm, our routing problem corresponds to the Token Distribution Problem defined by Peleg and Upfal in [28]. It consists in computing the optimal routing of identical tokens in any interconnection network so as to lead from a given initial distribution to a final distribution without creating contentions in the network. Meyer auf der Heide *et al.* showed in [23] that this problem can be solved in polynomial time for any k-ary d-cube network (torus of dimension d with k processors per ring) when each processor knows the tokens owned by every other processor.

In [5] we use a different approach to solve this problem trying to reduce the knowledge required to compute the token routing. Our algorithm is defined for the large class of k-ary d-cube interconnection networks. We showed that the routing can be computed from reduced information which involve a message passing complexity of $O(k\ d)$ instead of $O(k^d)$ for the total exchange. The token routing can of course not be optimal as it is based on partial information. But it can be shown that our routing has the same worst case complexity as the optimal routing.

We are currently applying our global load balancing approach to other problems than the branch and bound algorithm like the important family of minimax algorithms used in game playing [27]. Its best-first version, known under the name of SSS* [34], is in fact a generalization of the best-first branch and bound algorithm. Recent formulations of the SSS* algorithm due to [29] should make it amenable for a parallelization that is very close to the one we use for best-first branch and bound algorithm.

5 CONCLUSION

In this paper we discussed some needs for load balancing. We showed the properties needed to make static scheduling possible and defined two basic types of programs that need dynamic load balancing mechanisms. The first type consists in problems that generate independent dynamic tasks. Such problems are easily parallelized as all load balancing decisions may be taken locally considering only a neighborhood of the processor. Algorithms that generate depend dynamic tasks are, on the contrary, much more difficult to parallelize. The reason for this stems from the fact that the task scheduling now needs global information. Such problems ask for more sophisticated load balancing schemes which are typically global.

For the class of problems using dependent tasks we exhibited two kinds of solutions based either on local or on global decisions. Both schemes suffer from problems. The local strategies do not necessarily accurately reflect the global state and processors may take suboptimal decisions, possibly increasing the overall execution time. Global strategies, on the other hand, are based on perfect knowledge. They do not introduce useless computations but suffer from the drawback that a global synchronization is expensive.

Experiments using both techniques show that, under certain circumstances, the approaches are well suited to massive parallelism. More importantly, the hypotheses needed to obtain good efficiencies are basically the same. First, one needs large sets of independent tasks so as to have enough parallelism and keep all processors busy. Next, the number of sets of independent tasks should be limited so as to avoid too heavy usage of the load balancing mechanisms and improve the global efficiency.

REFERENCES

[1] T. Casavant, J. G. Kuhl: A Taxonomy of Scheduling in General-purpose Distributed Computing Systems. IEEE Trans. on Software Engineering, 14, pp. 141–154, 1988.

[2] A. Corradi, L. Leonardi, F. Zambonelli: Load balancing strategies for massively parallel architectures. Parallel Processing Letters, 2(2&3), pp. 139–148, 1992.

[3] G. Cybenko: Dynamic Load Balancing for Distributed Memory Multiprocessors. J. of Parallel and Distributed Computing, 7, pp. 279–301, 1989.

[4] C. G. Diderich, M. Gengler: Experiments with a Parallel Synchronized Branch and Bound Algorithm. In this volume.

[5] C. G. Diderich, M. Gengler, St. Ubéda: An Efficient Algorithm for Solving the Token Distribution Problem on k-ary d-cube Networks. (Extended abstract) Proceedings ISPAN 94, to appear, 1994.

[6] M. Dion, M. Gengler, St. Ubéda: Comparing two Probabilistic Models of the Computational Complexity of the Branch and Bound Algorithm. Proceedings CONPAR 94 – VAPP VI, LNCS 854, pp. 359–370, 1994.

[7] D. L. Eager, E. D. Lazowska, J. Zahorjan: Adaptive load sharing in homogeneous distributed systems. IEEE Trans. on Software Engineering, 12(5), pp. 662-675, 1986.

[8] M. J. Flynn: Very High Speed Computing Systems. Proceedings IEEE 54 (12), pp. 1901-1909, 1966.

[9] M. R. Garey, D. S. Johnson: *Computers and Intractability - A Guide to the Theory of NP-Completeness.* W.H. Freeman, 1979.

[10] M. Gengler, G. Coray: A Parallel Best-first B&B with Synchronization Phases. Proceedings CONPAR 92 - VAPP V, LNCS 634, pp. 515–526, 1992.

[11] M. Gengler, G. Coray: A Parallel Best-first B&B Algorithm and its Axiomatization. IEEE Proceedings HICSS–26, Vol. 2, pp. 263–272, 1993. Also in J. of Parallel Algorithms and Applications, Vol. 2, pp. 61–80, 1994.

[12] M. Held, R. M. Karp: The Traveling Salesman Problem and Minimum Spannig Trees. Operations Research 18, pp. 1138–1162, 1970.

[13] M. Held, R. M. Karp: The Traveling Salesman Problem and Minimum Spanning Trees: part II. Mathematical Programming 1, pp. 6–25, 1971.

[14] R. Jonker, T. Volgenant: Non-optimal Edges for the Symmetric Traveling Salesman Problem. Operations Research 32, pp. 837–846, 1984.

[15] S. Kirkpatrick, C. D. Gelatt, M. P. Vecchi: Optimization by Simulated Annealing. Science, 220(4598), pp. 671–680, 1983.

[16] T. H. Lai, S. Sahni: Anomalies in Parallel Branch and Bound Algorithms. Comm. ACM, 27(6), pp. 594–602, 1984.

[17] E. L. Lawler, J. K. Lenstra, A. H. G. Rinnooy Kan, D. B. Shmoys (eds.): *The Traveling Salesman Problem: a Guided Tour of Combinatorial Optimization.* Wiley and sons, Chichester (GB), 1985.

[18] M. R. Samatham, D. K. Pradhan: The de Bruijn Multiprocessor Network: a Versatile Parallel Processing and Sorting Network for VLSI. IEEE Trans. on Comp., 38(4), pp. 567–581, 1989.

[19] Th. Le Sergent, B. Berthomieu: Balancing Load under Large and Fast Load Changes in Distributed Computing Systems - A Case Study. Proceedings CONPAR 94 - VAPP VI, LNCS 854, pp. 854–866, 1994.

[20] R. Lüling, B. Monien, M. Räcke, S. Tschöke: Efficient Parallelization of a Branch& Bound Algorithm for the Symmetric Traveling Salesman Problem. Tech. Report, University of Padreborn, 1992.

[21] R. Lüling, B. Monien: Load Balancing for Distributed Branch and Bound Algorithms. Proceedings Int. Parallel Processing Symp. (IPPS), pp. 543–549, 1992.

[22] J. W. Meyer: Self-organizing Processes. Proceedings CONPAR 94 – VAPP VI, LNCS 854, pp. 842–853, 1994.

[23] F. Meyer auf der Heide, B. Oesterdiekhoff, R. Wanka: Strongly Adaptive Token Distribution. Proceedings 20th ICALP 93, LNCS 700, pp. 398–409, 1993.

[24] L. G. Mitten: Branch and Bound Methods: General Formulation and Properties. Operations Research 18, pp. 24–34, 1970.

[25] J. von Neumann: *John von Neumann – Collected Works, Volume 5.* A. H. Toub (ed.), Pergamon Press, 1961.

[26] C. H. Papadimitriou, K. Steiglitz: *Combinatorial Optimization, Algorithms and Complexity*. Prentice Hall, 1982.

[27] J. Pearl: *Heuristics*, Addison-Wesley, 1985.

[28] D. Peleg, E. Upfal: The Token Distribution Problem. SIAM J. Comput. 18(2), pp. 229–243, 1989.

[29] W. Pijls, A. de Bruin: Another View on the SSS* Algorithm. Proceedings SIGAL'90, 1990.

[30] G. Reinelt: *Tsplib v1.2 – Traveling Salesman Problems Library.* University of Augsburg, Germany, 1990. Anonymous ftp from `softlib.rice.edu`.

[31] A. H. G. Rinnooy Kan: On Mitten's Axioms for Branch and Bound. Graduate School of Management, Delft, Tech. Rep. W/74/45/03, 1974.

[32] C. Roucairol: Parallel Branch and Bound Algorithms: an Overview. In *Parallel and Distributed Algorithms*, Cosnard, Robert, Quinton, Raynal (eds.), pp. 153–163, Elsevier Science Publishers, 1988.

[33] M. Schwehm, Th. Walter: Mapping and Scheduling by Genetic Algorithms. Proceedings CONPAR 94 – VAPP VI, LNCS 854, pp. 832-841, 1994.

[34] G. C. Stockman: A Minimax Algorithm better than Alpha-beta? Artificial Intelligence, 12, pp. 179–196, 1979.

[35] J. D. Ullman: NP-complete scheduling problems. J. Comput. System Sci., 10, pp. 384–393, 1975.

[36] T. Volgenant, R. Jonker: The Symmetric Traveling Salesman Problem and Edge Exchanges in Minima 1-trees. European Journal of Operational Research 12, pp. 394–403, 1983.